AF878626

# Springer Proceedings in Mathematics & Statistics

Volume 80

For further volumes:
http://www.springer.com/series/10533

**Springer Proceedings in Mathematics & Statistics**

This book series features volumes composed of select contributions from workshops and conferences in all areas of current research in mathematics and statistics, including OR and optimization. In addition to an overall evaluation of the interest, scientific quality, and timeliness of each proposal at the hands of the publisher, individual contributions are all refereed to the high quality standards of leading journals in the field. Thus, this series provides the research community with well-edited, authoritative reports on developments in the most exciting areas of mathematical and statistical research today.

Juan Carlos Ferrando · Manuel López-Pellicer
Editors

# Descriptive Topology and Functional Analysis

## In Honour of Jerzy Kąkol's 60th Birthday

Springer

*Editors*
Juan Carlos Ferrando
The University Research Institute Center of Operations Research
Universidad Miguel Hernandez
Elche
Alicante
Spain

Manuel López-Pellicer
Matematica Aplicada
Universidad Politecnica de Valencia
Valencia
Spain

ISSN 2194-1009 ISSN 2194-1017 (electronic)
ISBN 978-3-319-05223-6 ISBN 978-3-319-05224-3 (eBook)
DOI 10.1007/978-3-319-05224-3
Springer Cham Heidelberg New York Dordrecht London

Library of Congress Control Number: 2014942526

© Springer International Publishing Switzerland 2014
This work is subject to copyright. All rights are reserved by the Publisher, whether the whole or part of the material is concerned, specifically the rights of translation, reprinting, reuse of illustrations, recitation, broadcasting, reproduction on microfilms or in any other physical way, and transmission or information storage and retrieval, electronic adaptation, computer software, or by similar or dissimilar methodology now known or hereafter developed. Exempted from this legal reservation are brief excerpts in connection with reviews or scholarly analysis or material supplied specifically for the purpose of being entered and executed on a computer system, for exclusive use by the purchaser of the work. Duplication of this publication or parts thereof is permitted only under the provisions of the Copyright Law of the Publisher's location, in its current version, and permission for use must always be obtained from Springer. Permissions for use may be obtained through RightsLink at the Copyright Clearance Center. Violations are liable to prosecution under the respective Copyright Law.
The use of general descriptive names, registered names, trademarks, service marks, etc. in this publication does not imply, even in the absence of a specific statement, that such names are exempt from the relevant protective laws and regulations and therefore free for general use.
While the advice and information in this book are believed to be true and accurate at the date of publication, neither the authors nor the editors nor the publisher can accept any legal responsibility for any errors or omissions that may be made. The publisher makes no warranty, express or implied, with respect to the material contained herein.

Printed on acid-free paper

Springer is part of Springer Science+Business Media (www.springer.com)

*In honour of Jerzy Kąkol's 60th birthday*

# Preface

Jerzy Kąkol is a Full Professor at the Adam Mickiewicz University, Poznan, Poland. He obtained his Ph.D. in 1980 with his dissertation, *On inductive limits of topological vector spaces and algebras*. In 1987 he got his Habilitation with his thesis, *On singularities of non-locally convex topological vector spaces*. His research interest is related mostly to Functional Analysis and applications, Function Spaces, General and Descriptive Topology, Topological Groups, and Non-Archimedean Functional Analysis.

Professor Kąkol has published over 120 research articles. He was a co-editor of two proceedings of the *International Conferences 4th and 5th on p-adic Functional Analysis*, by Marcel Dekker, and coauthor, jointly with W. Kubis and M. López-Pellicer, of the monograph *Descriptive Topology in Selected Topics of Functional Analysis*, Springer 2011.

He has been co-organizer (and organizer) of the fourth (and fifth) *International Conference on p-adic Functional Analysis*, Nijmegen, the Netherlands 1996 (and Poznan 1998), as well as co-organizer of the sessions *Topological methods in Functional Analysis* for annual Summer Topology Conferences, Brno (Czech 2009), Kielce (Poland 2010), New York (USA 2011 and 2014). He is a Corresponding Member of the Royal Academy of Sciences in Madrid. He has given several invited talks in Europe and in the United States of America.

In the years 1989–1991 and 1994 he was a scholar (Munich and Saarbrucken, Germany) of the prestigious A. von Humbold scholarship and was awarded with various short-period (2–5 months) research grants at several Universities in Belgium, Germany, Italy, Spain, the Netherlands, and the United States of America. He got a Visiting Professor position for the whole year 1997 at the University of North Dakota, Grand Forks. He was a Weisman Visiting Professor at Baruch College, CUNY, Fall term, 2002. He spent two Spring Terms in 2003 and 2004 as a Visiting Professor at the University of Florida, Gainesville, cooperating with Prof. S. A. Saxon.

In the years 1998–2001 and 2004–2012 he was the Dean for Science at the Department of Mathematics and Computer Science in the Adam Mickiewicz University of Poznan, Poland, where currently he is member of its Senate.

In 2013, September 27 and 28, we organized in Elche the First Meeting in Topology and Functional Analysis dedicated to Prof. Jerzy Kąkol on the occasion of his 60th birthday. We appreciate the invitation of Springer to publish the

Proceedings of this Meeting and thank very much all participants for their contributions and, particularly, to the invited speakers who are the authors of the 11 chapters of these Proceedings. The first chapter deals with some aspects in the Mathematical work of Jerzy Kąkol, the targets of the next five chapters are topological and metric questions in Functional Analysis, and the last five chapters are devoted to Topological Groups, Banach spaces, Banach algebras, and Operator Theory.

Certainly, it is challenging to summarize the wide and diverse mathematical work of Prof. Kąkol, but it is more difficult to present his excellent human qualities, which are partially captured in the following words intended for Prof. Kąkol at this conference. We thank their author Aaron R. Todd (Baruch College, CUNY) for permission to use them.

Jerzy Kąkol's 60th
A force of nature in his 60th year,
Surely worthy of a celebratory conference.
Yet a portion of the natural world!
Can a person be a force of nature?
Still arriving from any and all directions,
Laying waste to the virtual landscape.
Leaving still, some quiet corners,
For useful improvements yet.
Welcome to this side of a marker!
No "humph" from us for your arrival.
A force of kindness, of gentleness,
Of good humor, and much more,
All worthy of celebratory words.
To Jerzy Kąkol, an extraordinary human being.

A. R. Todd

We have all enjoyed and benefited from the friendship of Professor Kąkol.

Valencia, April 2014

Juan Carlos Ferrando
Manuel López-Pellicer

# Contents

# Contributors

**Domingo Alcaraz** Universidad Politécnica de Cartagena, Cartagena, Spain

**Carlos Angosto** Departamento de Matemática Aplicada y Estadistica, Universidad Politécnica de Cartagena, Cartagena, Spain

**Salud Bartoll** IUMPA, Universitat Politècnica de València, Valencia, Spain

**Dikran Dikranjan** Dipartimento de Matematica e Informatica, Università di Udine, Udine, Italy

**J. C. Ferrando** Centro de Investigación Operativa, Universidad Miguel Hernández, Elche, Alicante, Spain

**Domingo García** Departamento de Análisis Matemático, Universidad de Valencia, Burjasot (Valencia), Spain

**Antonio José Guirao** Instituto de Matemática Pura y Aplicada, Universitat Politècnica de València, Valencia, Spain

**Manuel López-Pellicer** Department of Applied Mathematics and IUMPA, Universitat Politècnica de València, València, Spain; Departamento de Matemática Aplicada and IMPA, Universitat Politècnica de València, Valencia, Spain

**Manuel Maestre** Departamento de Análisis Matemático, Universidad de Valencia, Burjasot (Valencia), Spain

**Félix Martínez-Giménez** IUMPA, Universitat Politècnica de València, Valencia, Spain

**Santiago Moll** Department of Applied Mathematics, Universitat Politècnica de València, València, Spain

**Vicente Montesinos** Instituto de Matemática Pura y Aplicada, Universitat Politècnica de València, Valencia, Spain

**Gaspar Mora** Department of Mathematical Analysis, University of Alicante, Alicante, Spain

**María Muñoz Guillermo** Departamento de Matemática Aplicada y Estadística, Universidad Politécnica de Cartagena, Cartagena, Murcia, Spain

**Marina Murillo-Arcila** IUMPA, Universitat Politècnica de València, Valencia, Spain

**José Manuel Díaz Nieto** Departamento de Geometría y Topología, Facultad de Matemáticas, Universidad Complutense de Madrid, Madrid, Spain

**Elena Martín Peinador** Departamento de Geometría y Topología, Facultad de Matemáticas, Universidad Complutense de Madrid, Madrid, Spain

**Alfredo Peris** IUMPA, Universitat Politècnica de València, Valencia, Spain

**Manuel Sanchis** Institut de Matemàtiques i Aplicacions de Castelló (IMAC), Universitat Jaume I de Castelló, Castelló, Spain

**Stephen A. Saxon** Department of Mathematics, University of Florida, Gainesville, FL, USA

**Václav Zizler** Department of Mathematics, University of Alberta, Edmonton, AB, Canada

# Chapter 1
# Some Aspects in the Mathematical Work of Jerzy Kąkol

**Manuel López-Pellicer and Santiago Moll**

**Abstract** We present some selected topics of the research developed by Professor Jerzy Kąkol with his collaborators. After some comments on Kąkol's Ph.D. dissertation we group together the chosen contributions in three sections: Topological vector spaces, Descriptive Topology and Functional Analysis and Other aspects in Kąkol's work. For the sake of clarity the sections have been divided in subsections.

**Keywords** Analytic space · Baire type condition · Compactness · Descriptive topology · Fredholm operator · Functional analysis · Hahn–Banach extension property · Hewitt spaces · *K*-analytic · Locally convex space · Non-archimedean functional analysis · Resolution · Space of the continuous functions · Sequential condition · Topological vector space

**MSC 2010 Classification:** Primary 46A04, 46A22, 54A20, 54E52, Secondary 46A50, 46E10, 46S10, 47A53, 54D60, 54H05

## 1.1 Kąkol's Ph.D. Dissertation

Professor Kąkol got his Ph.D. in 1980, 2 years after his graduation in the A. Mickiewicz University. In his fundamental work on locally $m$-convex algebras [80, 81], E. Michael gave sufficient conditions for the local $m$-convexity of an algebra

---

M. López-Pellicer
Department of Applied Mathematics and IUMPA, Universitat Politècnica de València, 46022 València, Spain
e-mail: mlopezpe@mat.upv.es

S. Moll (✉)
Department of Applied Mathematics, Universitat Politècnica de València, 46022 València, Spain
e-mail: sanmollp@mat.upv.es

J. C. Ferrando and M. López-Pellicer (eds.), *Descriptive Topology and Functional Analysis*, Springer Proceedings in Mathematics & Statistics 80, DOI: 10.1007/978-3-319-05224-3_1, 
© Springer International Publishing Switzerland 2014

$X$ endowed with the linear inductive limit topology associated with an increasing sequence $(X_n, \xi_n)_n$ of locally $m$-convex subalgebras of $X$, a line of research that was continued in the works [97, 98] of Warner. Kąkol's dissertation deals with a similar problem but in the frame of the so-called generalized inductive limits of topological algebras in the sense of Alexiewicz.

This approach provided more applications in the realm of equivalent inductive systems. The notion of a generalized inductive limit of locally convex spaces was introduced first by Garling and systematically studied by Adasch, Ernst, Keim, Roelcke, Ruess, Turpin and Wiweger with the theory of the so-called two-normed spaces in the sense of Alexiewicz [1].

The essential result of Kąkol Ph.D. dissertation was included in the following main result:

**Theorem 1.1** (Kąkol) *If $\Gamma = \{(S_n, \xi_n) : n \in \mathbb{N}\}$ is an m-bounded inductive system on the algebra X, then $(X, \xi_\Gamma)$ is a topological algebra, where $\xi_\Gamma$ is the finest linear topology $\xi_\Gamma$ on X such that $\xi_\Gamma|S_n \leq \xi_n$ for all $n \in \mathbb{N}$.*

This theorem provides several applications, yielding the following applicable fact which essentially extends some results of Michael.

**Corollary 1.1** *If $\Gamma$ is the usual inductive system of commutative Banach algebras on the algebra X with a unit, then $(X, \xi_\Gamma)$ is a locally m-convex algebra.*

## 1.2 Topological Vector Spaces

### *1.2.1 Hahn–Banach Extension Property*

A number of results published by Kąkol deal with non-locally convex spaces and Hahn–Banach type extension theorems. Recall that a topological vector space $E$ is said to have the Hahn–Banach extension property if for every subspace $F$ of $E$ every continuous linear functional on $F$ can be extended to a continuous linear functional on $E$. In this case the topological dual $E'$ of $E$ is separating, i.e. $E'$ separates points of $E$ from zero. Every locally convex space has the Hahn–Banach extension property. In [25], Duren, Romberg and Shields gave an example of a dual separating $F$-space $E$, i.e. a metrizable and complete topological vector space, such that $E'$ is separating and $E$ has a subspace $M$ such that $E/M$ has a closed weakly dense subspace, so that $E$ does not have the Hahn–Banach extension property. Shapiro showed in [89] that every sequence space $\ell^p$, $0 < p < 1$, also contains such subspaces. Finally, in 1974 Kalton, developing basic sequence techniques for $F$-spaces, proved that all $F$-spaces with the Hahn–Banach extension property are locally convex [70].

In 1992, Kąkol presented in [47] an elementary construction for an abundance of vector topologies $\xi$ on a fixed infinite dimensional vector space $E$ such that $(E, \xi)$ does not have the Hahn–Banach extension property but the topological dual $(E, \xi)'$

separates points of $E$. This construction might be placed in any academic book devoted to the classical Hahn–Banach theorem.

**Theorem 1.2** (Kąkol) *Let $E$ be an infinite dimensional vector space and $\Delta := \{(x, x) : x \in E\}$ the diagonal of $E \times E$. For every non-zero linear functional $f$ on $\Delta$ there always exist two vector topologies $\xi_1$ and $\xi_2$ on $E$ such that $f$ is $\xi_1 \times \xi_2$-continuous on $\Delta$ but $f$ cannot be extended to a continuous linear functional on $(E, \xi_1) \times (E, \xi_2)$.*

## 1.2.2 The Mackey–Arens Theorem

The classical Mackey–Arens theorem states that for any locally convex space $E = (E, \xi)$ there exists always the *finest locally convex topology* $\mu(E, E')$ on $E$ compatible with $\xi$, i.e. both $\xi$ and $\mu(E, E')$ have the same continuous linear functionals. It is natural to ask whether for a topological vector space $E$ there exists the finest vector topology compatible with the original one. In papers [44, 45] Kąkol deals with the problem of the existence on a topological vector space $E$ of not locally convex vector topologies producing the same topological dual as the original topology of $E$. The author proved, among others, the following negative result concerning the Mackey–Arens theorem.

**Theorem 1.3** (Kąkol) *Let $\xi$ be a vector topology on a vector space $X$ different from the finest one on $X$ and compatible with a dual pair $(X, Y)$, i.e. $Y$ being the topological dual of $(X, \xi)$. If $(X, \xi)$ contains a dense subspace of infinite codimension, then $X$ does not admit the finest vector topology compatible with $(X, Y)$.*

Kąkol results on the Mackey–Arens theorem have been used by Chasco et al. in [14], Perez García and Schikhof in [85] and Khan in [72].

## 1.2.3 Basic Sequences and the Hahn–Banach Theorem

In the paper [56], Kąkol and Sorjonen being motivated by the fundamental 1974 Kalton theorem of [70] provided necessary and sufficient conditions for a metrizable topological vector space with a generalized form of the Hahn–Banach Extension Property to be locally convex. The main theorem lead to the following extension of Kalton's theorem.

**Corollary 1.2** (Kąkol–Sorjonen) *A metrizable [and separable] topological vector space with the Markushevich–Hahn–Banach Extension Property [Hahn–Banach Extension property] is either locally convex or its completion is not dual-separating.*

It is interesting to mention here that, having in mind the above corollary, Kalton constructed in [71] a quasi-Banach space which contains no basic sequence. This led to provide a metrizable locally bounded (non-complete) and non-locally convex space with the Hahn–Banach Extension property.

### 1.2.4 Fredholm Operators

Professor Kąkol also published a few articles connected with the Theory of Operators. Let $(X, \tau)$ and $(Y, \xi)$ be $F$-spaces, i.e. metrizable and complete topological vector spaces, and let $T : X \to Y$ be a continuous linear operator. Consider the following condition:

(i) *There exists a regular basic sequence $(x_n)_n$ in $X$ such that $T(x_n) \to 0$.*

In [23] Drewnowski proved that a continuous linear operator $T$ mapping a Banach space into another one is semi-Fredholm if and only if $T$ fails to have property (i).

The following two results of Kąkol in [46] essentially extend known results, including Drewnowski's result, to semi-Fredholm operators acting between $F$-spaces, i.e. metrizable and complete topological vector spaces.

**Theorem 1.4** (Kąkol) *Let $X$ and $Y$ be $F$-spaces and $T \in L(X, Y)$. Consider the following conditions:*

(a) *$T$ is m-semi-Fredholm, i.e. the range of $T$ is closed and its kernel is a minimal space.*
(b) *Every closed non-minimal subspace of $X$ contains a closed non-minimal subspace $G$ such that $T|G$ is an isomorphism.*
(c) *Does not exist a non-minimal closed subspace $G$ of $X$ such that $T|G$ is compact.*
(d) *Does not exist a strongly regular M-basic sequence $(x_n)_n$ in $X$ such that $T(x_n) \to 0$.*

*Then (a) $\Rightarrow$ (b) $\Rightarrow$ (c) $\Rightarrow$ (b). Moreover, if every minimal subspace of $X$ is q-minimal, then (b) $\Rightarrow$ (a).*

Additionally to (i) Kąkol consider the following conditions on the $F$-spaces $(X, \tau)$ and $(Y, \xi)$ and on the continuous linear operator $T : X \to Y$:

(ii) *There exists on $X$ a strictly weaker metrizable vector topology $\gamma$ such that $\tau$ is $\gamma$-polar, i.e. $\tau$ has a base of $\gamma$-closed neighborhoods of zero, and $T : (X, \gamma) \to (Y, \xi)$ is continuous.*
(iii) *There exists a closed non-minimal subspace $G \subset X$ such that $T|G$ is compact.* With these two conditions Kąkol obtain the following corollary with two interesting equivalences.

**Corollary 1.3** (Kąkol) *Let $X$, $Y$ be $F$-spaces and $T \in L(X, Y)$. Then conditions (i) and (ii) are equivalent. Moreover, if $X$ or $Y$ is locally bounded, then (i) and (iii) are equivalent.*

### *1.2.5 Baire Type Conditions and Baire-Like Spaces in Saxon' Sense*

The classical *Baire category theorem* says that if $X$ is either a complete metric space or a locally compact Hausdorff space, then the intersection of countable many dense, open subsets of $X$ is a dense subset of $X$. This result provides many applications as the Closed Graph Theorem, the Open Mapping Theorem and the Uniform Boundedness Theorem. Spaces having the topological property stated in the conclusion of the Baire theorem are called Baire spaces.

A new line of research concerning Baire-type conditions started with Saxon in [87]. Motivated by earlier results of Amemiya and Kōmura in [2], Saxon introduced the class of *Baire-like spaces* as those locally convex spaces $X$ for which every increasing sequence of closed absolutely convex subsets of $X$ covering $X$ contains one member with a non-empty interior (equivalently, that it is a neighborhood of zero). It turns out that any metrizable locally convex space is Baire-like if and only if it is barrelled. Saxon proved that even more is true: *A barrelled locally convex space $X$ is Baire-like provided $X$ does not contain a copy of $\varphi$,* i.e. the $\aleph_0$-dimensional vector space with the finest locally convex topology. Saxon also obtained the following deep Closed Graph Theorem for Baire-like spaces: *A linear map with closed graph from a Baire-like space into an $(LB)$-space is continuous.*

In [22] Dierolf and Kąkol, being motivated by works of De Wilde, introduced, characterized and studied the locally convex spaces $E$ (under the name *s-barrelled spaces*) for which one has that *every linear map from $E$ into a Fréchet space* (i.e. a locally convex $F$ space) *with sequentially closed graph is continuous.*

They provided a general construction which led to concrete examples of linear maps into Fréchet spaces with non-closed but sequentially closed graphs, and this approach was applied to show that the De Wilde's closed graph theorem fails if $X$ is the inductive limit of a family of Baire bornological locally convex spaces. This result was used to answer in the negative a 20 years old question posed by W. Roelcke.

A number of papers related to this subject were jointly published with Roelcke, where some results of Lurje in [73, 74] have been extended. For more details see the Kąkol and Roelcke papers [53–55].

Many historical facts about this line of research with a long list of contributors can be found in the report made by Kąkol in [59].

### *1.2.6 The Spaces $C_p(X, E)$ and $C_c(X, E)$*

If $X$ is a completely regular Hausdorff space and $E$ is a locally convex space, then $C_p(X, E)$ and $C_c(X, E)$ are the spaces of all continuous $E$-valued maps endowed with the pointwise and compact-open topology, respectively.

A characterization of the barrelledness of $C_c(X, E)$ is due to Mendoza [79]. On the other hand, the space $C_p(X, E)$ *is barrelled if and only if $C_p(X)$ and $E$ are barrelled*

[88, Theorem IV.7.2], while Buchwalter and Schmets proved in [10] that $C_p(X)$ *is barrelled if and only if every topologically bounded set in X is finite.*

A good sufficient condition for $C_p(X)$ or $C_c(X)$ to be a Baire space seems to be difficult to find. Earlier results by Gruenhage and Ma in [42], showed that if $X$ is a locally compact space or a first-countable space (or more generally, a $k$-space) then $C_c(X)$ *is Baire if and only if X has the moving-off property.* Granado and Gruenhage in [39] show that this characterization holds for generalized ordered spaces. As a corollary they prove that if $X$ is a locally compact generalized ordered space, $C_c(X)$ is a Baire space if and only if $X$ is paracompact.

In [86] Pytkeev, slightly modifying the game $\Gamma(X)$ of Lutzer and McCoy in [75] obtained that $C_p(X)$ *is a first category space if and only if player I has a winning strategy in the game* $\Gamma_1(X)$ *and if and only if there exists a disjoint sequence of finite nonempty subsets* $(\Delta_n)_n$ *of Xsuch that* $\sup_n \min\{|f(x)| : x \in \Delta_n\} < \infty$ *for any* $f \in C(X)$. In this paper it is also proved that *for a locally convex topological space X the space* $C_p(X)$ *is barrelled if and only if X is monotonically-convex Baire.* Let us recall that a locally convex space $X$ is monotonically-convex Baire if it is not the union of an increasing sequence of closed convex nowhere dense subsets of $X$.

Kąkol has obtained nice results in $c_0(E)$, $\ell_\infty(E)$ and $C(X, E)$ related to Baireness and Baire-likeness. For instance in [61] it is proved the following theorem:

**Theorem 1.5** (Kąkol–Gilsdorff–Sánchez-Ruiz)

(a) $c_0(E)$ *is Baire-like if and only if E is barrelled and the strong dual* $E'_\beta$ *of E is strong fundamentally* $\ell_1$*-bounded.*
(b) *If* $\ell_\infty(E)$ *is Baire-like, then* $c_0(E)$ *is Baire-like. If every bounded subset of E is precompact and* $c_0(E)$ *is Baire-like, then* $\ell_\infty(E)$ *is Baire-like.*

Recall that a topological space $X$ is called Fréchet-Urysohn if for each subset $A \subset X$ and each $x \in \overline{A}$ there exists a sequence in $A$ which converges to $x$. The following relation between Fréchet-Urysohn and Baire properties was obtained by Kąkol and Sánchez Ruiz in [58]:

**Theorem 1.6** (Kąkol–Sánchez-Ruiz) *Every sequentially complete Fréchet-Urysohn topological vector space is Baire. Consequently, if X is a Lindelöf P-space and E is a Fréchet locally convex space, then every sequentially closed linear subspace of* $C_p(X, E)$ *is Baire and ultrabornological.*

This provides another simple proof of the classical Baire theorem: *Any product of metrizable and complete locally convex space is Baire.*

### *1.2.7 Strongly Hewitt Spaces*

This line of research was continued by Kąkol and Śliwa in the paper [62]. In the class of realcompact or Hewitt spaces they introduced a new class of completely regular Hausdorff spaces, named *strongly Hewitt spaces*, as those spaces $X$ such that for each sequence $D := (x_n)_n$ in $X^* := \beta X \setminus X$ there exists $f \in C(\beta X)$ which is positive on $X$ and vanishes on a subsequence of $D$. If $X$ is strongly Hewitt, then every infinite set $A \subset X^*$ contains an infinite subset $S$ which is relatively compact in $X^*$ and $C^*$-embedded in $\beta X$. In the following theorem they show that strongly Hewitt spaces are related also to Baire properties.

**Theorem 1.7** (Kąkol–Śliwa)

(a) *A space $X$ is strongly Hewitt if and only if it is Hewitt and $X^*$ is countably compact. Hence every locally compact Hewitt space is strongly Hewitt.*
(b) *Every strongly Hewitt space of pointwise countable type is locally compact.*
(c) *Let $X$ be strongly Hewitt and $E$ a metrizable locally convex space. Then $C_c(X, E)$ is Baire-like if and only if $E$ is barrelled. Hence, if $X$ is locally compact, then $C_c(X, E)$ is Baire-like if and only if $E$ is barrelled and $X$ is Hewitt.*
(d) *There exists a locally compact strongly Hewitt space $X$ such that $C_c(X)$ is not Baire.*

### 1.2.8 Jarchow Problem: (*df*)-spaces $C_c(X)$

It is well known that the strong dual $E'_\beta$ of any bornological locally convex space $E$ with a fundamental sequence of bounded sets is a Fréchet space. A locally convex space $E$ is a *df-space* if it contains a fundamental sequence of bounded sets and every null sequence in $E'_\beta$ is equicontinuous. The class of *df*-spaces contains Grothendieck's class of *DF*-spaces. In page 270 of Jarchow's book [43] was stated that "*nothing seems to be known of when precisely $C_c(X)$ is a df-space*", while in the proof of Theorem 12.4.1 it is shown that for an arbitrary locally convex space $E$ the strong dual $E'_\beta$ is Fréchet if and only if $E$ equipped with its Mackey topology becomes a *df*-space. For $E = C_c(X)$, Mazón adds in [78] the equivalent condition that each regular Borel measure on $X$ has compact support. In the paper [77] McCoy and Todd characterize those $X$ for which the uniform dual of $C_c(X)$ is a Banach space. These three authors unearthed that the two conditions where equivalent to the Jarchow's ancient query.

In the remarkable paper [65] Kąkol, Saxon and Todd prove that the Jarchow problem was equivalent to the two mentioned conditions and also to the eight additional conditions given in the next main theorem.

**Theorem 1.8** (Kąkol–Saxon–Todd) *The following statements about $E := C_c(X)$ are equivalent:*

(i) *$E$ is a (df)-space.*
(ii) *$E$ has a fundamental sequence of bounded sets and is $\ell^\infty$-barrelled.*
(iii) *$(E', \beta(E', E))$ is a Fréchet space.*
(iv) *$(E', \beta(E', E))$ is a Banach space and equals $(E', n(E', E))$.*

(v) $(E', n(E', E))$ *is a Banach space.*
(vi) $(E', \beta(E', E))$ *is docile and locally complete.*
(vii) $(E', \sigma(E', E))$ *is docile and locally complete.*
(viii) *For each sequence* $(\mu_n)_n$ *in* $E'$ *there exists a sequence* $(t_n)_n \subset (0, 1]$ *such that* $\{t_n\mu_n\}_n$ *is equicontinuous.*
(ix) $X$ *is pseudocompact and* $(E', \sigma(E', E))$ *is locally complete.*
(x) *Each regular Borel measure on* $X$ *has compact support.*
(xi) *Every countable union of supports sets in* $X$ *is relatively compact.*

They apply this equivalence to provide a space $X$ for which $C_c(X)$ is a *df*-space but not a $DF$-space. This concrete space $C_c(X)$ provides an example of an $\ell^\infty$-barrelled space $C_c(X)$ which is not $\aleph_0$-quasibarrelled, answering an old Buchwalter–Schmets question posed in [10, Remark 1].

## 1.3 Descriptive Topology and Functional Analysis

### *1.3.1 Sequential Conditions for Topological Groups and Montel (DF) and (LM)-Spaces*

A part of the research of Professor Kąkol deals with sequential conditions of topological spaces and topological groups, a field of topology intensively developed from many years. We analyze some of results from this area, which were jointly obtained by Kąkol and Saxon in their celebrated paper [63].

Following Webb [99] a topological space $X$ is *sequential* if every sequentially closed subset is closed and $X$ verifies property $C_3$ if the sequential closure of every subset is sequentially closed. Clearly if $X$ is sequential and verifies $C_3$ then $X$ is Fréchet-Urysohn. It is well known that the strong dual $\varphi$ of the Fréchet space $\mathbb{K}^{\mathbb{N}}$ of all scalar sequences is an $\aleph_0$-dimensional, Montel, (*LB*) and (*DF*)-space. Nyikos showed in [83, Example 1] that the space $\varphi$ is sequential but it is not Fréchet-Urysohn.

Yoshinaga proved in [100] that the strong dual of a Fréchet–Schwartz space is sequential and Webb, in [99], extended Yoshinaga result to strong duals of Fréchet–Montel spaces (equivalently, to Montel (*DF*)-spaces).

A natural question is whether the Yoshinaga–Webb result extends to the class of (*LB*)-spaces, (*DF*)-spaces or (*LM*)-spaces. The answers are negative. Kąkol and Saxon proved the converse of Webb's result stating in [63, Theorems 4.5 and 4.6] that:

**Theorem 1.9** (Kąkol–Saxon)

(A) *For a dual metric space* $E$ *the following assertions are equivalent:*

1. $E$ *is sequential.*
2. $E$ *is normable or it is a Montel* (*DF*)*-space.*

(B) *For an (LM)-space E the following assertions are equivalent:*

1. *E is sequential.*
2. *E is metrizable or is a Montel (DF)-space.*

From these results it follows an easy answer to Nyikos topological group question in [83, Problem 1]. In the same frame Kąkol and Saxon prove that *every sequential proper (LB)-space is Montel* and that *a quasibarrelled (DF)-space E has property $C_3$ if and only if it is metrizable* [63, Corollary 4.3]. This result improves the analogous Webb's property for bornological spaces.

The topological property $C_3$ has been also used by Kąkol to characterize metrizability for (*LM*)-spaces in [57].

A topological vector space $X$ has property $C_3^-$ if the sequential closure of every linear subspace is sequentially closed. This definition follows from a nice results of Bonet and Defant in [9] showing that *the only infinite-dimensional Silva space with property $C_3^-$ is $\varphi$*. They also proved that *if E is an infinite dimensional nuclear (DF)-space different from $\varphi$, then E contains a subspace whose sequential closure is not sequentially closed.* Kąkol and Saxon in [63] also obtained that if $M$ is any metrizable space, then the product $M \times \varphi$ has property $C_3^-$ but not property $C_3$ and they got the following descriptive results for (*LB*) and (*LF*) spaces that have property $C_3^-$:

**Theorem 1.10** (Kąkol–Saxon) *An (LB)-space* $(E, \tau) = \lim(E_n, \tau_n)$ *has property $C_3^-$ if and only if E is isomorphic to some Banach step $E_q$, to $\varphi$, or to the product $E_q \times \varphi$.*

**Theorem 1.11** (Kąkol–Saxon) *An (LF)-space* $(E, \tau) = \lim(E_n, \tau_n)$ *has property $C_3^-$ if and only if E is isomorphic to some metrizable (LF)-space M, to $\varphi$, or to the product $M \times \varphi$.*

A topological space $X$ has countable tightness if for each subset $A$ and each point $a \in \overline{A}$ there exists a countable subset $B$ in $A$ such that $a \in \overline{B}$. Sequential spaces have countable tightness but the converse is not true. The following topological description of distinguished Fréchet spaces was obtained by Ferrando et al. in [26]:

**Theorem 1.12** (Ferrando–Kąkol–López-Pellicer–Saxon) *A Fréchet space E is distinguished if and only if its strong dual $E'_\beta$ has countable tightness.*

Based in this result, Ferrando et al. found in [27] that *a Fréchet space E is distinguished if and only if every bounded set in $E'_\beta$ has countable tightness.*

### 1.3.2 K-Analytic, Analytic Spaces, Compact Coverings (Resolutions)

A number of papers of Professor Kąkol deal with the *K-analyticity* of a topological (vector) space $E$ and the concept of a *resolution* on $E$, i.e. a family of sets

$\{K_\alpha : \alpha \in \mathbb{N}^\mathbb{N}\}$ such that $E = \cup_{\alpha\in\mathbb{N}^\mathbb{N}} K_\alpha$ and $K_\alpha \subset K_\beta$ if $\alpha \leq \beta$. Compact resolutions, i.e. resolutions $\{K_\alpha : \alpha \in \mathbb{N}^\mathbb{N}\}$ whose members are compact sets, naturally appear in many situations in Topology and Functional Analysis. A resolution $\{K_\alpha : \alpha \in \mathbb{N}^\mathbb{N}\}$ in $E$ swallows compact sets if for each compact subset $K$ in $E$ there exists $K_\alpha$ such that $K \subset K_\alpha$. It is an easy and elementary exercise to observe that any separable metric and complete space $E$ admits a compact resolution swallowing compact sets. Very far from triviality is the remarkable theorem due to Christensen [15, 68, Theorems 3.3 and 6.1, respectively], stating that a separable metric topological space $X$ is a Polish space if and only if $X$ admits a compact resolution swallowing compact sets.

Talagrand proved in [90] that any $K$-analytic space admits a compact resolution. It is worthwhile to mention two results of Cascales and Orihuela stating that for many topological spaces $X$ the existence of such a resolution is enough for $X$ to be $K$-analytic [11, Corollary 1.1] or analytic [12, Theorem 15, extending Talagrand's result that every metric K-analytic space is analytic].

Quasi-(*LB*)-spaces are locally convex spaces that admit special resolutions consisting of Banach discs. Quasi-(*LB*)-spaces and its relationships with the Closed Graph Theorem were defined and studied by Valdivia in [95], where it is proved that a Baire quasi-(*LB*)-space is a Fréchet space.

Tkachuk in his paper [93] proved that if $C_p(X)$ is $K$-analytic and Baire, the space $X$ is countable and discrete. Hence a $K$-analytic Baire space $C_p(X)$ is a separable Fréchet space. In fact Tkachuk's theorem follows from the following theorem, due to De Wilde and Sunyach in [21].

**Theorem 1.13** (De Wilde–Sunyach) *A Baire $K$-analytic locally convex space is a separable Fréchet space.*

Lutzer, van Mill and Pol complete this picture in [76] exhibiting a countable space $X$, having a unique non-isolated point, such that $C_p(X)$ is a separable, metrizable, non-complete Baire space. By Theorem 1.13 $X$ is not $K$-analytic.

This line of research was continued by Kąkol and López-Pellicer in the paper [66]. Their following result essentially extends De Wilde–Sunyach theorem.

**Theorem 1.14** (Kąkol–López-Pellicer) *A Baire locally convex space $F$ with a relatively countably compact resolution is a separable Fréchet space.*

The proof used the following technical result from the same paper.

**Proposition 1.1** (Kakol–López-Pellicer) *Let $E$ be a topological space which admits a weaker topology $\xi$ generated by a metric $d$. Let $F$ be a dense Baire subspace of $E$ having a resolution consisting of closed sets in $\xi$. Then $E \setminus F$ is of first Baire category.*

The concept of the angelicity introduced by Fremlin in [36] provided a strong motivation for several lines of research concerning to some variants of compactness. In [84] Orihuela introduced a large class of topological spaces $X$ (under the name *web-compact*) for which the space $C_p(X)$ is angelic. Orihuela's theorem covers many already known partial results of Eberlein–Šmulian type.

Kąkol, Ferrando, López-Pellicer and Śliwa studied the subclass of web-compact spaces formed by the spaces that admit a resolution of relatively countably compact sets, called *strongly web-compact spaces*, in their paper [32]. They found that the square of a strongly web-compact space need not be strongly web-compact. Their construction used some idea presented by Novák in [82] and they proved:

*Example 1.1 (Ferrando–Kakol–Lopez-Pellicer–Sliwa)* There exists a countably compact topological space $G$ such that the product $G \times G$ cannot be covered by a resolution of relatively countably compact sets. Hence, there exists a quasi-Suslin space $X$ such that $X \times X$ is not quasi-Suslin.

It turns out that *strongly web-compact spaces* can be used to extend some classical Closed Graph Theorems. In [94, Chap. 1, 4.2. (11)] Valdivia stated that a linear map with closed graph from a metrizable Baire locally convex space $E$ into a quasi-Suslin locally convex space $F$ is continuous. Drewnowski proved in [24] that every continuous linear map from a topological vector space having a compact resolution onto an $F$-space is open. The Theorem 1.15 from [30] extends Valdivia's and Drewnowski's results.

**Theorem 1.15** (Ferrando–Kąkol–López-Pellicer) *Let $E$ and $F$ be topological vector spaces such that $E$ is Baire and $F$ admits a relatively countably compact resolution. If $f : E \to F$ is a linear map with closed graph, there is a sequence $(U_n)_n$ in the family $\mathfrak{F}(E)$ of zero neighborhoods of $E$ such that for every $V \in \mathfrak{F}(F)$ there exists $m \in \mathbb{N}$ with $m^{-1}U_m \subset f^{-1}(V)$. If $E = F$, then $E$ is a separable $F$-space.*

From this theorem they get the Corollary 1.4.

**Corollary 1.4** (Ferrando–Kakol–López-Pellicer) *Every linear map $f$ from a Baire topological vector space $E$ into a topological vector space $F$ whose graph $G$ admits a relatively countably compact resolution is continuous. Hence, every linear map from a metrizable and complete topological vector space into a separable metrizable topological vector space whose graph admits a complete resolution is continuous.*

For spaces $C_p(X)$, Ferrando and Kąkol in [29] obtain the following equivalences:

**Theorem 1.16** (Ferrando–Kąkol) *The following conditions are equivalent:*

1. *$C_p(X)$ admits a bounded resolution.*
2. *$C_p(X)$ is $K$-analytic-framed in $\mathbb{R}^X$, i.e. there exists a $K$-analytic space $Z$ such that $C_p(X) \subset Z \subset \mathbb{R}^X$.*
3. *$C_p(X)$ is $K$-analytic-framed in $\mathbb{R}^X$ and $C_p(X)$ is angelic.*

This applies to provide simple, almost elementary, proofs of the following remarkable results due to Talagrand, Tkachuk and Velichko-Tkachuk-Shakhmatov [92], respectively:

(i) *Let $X$ be a compact space. Then $C_p(X)$ is $K$-analytic if and only if $C_c(X)$ is weakly $K$-analytic* [90, Théorème 3.4].

(ii) *For any completely regular space* $X$, *the space* $C_p(X)$ *is* $K$*-analytic if and only if* $C_p(X)$ *has a compact resolution* [93, Theorem 2.8].
(iii) $C_p(X)$ *is covered by a sequence of compact (relatively countably compact) sets if and only if* $X$ *is finite* [7, Theorem I.2.1 and Corollary I.2.4].

Ferrando gives the first contribution in [31] to the general problem of when analyticity or $K$-analyticity of the weak topology $\sigma(E, E')$ of a dual pair $(E, E')$ can be lifted to a stronger topology on $E$ compatible with the dual pair, by means of the following theorem.

**Theorem 1.17** (Ferrando) *The dual of* $C_p([0,1])$ *endowed with the Mackey topology is a weakly analytic space that it is not* $K$*-analytic.*

This nice result was complemented in [67] with the following theorem:

**Theorem 1.18** (Kąkol–López-Pellicer–Śliwa) *For a completely regular space* $X$, *the Mackey dual of* $C_p(X)$ *is analytic if and only if* $X$ *is countable.*

### *1.3.3 Class* $\mathfrak{G}$ *and K-Analyticity*

Let $E$ be an $(LM)$-space, i.e. the inductive limit of a sequence of metrizable locally convex spaces $(E_k)_k$, with a countable base of absolutely convex neighborhoods of zero $(U_n^k)_n$ in each $E_k$ such that $U_{n+1}^k \subset U_n^k$, for every $k, n \in \mathbb{N}$. For $\alpha = (n_k) \in \mathbb{N}^{\mathbb{N}}$ set $K_\alpha := \bigcap_k (U_{n_k}^k)^\circ$. The family $\{K_\alpha : \alpha \in \mathbb{N}^{\mathbb{N}}\}$ is a relatively countably compact resolution in $(E', \sigma(E', E))$ and each sequence in every $K_\alpha$ is equicontinuous.

Let $E$ be a $(DF)$-space and let $(S_n)_n$ be a fundamental sequence of bounded sets. For $\alpha = (n_k) \in \mathbb{N}^{\mathbb{N}}$ set $K_\alpha := \bigcap_k n_k S_k^\circ$. Then $\{K_\alpha : \alpha \in \mathbb{N}^{\mathbb{N}}\}$ is as above.

Let $F$ be a locally complete $(LF)$-space and let $E := (F', \beta(F', F))$ be its strong dual. Then $F$ has a resolution $\{A_\alpha : \alpha \in \mathbb{N}^{\mathbb{N}}\}$ consisting of Banach discs and each bounded set in $F$ is contained in some $A_\alpha$. For $\alpha \in \mathbb{N}^{\mathbb{N}}$ set $K_\alpha := A_\alpha^{\circ\circ}$ in $E'$ and $\{K_\alpha : \alpha \in \mathbb{N}^{\mathbb{N}}\}$ in $(E', \sigma(E', E))$ is as above. In particular this holds for the space of distributions $E := \mathfrak{D}'(\Omega)$ over an open set $\Omega \subset \mathbb{R}^n$, as well as for the space $A(\Omega)$ of real analytic functions on $\Omega$.

The common topological structure that appears in the dual $(E', \sigma(E', E))$ of the above examples motivated Cascales and Orihuela in [12] to introduce the class $\mathfrak{G}$ of locally convex spaces, such that a locally convex space $E$ belongs to class $\mathfrak{G}$ if $(E', \sigma(E', E))$ admits a resolution $\{A_\alpha : \alpha \in \mathbb{N}^{\mathbb{N}}\}$ with the property that each sequence in every $A_\alpha$ is equicontinuous. Therefore the locally convex spaces mentioned above belong to class $\mathfrak{G}$.

Class $\mathfrak{G}$ enjoys good permanence properties, being its precompact subsets metrizables [12, Theorem 2]. A general characterization for the metrization of precompacts subsets of a locally convex space was obtained in [28, Theorem 3]:

**Theorem 1.19** (Ferrando–Kąkol–López-Pellicer) *Precompact sets in a locally convex space $E$ are metrizable if and only if $E'$ endowed with the topology $\tau_p$ of uniform convergence on the precompact sets of $E$ is trans-separable.*

Nevertheless in [64, Corollary 2.8] Cascales, Kąkol and Saxon proved that $C_p(X) \in \mathfrak{G}$ if and only if $X$ is countable. Theorems 2.2 and 2.5 of this paper established the following characterization of global nature of metrizability in the class $\mathfrak{G}$ that extends a number of metrization theorems. Recall that $\varphi$ is the $\aleph_0$-dimensional vector space with the finest locally convex topology.

**Theorem 1.20** (Cascales-Kąkol–Saxon) *For a locally convex space $E$ in class $\mathfrak{G}$ the following conditions are equivalent:*

*(i) $E$ is metrizable.*
*(ii) $E$ is Fréchet-Urysohn.*

*If, additionally $E$ is barrelled, the preceding conditions are equivalent to the property that $E$ does not contain $\varphi$.*

The next two theorems are in [13, Theorems 4.6, 4.7 and 4.8]. Its motivation and proof comes in part from Cascales result that if $E \in \mathfrak{G}$ then $(E', \sigma(E', E))$ is quasi-Suslin [11, Proposition 1]. They are related to the inverse case of a following general problem due to Corson asking if it is true that the unit ball in $E'$ has $\sigma(E', E)$ countable tightness when $E$ is a weakly Lindelöf Banach space. This Corson problem seems to be still open.

**Theorem 1.21** (Cascales-Kąkol–Saxon) *If $E \in \mathfrak{G}$ the following assertions are equivalent:*

1. *$(E, \sigma(E, E'))$ has countable tightness.*
2. *$(E', \sigma(E', E))$ is realcompact.*
3. *$(E', \sigma(E', E))$ is Lindelöf.*
4. *$(E', \sigma(E', E))^n$ is Lindelöf for each $n \in \mathbb{N}$.*
5. *$(E', \sigma(E', E))$ is K-analytic.*

**Theorem 1.22** (Cascales-Kąkol–Saxon)

(1) *If $E \in \mathfrak{G}$ and has countable tightness then $(E, \sigma(E, E'))$ have countable tightness.*
(2) *If $E \in \mathfrak{G}$ and it is quasibarrelled then $E$ has countable tightness.*

The next result obtained by Cascales and Orihuela in [12, Theorem 13] extends a deep theorem of Amir-Lindestrauss for weakly compact generated Banach spaces. Let us notice that locally convex Lindelöf $\Sigma$-spaces, called also weakly countably determined, have attracted specialists for many years because of some important applications in the theory of $C_p(X)$ spaces.

**Theorem 1.23** (Cascales-Orihuela) *If the space $E$ belongs to the class $\mathfrak{G}$ and $(E, \sigma(E, E'))$ is a Lindelöf $\Sigma$-space then* $\mathrm{dens}(E, \sigma(E, E')) = \mathrm{dens}(E', \sigma(E', E))$.

The next characterization for duals Lindelöf $\Sigma$-spaces was obtained by Ferrando and Kąkol in [33].

**Theorem 1.24** (Ferrando–Kąkol) *Let $E$ be a locally convex space such that every weak*-bounded set in $E'$ is relatively weak*-compact. Then $(E, \sigma(E, E'))$ has a $\Sigma$-covering with limited envelope if and only if $\left(E', \sigma\left(E', E\right)\right)$ is a Lindelöf $\Sigma$-space.*

A particular case of a version of an Amir-Lindestrauss theorem-type for spaces $C_p(X)$ is given by Ferrando et al. in [35, Theorem 1]. With the use of Nagami index they obtain new topological cardinal inequalities for spaces $C_p(\upsilon_\lambda X)$. For instance it is stated that:

**Theorem 1.25** (Ferrando–Kąkol–López-Pellicer–Muñoz) *If $L \subset C_p(X)$ is a Lindelöf $\Sigma$-space and the Nagami index $Nag(X)$ of $X$ is less or equal than the density* dens$(L)$ *of $L$ (this holds for example if $X$ is a Lindelöf $\Sigma$-space), then:*

(i) *There exists a completely regular Hausdorff space $Y$ such that $Nag(Y) \leq Nag(X)$, $L \subset C_p(Y)$ and* $\text{dens}(Y) = \text{dens}(L)$.
(ii) *$Y$ admits a weaker completely regular Hausdorff topology $\tau'$ such that $w(Y, \tau') \leq d(Y)$.*

A locally convex space $E$ has a $\mathfrak{G}$-base if it admits a base of neighborhoods of zero $\{U_\alpha : \alpha \in \mathbb{N}^{\mathbb{N}}\}$ satisfying the decreasing condition $U_\beta \subset U_\alpha$ whenever $\alpha \leqslant \beta$ in $\mathbb{N}^{\mathbb{N}}$ [68, Lemma 15.2 (iii)]. If $E$ has a $\mathfrak{G}$ -base then $E \in \mathfrak{G}$ but, in general, the converse is not true. Concerning to $\mathfrak{G}$-bases is worthwhile to mention the following result of Ferrando and Kąkol in [34].

**Theorem 1.26** (Ferrando–Kąkol) *For a completely regular topological space $X$, the space $C_c(X)$ has a $\mathfrak{G}$-base if and only if $X$ contains a compact resolution swallowing the compact subsets.*

Generalizing the notion of metrizability, Tsaban and Zdomskyy introduced in [91] the strong Pytkeev property in relation with some deep metrization problems. They proved that the space $C_c(X)$ has the strong Pytkeev property for any Polish space $X$. This theorem has been recently extended by Gabriyelyan et al. in their preprint [38] where it is proved:

**Theorem 1.27** (Gabriyelyan–Kąkol–Leiderman) *Let $X$ be a space admitting a compact resolution swallowing the compact sets. The following are equivalent:*

(i) *$C_c(X)$ has the strong Pytkeev property.*
(ii) *$C_c(X)$ has countable tightness.*
(iii) *$C_p(X)$ has countable tightness.*
(iv) *$X$ is a $\mu$-space (this condition is equivalent to the barrelledness of $C_c(X)$).*

In the same paper [38] Gabriyelyan, Kąkol and Leiderman found the following interesting relations between $\mathfrak{G}$-bases and strong Pytkeev property in topological groups.

**Theorem 1.28** (Gabriyelyan–Kąkol–Leiderman) *A sequential topological group with a $\mathfrak{G}$-base has the strong Pytkeev property. Consequently, a Baire topological group G is metrizable if and only if G has a $\mathfrak{G}$-base and is sequential. Conversely, any topological group which has the strong Pytkeev property admits a base of neighborhoods of the unit of the form $\{U_\alpha : \alpha \in \mathbb{M}\}$, where $\mathbb{M}$ is a subset of the partially ordered set $\mathbb{N}^{\mathbb{N}}$.*

### 1.3.4 $\ell_c$-Equivalence and Metrizability

Two spaces $X$ and $Y$ are said to be *$\ell_p$-equivalent* if the spaces $C_p(X)$ and $C_p(Y)$ are linearly homeomorphic and, analogously, $X$ and $Y$ are *$\ell_c$-equivalent* if the spaces $C_c(X)$ and $C_c(Y)$ are linearly homeomorphic.

Arhangel'skiĭ asked in [6] *if metrizability is preserved by $\ell_p$-equivalence in the class of first countable spaces.* Baars et al. proved in [8, Theorem 3.3] that *complete metrizability is preserved by $\ell_p$-equivalence in the class of metrizable spaces*, given an alternative proof in the class of separable metrizable spaces by using a classical result of Christensen [15, 68, Theorems 3.3 and 6.1, respectively]. Later on, Valov in [96, Corollary 4.6] proved that a Čech-complete and first countable space $Y$ is metrizable when it is $\ell_p$-equivalent to a metrizable space $X$.

In the class of first countable spaces the preservation by *$\ell_c$-equivalence of separable metrizability as well as the $\ell_c$*-invariance of Polish spaces have been considered very recently by Kąkol et al. in [69, Corollaries 22 and 23]. These two $\ell_c$-invariant properties may be extended to the class of spaces of pointwise countable type, that contains also the class of locally compact spaces (see *op. cit.* Corollary 24 for the $\ell_c$-invariance of Polish spaces).

## 1.4 Other Aspects in Kąkol Work

### 1.4.1 Quantitative Outlook of Compactness

In recent years, several quantitative counterparts of some classical theorems such as Krein-Šmulyan, Eberlein-Šmulyan and Grothendieck theorems have been proved by different authors. These new versions also are in the Kąkol's work, always focused on new problems and its applications in topology and analysis. In [3, 4] we may find among others the next results:

**Theorem 1.29** (Angosto–Kąkol–López-Pellicer) *Let $X$ be a web-compact space and $(Z, d)$ a separable metric space, let $H \subset Z^X$ be a $\tau_p$-relatively compact set and let $C(X, Z)$ be the space of $Z$-valued continuous functions defined on $X$. Then $ck(H) \leq \hat{d}(\overline{H}, C(X, Z)) \leq 17ck(H)$, where $ck(H)$ is the "worst" distance to $C(X, Z)$ of the* *set of cluster points in $Z^X$ of sequences in $H$. Therefore the following conditions are equivalent:*

1. $ck(H) = 0$.
2. *$H$ is a relatively countably compact subset of $C(X, Z)$.*
3. *$H$ is a relatively compact subset of $C(X, Z)$.*

A quantitative sequential approximation of closure points [3, Theorem 15] enables us to give a nice proof of the well known Orihuela result about the angelicity of $C_p(X, Z)$ for web-compact $X$ and $Z$ metric [84, Corollary 17].

For a Fréchet space $E$ we define $k(H) := \sup\{d(h, E) : h \in \overline{H}^{\sigma(E'',E')}\}$, where $d(h, E)$ is the natural distance from $h$ to $E$ in the bidual $E''$. As usual $coH$ is the convex hull of $H$.

**Theorem 1.30** (Angosto–Kąkol–Kubzdela–López-Pellicer) *For a bounded set $H$ in a Fréchet space $E$ the following inequality holds $k(coH) < (2^{n+1} - 2)k(H) + \frac{1}{2^n}$ for all $n \in \mathbb{N}$. Consequently, this yields also the following formula $k(coH) \leq \sqrt{k(H)}(3 - 2\sqrt{k(H)})$.*

Hence $coH$ is weakly relatively compact provided $H$ is weakly relatively compact in $E$. This theorem extends the quantitative version of Krein's theorem for Banach spaces obtained by Granero, Hájek and Montesinos Santalucía in [40, 41].

Two another measures of weak non-compactness $lk(H)$ and $k'(H)$ for a Fréchet space were discussed in [4] providing two quantitative versions of Krein's theorem for the both functions.

### *1.4.2 A Glance on Non-archimedean Functional Analysis in Kąkol Work*

An essential part of the research of Professor Kąkol deals also with non-archimedean Functional Analysis, including some results about tensor products, Hahn–Banach type theorems and Dugundji Extension type theorems among others. Kąkol provided an essential application of so-called $t$-frames to the study of compactoids. His results may be found in [16–20, 60].

Since this line of research for non-specialists might provide some extra technical and theoretical difficulties, we refer to a recent monograph [85] by Perez Garcia and Schikhof, published in 2010, where many results of Kąkol have been presented. Kąkol results after 2010 may be found in his papers [5, 48–52].

It must be noted how far some non-archimedean cases might be from the classical ones. For instance Kąkol proved that a non-archimedean field $\mathbb{K}$ is spherically complete if and only if every locally convex space over $\mathbb{K}$ has the Hahn–Banach Extension Property.

**Acknowledgments** The first named author has been supported by Generalitat Valenciana, Conselleria d'Educació Cultura i Esport, Spain, Grant PROMETEO/2013/058.

# References


1. Alexiewicz, A.: The two-norm spaces. Stud. Math. (Ser. Specjalna) Zeszyt, **1**, 17–20 (1963)
2. Amemiya, I., Ichiro, Kōmura, Y.: Über nicht-vollständige Montelräume. Math. Ann. **177**, 273–277 (1968)
3. Angosto, C., Kąkol, J., López-Pellicer, M.: A quantitative approach to weak compactnesss in Fréchet spaces and spaces $C(X)$. J. Math. Anal. Appl. **403**, 13–22 (2013)
4. Angosto, C., Kąkol, J., Kubzdela, A.: A quantitative version of Krein's theorems for Fréchet spaces. Arch. Math. **101**, 65–77 (2013)
5. Angosto, C., Kąkol, J., Kubzdela, A.: Measures of weak noncompactness in non-archimedean Banach spaces. J. Convex Anal. **21**(3) (2014)
6. Arkhangel'skiĭ, A.V.: A survey of $C_p$-theory. Questions Answers Gen. Topology **5**, 1–109 (1987)
7. Arkhangel'skiĭ, A.V.: Topological function spaces. Mathematics and Its Applications, vol. 78. Kluwer, Dordrecht (1992)
8. Baars, J., de Groot, J., Pelant, J.: Function spaces of completely metrizable spaces. Trans. Am. Math. Soc. **340**, 871–883 (1993)
9. Bonet, J., Defant, A.: The Levy-Steinitz rearrangement theorem for duals of metrizable spaces. Isr. J. Math. **117**, 131–156 (2000)
10. Buchwalter, H., Schmets, J.: Sur quelques propriétés de l'espace $C_s(T)$. J. Math. Pures Appl. **52**, 337–352 (1973)
11. Cascales, B.: On $K$-analytic locally convex spaces. Arch. Math. **49**, 232–244 (1987)
12. Cascales, B., Orihuela, J.: On compactness in locally convex spaces. Math. Z. **195**, 365–381 (1987)
13. Cascales, B., Kąkol, J., Saxon, S.: Weight of precompact subsets and tightness. J. Math. Anal. Appl. **269**, 500–518 (2002)
14. Chasco, M.J., Peinador, E., Tarieladze, V.: On Mackey topology for groups. Stud. Math. **132**(3), 257–284 (1999)
15. Christensen, J.P.R.: Topology and Borel structure. North-Holland Mathematics Studies, vol. 10. North-Holland, Amsterdam (1974)
16. De Grande-De Kimpe, N., Kąkol, J., Pérez-García, C.: $p$-adic locally convex inductive limits. In: Kąkol, J., Pérez-García, C., Schikhof, W.H. (eds.) Lectures Notes in Pure and Applied Mathematics. Proceedings of the Fourth Conference in P-adic Analysis, pp. 159–222. Marcel Dekker, New York (1997)
17. De Grande-De Kimpe, N., Kąkol, J., Pérez-García, C.: On compactoides in ($LB$)-spaces. Bull. Acad. Polon. Sci. **45**(4), 313–321 (1997)
18. De Grande-De Kimpe, N., Kąkol, J., Pérez-García, C., Schikhof, W.: Orthogonal sequences in non-archimedean locally convex spaces. Indag. Math. (N.S.) **11**(2), 187–195 (2000)
19. De Grande-De Kimpe, N., Kąkol, J., Pérez-García, C., Schikhof, W.: Weak bases in $p$-adic spaces. Boll. Unione Mat. Ital. **85-B**, 667–676 (2002)
20. Kimpe, De Grande-De: N., Kąkol, J., Pérez-García, C., Schikhof, W.: Metrizability of compactoid sets in non archimedean Hausdorff ($LM$)-spaces. In: Schikhof, W., Perez-Garcia, C., Escassut, A. (eds.) Proceedings of the Ultrametric Functional Analysis (Nijmegen 2002) Contemporary Mathematics, vol. 319. American Mathematical Society, Providence (2003)
21. De Wilde, M., Sunyach, C.: Un théorèm de selection pour des applications à graphe borelién. C. R. Acad. Sci. Paris **269**, 273–274 (1969)
22. Dierolf, S., Kąkol, J.: On $s$-barrelled spaces. Results Math. **28**(1 and 2), 40–48 (1995)
23. Drewnowski, L.: Some characterizations of semi-Fredholm operators. Comment. Math. Prace Mat. **24**(2), 215–218 (1984)
24. Drewnowski, L.: Resolutions of topological linear spaces and continuituy of linear maps. J. Math. Anal. Appl. **335**, 1177–1195 (2007)
25. Duren, P.L., Romberg, B.W., Shields, A.L.: Linear functionals on $H^p$ spaces with $0 < p < 1$. J. Reine Angew. Math. **238**, 32–60 (1969)

26. Ferrando, J.C., Kąkol, J., López-Pellicer, M., Saxon, S.A.: Tightness and distinguished Fréchet spaces. J. Math. Anal. Appl. **324**, 862–881 (2006)
27. Ferrando, J.C., Kąkol, J., López-Pellicer, M.: A characterization of distinguished Fréchet spaces. Math. Nachr. **279**, 1783–1786 (2006)
28. Ferrando, J.C., Kąkol, J., López-Pellicer, M.: Necessary and sufficient conditions for precompact sets to be metrizable. Bull. Aust. Math. Soc. **74**, 7–13 (2006)
29. Ferrando, J.C., Kąkol, J.: A note on spaces $C_p(X)$ $K$-analytic-framed in $\mathbb{R}^X$. Bull. Aust. Math. Soc. **78**, 141–146 (2008)
30. Ferrando, J.C., Kąkol, J., López-Pellicer, M.: A revised closed graph theorem for quasi-Suslin spaces. Czech. Math. J. **59**, 1115–1122 (2009)
31. Ferrando, J.C.: A weakly analytic space which is not $K$-analytic. Bull. Aust. Math. Soc. **79**, 31–35 (2009)
32. Ferrando, J.C., Kąkol, J., López-Pellicer, M., Śliwa, W.: Web-compact spaces, Fréchet-Urysohn groups and a Suslin type closed graph theorem. Math. Nachr. **283**(5), 704–711 (2010)
33. Ferrando, J.C., Kąkol, J.: On locally convex weakly Lindelöf $\Sigma$-spaces. Bull. Belg. Math. Soc. Simon Stevin **20**(4), 753–763 (2013)
34. Ferrando, J.C., Kąkol, J.: On precompact sets in spaces $C_c(X)$. Geogian Math. J. **20**(2), 247–254 (2013)
35. Ferrando, J.C., Kąkol, J., López-Pellicer, M., Muñoz, M.: Some topological cardinal inequalities for spaces $Cp(X)$. Topology Appl. **160**(10), 1102–1107 (2013)
36. Fremlin, D.H.: $K$-analytic spaces with metrizable compacts. Mathematika **24**, 257–261 (1977)
37. Garling, D.J.H.: A generalized form of inductive-limit topology for vector spaces. Proc. Lond. Math. Soc. **3**(14), 1–28 (1964)
38. Gabriyelyan, S.S., Kąkol, J., Leiderman, A.: The strong Pytkeev property for topological groups and topological vector spaces. (Preprint)
39. Granado, M., Gruenhage, G.F.: Baireness of $C_c(X)$ for ordered $X$. Comm. Math. Univ. Carolin. **47**(1), 103–111 (2006)
40. Granero, S., Hájek, P., Montesinos Santalucía, V.: Convexity and $\omega^*$-compactness in Banach spaces. Math. Ann. **328**, 625–631 (2004)
41. Granero, A.S.: An extension of the Krein-Smulian theorem. Rev. Mat. Iberoam. **22**, 93–110 (2006)
42. Gruenhage, G.F., Ma, D.K.: Baireness of $C_k(X)$ for locally compact $X$. Topology Appl. **80** (1–2), 131–139 (1997)
43. Jarchow, H.: Locally Convex Spaces. B.G. Teubner, Stuttgart (1981)
44. Kąkol, J.: Note on compatible vector topologies. Proc. Am. Math. Soc. **99**(4), 690–692 (1987)
45. Kąkol, J.: Compatible vector topologies and quotients of $F$-spaces. In: Function spaces (Poznań 1986), pp. 164–169. Teubner-Texte Mathematik, vol. 103. Teubner, Leipzig (1988)
46. Kąkol, J.: Note on semi-Fredholm and strictly singular operators acting between $F$-spaces. Math. Nachr. **143**, 69–74 (1989)
47. Kąkol, J.: Simple construction of spaces without the Hanh-Banach extension property. Comm. Math. Univ. Carolinae. **33**(4), 623–624 (1992)
48. Kąkol, J., Śliwa, W.: On the weak topology of Banach spaces over non-archimedian fields. Topology Appl. **158**, 1131–1135 (2011)
49. Kąkol, J., Pérez-García, C., Śliwa, W.: Non-archimedean function spaces and the Lebesgue dominated convergence theorem. Bull. Belg. Math. Soc. Simon Stevin **19**(1), 173–184 (2012)
50. Kąkol, J., Śliwa, W.: Descriptive topology in non-archimedean function spaces, part 1. Bull. Lond. Math. Soc. **44**(5), 899–912 (2012)
51. Kąkol, J., Kubzdela, A.J.: Non-archimedean quantitative Grothendieck and Krein's theorems. Convex Anal. **20**, 233–242 (2013)
52. Kąkol, J., Kubzdela, A., Śliwa, W.: A non-archimedean Dugundji extension theorem. Czech. Math. J. **63**, 157–164 (2013)
53. Kąkol, J., Roelcke, W.: Unordered Baire-like spaces without local convexity. Collect. Math. **43**(1), 43–53 (1992)

54. Kąkol, J., Roelcke, W.: Topological vector spaces with some Baire-type properties. Note di Mat. (special sigue to memory of Prof. Köthe) **12**, 77–92 (1992)
55. Kąkol, J., Roelcke, W.: On the barrelledness of $l^p$-direct sums of seminormed spaces for $1 \leqslant p \leqslant \infty$. Arch. Math. **36**(4), 331–334 (1994)
56. Kąkol, J., Sorjonen, P.: Basic sequences and the Hahn-Banach extension property. Acta Sci. Math. (Szeged), **59**(1–2), 161–171 (1994)
57. Kąkol, J.: Strongly Lindelöf spaces, Baire-type property and sequential closure conditions for inductive limits of metrizable spaces. In: Dierolf, S., Dineen, S., Domanski, P. (eds.) Functional Analysis, pp. 227–239. Walter de Gruyter, Berlin (1996)
58. Kąkol, J., Sánchez Ruiz, L.: A note on the Baire and the ultrabornological property of spaces $C_p(X, E)$. Arch. Math. (Basel) **67**, 493–499 (1996)
59. Kąkol, J.: Report on the book J. C. Ferrando, M. López-Pellicer, L. M. Sanchez-Ruiz, Metrizable barrelled spaces. Pitman Research Notes in Mathematics Series, vol. 332. Longman, New York. Bull. Am. Math. Soc. **34**, 101–105 (1997)
60. Kąkol, J., Gildsdorf, T.E.: On the weak bases theorems for $p$-adic locally convex spaces. In: De Grande-de Kimpe, N., Kąkol, J., Pérez-García, C. (eds.) Proceedings of the Fifth Conference in P-adic Analysis (Poznan 1998), pp. 149–167. Lectures Notes in Pure and Applied Mathematics, vol. 192. Marcel Dekker, New York (1999)
61. Kąkol, J., Gildsdorf, T., Sánchez Ruiz, L.: Baire-likeness of spaces $\ell_\infty(E)$ and $c_0(E)$. Period. Math. Hungar. **40**, 31–35 (2000)
62. Kąkol, J., Śliwa, W.: Strongly Hewitt spaces. Topology Appl. **119**, 219–227 (2002)
63. Kąkol, J., Saxon, S.A.: Montel $(DF)$-spaces, sequential $(LM)$-spaces and the strongest locally convex topology. J. Lond. Math. Soc. **66**(2), 388–406 (2002)
64. Kąkol, J., Saxon, S.A.: Metrizability versus Fréchet-Urysohn property. Proc. Am. Math. Soc. **131**(11), 3623–3631 (2003)
65. Kąkol, J., Saxon, S.A., Todd, A.R.: Pseudocompact spaces $X$ and $(df)$-spaces $C_c(X)$. Proc. Am. Math. Soc. **132**(6), 1703–1712 (2004)
66. Kąkol, J., López-Pellicer, M.: Compact coverings for Baire locally convex spaces. Math. Anal. Appl. **332**, 965–974 (2007)
67. Kąkol, J., López-Pellicer, M., Śliwa, W.: Weakly $K$-analytic spaces and the three-space property for analyticity. J. Math. Anal. Appl. **362**, 90–99 (2010)
68. Kąkol, J., Kubiś, W., López-Pellicer, M.: Descriptive Topology in Selected Topics of Functional Analysis. Developments in Mathematics, vol. 24. Springer, New York (2011)
69. Kąkol, J., López-Pellicer, M., Okunev, O.J.: Compact covers and functions spaces. Math. Anal. Appl. **411**, 372–380 (2014)
70. Kalton, N.J.: Basic sequences in $F$-spaces and their applications. Proc. Edinb. Math. Soc. **19**(2), 151–167 (1974)
71. Kalton, N.J.: Basic sequence problem. Stud. Math. **116**(2), 167–187 (1995)
72. Khan, L.A.: Linear Topological Spaces of Continuous Vector-Valued Functions. Academic, London (2013)
73. Lurje, P.: Tonneliertheit in lokalkonvexen Vektorgruppen. Manuscripta Math. **14**, 107–121 (1974)
74. Lurje, P.: Vollständige und B-vollständige topologische Vektorgruppen. Graphensätze und open-mapping theorems. Manuscripta Math. **15**, 109–137 (1975)
75. Lutzer, D.J., McCoy, R.A.: Category in function spaces. I. Pac. J. Math. **90**(1), 145–168 (1980)
76. Lutzer, D.J., van Mill, J., Pol, R.: Descriptive complexity of function spaces. Trans. Am. Math. Soc. **291**, 121–128 (1985)
77. McCoy, R.A., Todd, A.R.: Support spaces and $P(K)$-points in the Stone-Čech compactification of integers. Southeast Asian Bull. Math. **20**(1), 11–18 (1996)
78. Mazón, J.M.: On the $df$-spaces of continuous functions. Bull. Soc. Roy. Sci. Liège **53**, 225–232 (1984)
79. Mendoza, J.: Necessary and sufficient conditions for $C(X; E)$ to be barrelled or infrabarrelled. Simon Stevin **57**(1–2), 103–123 (1983)
80. Michael, E.A.: Locally $M$-convex algebras. Thesis (Ph.D.), The University of Chicago (1951)

81. Michael, E.A.: Locally multiplicatively-convex topological algebras. Am. Math. Soc. 11 (1952)
82. Novák, J.: On the cartesian product of two compact spaces. Fundam. Math. **40**, 106–112 (1953)
83. Nyikos, P.: Metrizability and the Fréchet-Urysohn property in topological groups. Proc. Am. Math. Soc. **83**, 793–801 (1981)
84. Orihuela, J.: Pointwise compactness in spaces of continuous functions. J. Lond. Math. Soc. **36**, 143–152 (1987)
85. Pérez García, C., Schikhof, W.H.: Locally convex spaces over non-archimedean valued fields. Cambridge Studies in Advanced Mathematics, vol. 119. Cambridge University Press, Cambridge (2010)
86. Pytkeev, E.G.: The Baire property of spaces of continuous functions. Mat. Zametki **38**(5), 726–740, 797 (1985)
87. Saxon, S.A.: Nuclear and product spaces, Baire-like spaces, and the strongest locally convex topology. Math. Ann. **197**, 87–106 (1972)
88. Schmets, J.: Spaces of Vector-Valued Continuous Functions. Lecture Notes in Mathematics, vol. 1003. Springer, Berlin (1983)
89. Shapiro, J.H.: Extension of linear functionals on $F$-spaces with bases. Duke Math. J. **37**, 639–645 (1970)
90. Talagrand, M.: Espaces de Banach faiblement $K$-analytiques. Ann. Math. **110**, 407–438 (1979)
91. Tsaban, B., Zdomskyy, L.: On the Pytkeev property in spaces of continuous functions (II). Houston J. Math. **35**, 563–571 (2009)
92. Tkachuk, V.V., Shakhmatov, D.B.: When is space $C_p(X)$ $\sigma$-countably compact? Mosc. Univ. Math. Bull. **42**, 23–27 (1987)
93. Tkachuk, V.V.: A space $C_p(X)$ is dominated by irrationals if and only if it is K-analytic. Acta Math. Hung. **107**, 253–265 (2005)
94. Valdivia, M.: Topics in Locally Convex Spaces. North-Holland Mathematics Studies, vol. 67. North-Holland, Amsterdam (1982)
95. Valdivia, M.: Quasi-LB-spaces. J. Lond. Math. Soc. **35**, 149–168 (1987)
96. Valov, V.: Function spaces. Topology Appl. **81**, 1–22 (1997)
97. Warner, S., Blair, A.: On symmetry in convex topological vector spaces. Proc. Am. Math. Soc. **6**, 301–304 (1955)
98. Warner, S.: Weak locally multiplicatively-convex algebras. Pac. J. Math. **5**, 1025–1032 (1955)
99. Webb, J.H.: Sequential convergence in locally convex spaces. Proc. Camb. Philos. Soc. **64**, 341–364 (1968)
100. Yoshinaga, K.: On a locally convex space introduced by J. S. e Silva. J. Sci. Hiroshima Univ. Ser. A-I Math. **21**, 89–98 (1957)

## List of Kakol Publications in MathSciNet

**MR3118492** Kąkol, J., López-Pellicer, M., Okunev, O.: Compact covers and function spaces. J. Math. Anal. Appl. **411**(1), 372–380 (2014) (54Cxx)

**MR3129072** Ferrando, J.C., Kąkol, J.: On locally convex weakly Lindelöf $\Sigma$-spaces. Bull. Belg. Math. Soc. Simon Stevin **20**(4), 753–763 (2013) (54D20) (46Axx)

**MR3073666** Angosto, C., Kąkol, J., Kubzdela, A., López-Pellicer, M.: A quantitative version of Krein's theorems for Fréchet spaces. Arch. Math. (Basel) **101**(1), 65–77 (2013) (46A50) (54C35)

**MR3086450** Kąkol, J., Kubzdela, A.: Non-archimedean quantitative Grothendieck and Krein's theorems. J. Convex Anal. **20**(1), 233–242 (2013) (46S10)

**MR3063343** Ferrando, J.C., Kąkol, J.: On precompact sets in spaces $C_c(X)$. Georgian Math. J. **20**(2), 247–254 (2013) (54D35)

**MR3056000** Ferrando, J.C., Kąkol, J., López-Pellicer, M., Muñoz, M.: Some topological cardinal inequalities for spaces $C_p(X)$. Topology Appl. **160**(10), 1102–1107 (2013) (54A25) (54C35)
**MR3035503** Kąkol, J., Kubzdela, A., Śliwa, W.: A non-archimedean Dugundji extension theorem. Czech. Math. J. **63**(138), no. 1, 157–164 (2013) (46S10) (54C35)
**MR3035068** Angosto, C., Kąkol, J., López-Pellicer, M.: A quantitative approach to weak compactness in Fréchet spaces and spaces $C(X)$. J. Math. Anal. Appl. **403**(1), 13–22 (2013) (46A50) (46E15)
**MR2975150** Kąkol, J., Śliwa, W.: Descriptive topology in non-archimedean function spaces $C_p(X, K)$. Part I. Bull. Lond. Math. Soc. **44**(5), 899–912 (2012) (46S10) (46A50 46E10 54C35 54H05)
**MR2952804** Kąkol, J., Perez-Garcia, C., Śliwa, W.: Non-archimedean function spaces and the Lebesgue dominated convergence theorem. Bull. Belg. Math. Soc. Simon Stevin **19**(1), 173–184 (2012) (46S10) (54C35)
**MR2862316** Kąkol, J., López-Pellicer, M.: Note about Lindelöf $\Sigma$-spaces $\upsilon X$. Bull. Aust. Math. Soc. **85**(1), 114–120 (2012) (54H05) (28A05 46A50)
**MR2953769** Kąkol, J., Kubiś, W., López-Pellicer, M.: Descriptive topology in selected topics of functional analysis. Developments in Mathematics, vol. 24. Springer, New York (2011) (xii+493 pp. 46–02, ISBN: 978-1-4614-0528-3) (54-02)
**MR2841188** Ferrer, J., Kąkol, J., López-Pellicer, M., Wójtowicz, M.: On a three-space property for Lindelöf $\Sigma$-spaces, $(WCG)$-spaces and the Sobczyk property. Funct. Approx. Comment. Math. **44**(2), 289–306 (2011) (46A30) (46A03)
**MR2794468** Kąkol, J., Śliwa, W.: On the weak topology of Banach spaces over non-archimedean fields. Topology Appl. **158**(9), 1131–1135 (2011) (46S10) (46B26 54C35)
**MR2783795** Kąkol, J., López-Pellicer, M.: On realcompact topological vector spaces. Rev. R. Acad. Cienc. Exactas Fís. Nat. Ser. A Math. RACSAM **105**(1), 39–70 (2011) (46A04) (46A50 54H05)
**MR2666299** Ferrando, J.C., Kąkol, J., López-Pellicer, M., Śliwa, W.: Web-compact spaces, Fréchet-Urysohn groups and a Suslin closed graph theorem. Math. Nachr. **283**(5), 704–711 (2010) (46A30) (46A08)
**MR2557670** Kąkol, J., López-Pellicer, M., Śliwa, W.: Weakly K-analytic spaces and the three-space property for analyticity. J. Math. Anal. Appl. **362**(1), 90–99 (2010) (46A50) (46A20)
**MR2563582** Ferrando, J.C., Kąkol, J., Lopez-Pellicer, M.: A revised closed graph theorem for quasi-Suslin spaces. Czech. Math. J. **59**(134), no. 4, 1115–1122 (2009) (46A30) (54C05 54D10)
**MR2541044** Kąkol, J., López-Pellicer, M., Todd, A.R.: A topological vector space is Fréchet-Urysohn if and only if it has bounded tightness. Bull. Belg. Math. Soc. Simon Stevin **16**(2), 313–317 (2009) (46A50) (46A30 54C35)
**MR2535274** Kąkol, J., López-Pellicer, M.: About an example of a Banach space not weakly $K$-analytic. Rev. R. Acad. Cienc. Exactas Fís. Nat. Ser. A Math. RACSAM **103**(1), 87–90 (2009) (46B99) (46A03 46E15 54C30 54H05)
**MR2667152** Kąkol, J., Martín-Peinador, E., Moll, S.: A characterization of K-analyticity of groups of continuous homomorphisms. Bol. Soc. Mat. Mexicana (3) **14**(1), 15–19 (2008) (22B05) (54H11)
**MR2546832** Kąkol, J., Śliwa, W.: On metrizability of compactoid sets in non-archimedean locally convex spaces. Indag. Math. (N.S.) **19**(4), 563–578 (2008) (46S10) (46A13)
**MR2458306** Ferrando, J.C., Kąkol, J.: A note on spaces $C_p(X)$ $K$-analytic-framed in $R^X$. Bull. Aust. Math. Soc. **78**(1), 141–146 (2008) (54C35) (46B99 46E15 54H05)
**MR2394919** Kąkol, J., López-Pellicer, M., Martín-Peinador, E., Tarieladze, V.: Lindelöf spaces $C(X)$ over topological groups. Forum Math. **20**(2), 201–212 (2008) (54H11) (46E10 54C35)
**MR2377082** Ferrando, J.C., Kąkol, J., López-Pellicer, M., Saxon, S.A.: Quasi-Suslin weak duals. J. Math. Anal. Appl. **339**(2), 1253–1263 (2008) (46A20) (46A03 46A08)

**MR2667463** Kąkol, J., López-Pellicer, M.: A note on Fréchet-Urysohn locally convex spaces. RACSAM. Rev. R. Acad. Cienc. Exactas Fís. Nat. Ser. A Mat. **101**(2), 127–131 (2007) (46A03) (46A04 54E40)
**MR2387046** Ferrando, J.C., Kąkol, J., López-Pellicer, M.: A characterization of trans-separable spaces. Bull. Belg. Math. Soc. Simon Stevin **14**(3), 493–498 (2007) (54E15) (46A50 54C35)
**MR2324313** Kąkol, J., López-Pellicer, M.: Compact coverings for Baire locally convex spaces. J. Math. Anal. Appl. **332**(2), 965–974 (2007) (46A04) (54C35)
**MR2274833** Ferrando, J.C., Kąkol, J., López-Pellicer, M.: A characterization of distinguished Fréchet spaces. Math. Nachr. **279**(16), 1783–1786 (2006) (46A04) (46A30 54C35)
**MR2265087** Ferrando, J.C., Kąkol, J., López-Pellicer, M., Saxon, S.A.: Tightness and distinguished Fréchet spaces. J. Math. Anal. Appl. **324**(2), 862–881 (2006) 46A04 (46A03 46A08 46E10)
**MR2248802** Ferrando, J.C., Kąkol, J., López-Pellicer, M.: Necessary and sufficient conditions for precompact sets to be metrisable. Bull. Aust. Math. Soc. **74**(1), 7–13 (2006) (46A50) (54C35)
**MR2216100** Ferrando, J.C., Kąkol, J., López-Pellicer, M.: Metrizability of precompact sets: an elementary proof. RACSAM. Rev. R. Acad. Cienc. Exactas Fís. Nat. Ser. A Mat. **99**(2), 183–185 (2005) (46A50) (46A03)
**MR2097264** Kąkol, J., Saxon, S.A., Todd, A.R.: The analysis of Warner boundedness. Proc. Edinb. Math. Soc. (2) **47**(3), 625–631 (2004) (46A08) (46A30 46E10)
**MR2088677** Ferrando, J.C., Kąkol, J., López-Pellicer, M.: Bounded tightness for locally convex spaces and spaces $C(X)$. Special issue dedicated to John Horváth. J. Math. Anal. Appl. **297**(2), 518–526 (2004) (Reviewer: Matias Raja) (46A03) (46E10 54F65)
**MR2088675** Kąkol, J., Saxon, S.A., Todd, A.R.: Weak barrelledness for $C(X)$ spaces. Special issue dedicated to John Horváth. J. Math. Anal. Appl. **297**(2), 495–505 (2004) (46E10) (46A08)
**MR2088033** Kąkol, J., López-Pellicer, M., Moll, S.: Banach disks and barrelledness properties of metrizable $c_0(\Omega, X)$. Mediterr. J. Math. **1**(1), 81–91 (2004) (46A08) (46B25 46E40)
**MR2059181** Ferrando, J.C., Kąkol, J., López-Pellicer, M.: On a problem of Horváth concerning barrelled spaces of vector valued continuous functions vanishing at infinity. Bull. Belg. Math. Soc. Simon Stevin **11**(1), 127–132 (2004) (46E40) (46A08 46B25)
**MR2051131** Kąkol, J., Saxon, S.A., Todd, A.R.: Pseudocompact spaces $X$ and $df$-spaces $C_c(X)$. Proc. Am. Math. Soc. **132**(6), 1703–1712 (2004) (46A08) (46E10 54C35)
**MR2040017** Kąkol, J., Saxon, S.A., Todd, A.R.: Docile locally convex spaces. Topological algebras and their applications, pp. 73–78. Contemporary Mathematics, vol. 341. American Mathematical Society, Providence (2004) (46A03) (03E10 46A08 46E10 54C35)
**MR1997755** Kąkol, J., Saxon, S.A.: The Fréchet-Urysohn property, $(LM)$-spaces and the strongest locally convex topology. Math. Proc. R. Ir. Acad. **103A**(1), 1–8 (2003) (46A03)
**MR1991777** Cascales, B., Kąkol, J., Saxon, S.A.: Metrizability versus Fréchet-Urysohn property. Proc. Am. Math. Soc. **131**(11), 3623–3631 (2003) (electronic) (54D10) (46A50 54E15 54E35)
**MR1977440** De Grande-De Kimpe, N., Kąkol, J., Perez-Garcia, C.: Metrizability of compactoid sets in non-archimedean Hausdorff $(LM)$-spaces. Ultrametric functional analysis (Nijmegen 2002), pp. 99–107. Contemporary Mathematics, vol. 319. American Mathematical Society, Providence (2003) (46S10) (46A13)
**MR1972199** Kąkol, J., López-Pellicer, M.: On countable bounded tightness for spaces $C_p(X)$. J. Math. Anal. Appl. **280**(1), 155–162 (2003) (54C35)
**MR2232643** Kąkol, J., Tweddle, I.: Spaces of continuous functions $C_p(X, E)$ as $(LM)$-spaces. Bull. Belg. Math. Soc. Simon Stevin **9**(Suppl.) 109–116, (2002) (46A13) (46A16 46E40)
**MR2232642** Kąkol, J., Saxon, S.A.: The Fréchet-Urysohn view of weak and $s$-barrelledness. Bull. Belg. Math. Soc. Simon Stevin **9**(Suppl.) 101–108, (2002) (46A30) (46A08 54C35)
**MR1934373** De Grande-De Kimpe, N., Kąkol, J., Perez-Garcia, C., Schikhof, W.H.: Weak bases in $p$-adic spaces. Boll. Unione Mat. Ital. Sez. B Artic. Ric. Mat. (8) **5**(3), 667–676 (2002) (46S10) (46A35)

**MR1920410** Kąkol, J., Saxon, S.A.: Montel (*DF*)-spaces, sequential (*LM*)-spaces and the strongest locally convex topology. J. Lond. Math. Soc. (2) **66**(2), 388–406 (2002) (46A50) (46A04 46A08 46A11 46A13)
**MR1907128** Cascales, B., Kąkol, J., Saxon, S.A.: Weight of precompact subsets and tightness. J. Math. Anal. Appl. **269**(2), 500–518 (2002) (54A25) (46A04 54C60)
**MR1886096** Kąkol, J.: Śliwa, Wiesław Strongly Hewitt spaces. Topology Appl. **119**(2), 219–227 (2002) (54C35) (46E40)
**MR1838285** De Grande-De Kimpe, N., Kąkol, J., Perez-Garcia, C., Schikhof, W.H.: Orthogonal and Schauder bases in non-archimedean locally convex spaces. *P*-adic functional analysis (Ioannina 2000), pp. 103–126. Lecture Notes in Pure and Applied Mathematics, vol. 222. Dekker, New York (2001) (46A35) (46S10)
**MR1912762** Kąkol, J.: Remarks on linear homeomorphisms of function spaces. Acta Math. Hung. **89**(4), 315–319 (2000) (46E10) (54C35)
**MR1822118** Kąkol, J.: Baire-like spaces $C(X, E)$. Rev. Mat. Complut. **13**(2), 341–347 (2000) (46E40) (46A45)
**MR1813159** De Grande-De Kimpe, N., Kąkol, J., Perez-Garcia, C., Schikhof, W.: Orthogonal sequences in non-archimedean locally convex spaces. Indag. Math. (N.S.) **11**(2), 187–195 (2000) (46S10) (46A35)
**MR1772130** Kąkol, J.: Baire property and related concepts for spaces $C(X, E)$. Function spaces (Poznań 1998), pp. 255–268. Lecture Notes in Pure and Applied Mathematics, vol. 213. Dekker, New York (2000) (46E40) (46A08 46E10 54C35)
**MR1774932** Kąkol, J., Gilsdorf, T., Sánchez Ruiz, L.: Baire-likeness of spaces $l_\infty(E)$ and $c_0(E)$. Period. Math. Hung. **40**(1), 31–35 (2000) (46A08) (54E52)
**MR1773194** Kąkol, J., Saxon, S.A.: The semi-Baire-like three-space problems. Adv. Appl. Math. **25**(1), 57–64 (2000) (46A08)
**MR1702053** Kąkol, J., Gilsdorf, T.: On the weak bases theorems for $p$-adic locally convex spaces. *P*-adic functional analysis (Poznań 1998), pp. 149–165. Lecture Notes in Pure and Applied Mathematics, vol. 207. Dekker, New York (1999) (46S10) (46A35)
**MR1668786** Kąkol, J.: Baire-Type Properties of Spaces of Continuous Functions. Functional Analysis, pp. 9–22. Narosa, New Delhi (1998) (46E10) (46A03 46E40)
**MR1602398** Kąkol, J., Perez-Garcia, C., Śliwa, W.: Remarks on quasi-Banach spaces over valued fields. Dedicated to Roman Taberski on the occasion of his 70th birthday. Funct. Approx. Comment. Math. 25, 173–179 (1997) (46S10) (46A20 46A22)
**MR1489873** De Grande-De Kimpe, N., Kąkol, J., Pérez-Garcia, C.: On compactoids in (*LB*)-spaces. Bull. Polish Acad. Sci. Math. **45**(4), 313–321 (1997) (46S10) (46A13)
**MR1459211** De Grande-De Kimpe, N., Kąkol, J., Perez-Garcia, C., Schikhof, W.H.: *P*-adic locally convex inductive limits. *P*-adic functional analysis (Nijmegen 1996), pp. 159–222. Lecture Notes in Pure and Applied Mathematics, vol. 192. Dekker, New York (1997) (46S10) (46A13)
**MR1459210** Gilsdorf, T.E., Kąkol, J.: On some non-archimedean closed graph theorems. *P*-adic functional analysis (Nijmegen 1996), pp. 153–158. Lecture Notes in Pure and Applied Mathematics, vol. 192. Dekker, New York (1997) (46S10) (46A30)
**MR1440245** Kąkol, J.: On Baire-like spaces of spaces of continuous vector-valued functions. Acta Math. Hung. **74**(3), 203–209 (1997) (46E40) (46A08)
**MR1420451** Kąkol, J.: Strongly Lindelöf spaces, Baire-type property and sequential closure conditions for inductive limits of metrizable spaces. Functional Analysis (Trier 1994), pp. 227–239. de Gruyter, Berlin (1996) (46A13)
**MR1418911** Kąkol, J., Sánchez Ruiz, L.: A note on the Baire and ultrabornological property of spaces $C_p(X, E)$. Arch. Math. (Basel) **67**(6), 493–499 (1996) (46E40) (46A08 46E10)
**MR1416418** Kąkol, J., Perez-Garcia, C., Schikhof, W.: Cardinality and Mackey topologies of non-archimedean Banach and Fréchet spaces. Bull. Polish Acad. Sci. Math. **44**(2), 131–141 (1996) (46S10) (46A03)

**MR1357524** Kąkol, J.: Remarks on bounded sets in (*LF*) topological vectors spaces. Comm. Math. Univ. Carolin. **36**(2), 239–244 (1995) (46A16) (46A13)
**MR1345353** Dierolf, S., Kąkol, J.: On *S*-barrelled spaces. Results Math. **28**(1–2), 40–48 (1995) (46A08) (46E10)
**MR1342811** Kąkol, J.: The Mackey-Arens and Hahn-Banach theorems for spaces over valued fields. In: Proceedings of the 3rd International Conference on p-adic Functional Analysis (Aubière 1994). Ann. Math. Blaise Pascal **2**(1), 147–153 (1995) (46S10) (46A03)
**MR1810694** Indexed Kąkol, J.: The Mackey-Arens property for spaces over valued fields. Bull. Polish Acad. Sci. Math. **42**(2), 97–101 (1994) (46S10) (46A32 46B28)
**MR1626419** Abul-Ez, M., Kąkol, J.: Mode of increase of power similar sets of polynomials. Bull. Fac. Sci. Assiut Univ. C **23**(2), 55–65 (1994) (30C10) (30A10 30D15)
**MR1457218** Kąkol, J.: A strong barrelledness property for spaces $C(X, E)$. Note Mat. **14**(2), 199–202 (1994, 1997) (46E40) (46A08)
**MR1336293** Kąkol, J., Śliwa, W., Wójtowicz, M.: *CS*-barrelled spaces. Collect. Math. **45**(3), 271–276 (1994) (46A08)
**MR1313172** Kąkol, J.: The weak and strong convergence in metrizable spaces over valued fields. Portugal. Math. **51**(4), 613–615 (1994) (46S10) (46A35)
**MR1311875** Kąkol, J.: Hahn-Banach extension property for Banach spaces over non-archimedean fields. Period. Math. Hung. **29**(2), 159–162 (1994) (46S10) (46A22 46B20)
**MR1298778** Kąkol, J.: Remarks on spherical completeness of non-archimedean valued fields. Indag. Math. (N.S.) **5**(3), 321–323 (1994) (46S10) (12J25)
**MR1285437** Kąkol, J., Sorjonen, P.: Basic sequences and the Hahn-Banach extension property. Acta Sci. Math. (Szeged) **59**(1–2), 161–171 (1994) (46A16) (46A03 46A35)
**MR1264704** Kąkol, J., Roelcke, W.: On the barrelledness of $l^p$-direct sums of seminormed spaces for $1 \leq p \leq \infty$. Arch. Math. (Basel) **62**(4), 331–334 (1994) (46A08) (46A13 46A16)
**MR1366717** Abul-Ez, M.A., Kąkol, J.: Mode of increase of power similar set of polynomials. Carib. J. Math. Comput. Sci. **3**(1–2), 27–33 (1993) (30E10) (30D15 41A10)
**MR1324070** Kąkol, J., Śliwa, W.: Remarks concerning the separable quotient problem. Note Mat. **13**(2), 277–282 (1993) (46A16)
**MR1314928** Kąkol, J.: The weak bases theorem for *K*-Banach spaces. Bull. Soc. Math. Belg. Sér. B **45**(1), 1–4 (1993) (46S10) (46B15)
**MR1283356** Kąkol, J.: Remarks on regular (*LF*)-spaces. Rend. Circ. Mat. Palermo (2) **42**(3), 453–458 (1993, 1994) (46A13) (46A16 46A35)
**MR1240205** Kąkol, J.: A note on a theorem of Klee. Comm. Math. Univ. Carolin. **34**(1), 79–80 (1993) (46A20) (46B10)
**MR1201642** Kąkol, J.: Sequential closure conditions and Baire-like spaces. Arch. Math. (Basel) **60**(3), 277–282 (1993) (46A08)
**MR1258565** Kąkol, J., Roelcke, W.: Topological vector spaces with some Baire-type properties. Dedicated to the memory of Professor Gottfried Köthe. Note Mat. 12, 77–92 (1992) (46A08) (46A03 46A30)
**MR1240183** Kąkol, J.: Simple construction of spaces without the Hahn-Banach extension property. Comm. Math. Univ. Carolin. **33**(4), 623–624 (1992) (46A22)
**MR1233048** Kąkol, J.: Sequentially barrelled spaces and the strongest locally convex topology. Math. Nachr. 157, 77–80 (1992) (46A08)
**MR1214222** Kąkol, J., Roelcke, W.: Unordered Baire-like spaces without local convexity. Collect. Math. **43**(1), 43–53 (1992) (46A16)
**MR1118287** Kąkol, J., Sorjonen, P.: Locally convex topologies in linear orthogonality spaces. Comm. Math. Univ. Carolin. **32**(1), 33–37 (1991) (46A16) (46C99)
**MR1094425** Kąkol, J.: Remarks and examples concerning the weak basis theorem. Arch. Math. (Basel) **56**(4), 376–379 (1991) (46A35) (46A04)

**MR1165080** Kąkol, J.: Ultrabarrelled spaces and dense vector subspaces. Rev. Real Acad. Cienc. Exact. Fís. Natur. Madrid **84**(3), 435–440 (1990) (46A16)
**MR1149649** Kąkol, J.: The Mackey-Arens theorem for non-locally convex spaces. Collect. Math. **41**(2), 129–132 (1990, 1991) (46A16)
**MR1100776** Kąkol, J.: On some dense compact operator range subspaces in separable Fréchet spaces. Acta Math. Hung. **56**(1–2), 53–55 (1990) (46A04) (47A05)
**MR1022750** Kąkol, J.: On the weak basis theorem. Bull. Soc. Math. Belg. Sér. B **41**(2), 247–260 (1989) (46A35)
**MR1018233** Kąkol, J.: Note on semi-Fredholm and strictly singular operators acting between $F$-spaces. Math. Nachr. 143, 69–74 (1989) (47A53) (47A55)
**MR1066531** Kąkol, J.: Compatible vector topologies and quotients of $F$-spaces. Function spaces (Poznań 1986), pp. 164–169. Teubner-Texte Mathematics, vol. 103. Teubner, Leipzig (1988) (46A16) (46A20)
**MR0962609** Kąkol, J.: On the cardinal number of some families of separable topological vector spaces. Rev. Roumaine Math. Pures Appl. **33**(7), 595–599 (1988) (46A06)
**MR0944650** Kąkol, J.: Basic sequences and nonlocally convex topological vector spaces. Rend. Circ. Mat. Palermo (2) **36**(1), 95–102 (1987) (46A15) (46A06)
**MR0937732** Kąkol, J.: On some dense dual-separating subspaces in $F$-spaces. Bull. Polish Acad. Sci. Math. **35**(9–10), 557–560 (1987) (46A15) (46B20)
**MR0877041** Kąkol, J.: Note on compatible vector topologies. Proc. Am. Math. Soc. **99**(4), 690–692 (1987) (46A15)
**MR0880146** Kąkol, J.: A note on dense subspaces in separable Fréchet spaces. Funct. Approx. Comment. Math. 15, 167–170 (1986) (46A07) (46A06)
**MR0874887** Kąkol, J.: A note on dual-separating $F$-spaces. Bull. Polish Acad. Sci. Math. **34**(7–8), 423–426 (1986) (46A06)
**MR0866623** Kąkol, J.: Some remarks on $^*$-Baire-like and on $b$-$^*$-Baire-like spaces. Math. Slovaca **36**(3), 233–239 (1986) (46A15) (46A05)
**MR0850431** Kąkol, J.: On nonlocally convex and on dual-less separable topological vector spaces. Simon Stevin **60**(1), 15–27 (1986) (46A15)
**MR0847764** Kąkol, J.: On suprema of metrizable vector topologies with trivial dual. Czech. Math. J. **36**(111), no. 3, 343–350 (1986) (46A15) (46A06 46A20)
**MR0847357** Kąkol, J.: On bounded multiplier summable sequences in topological vector spaces. Math. Nachr. **125**, 175–178 (1986) (46A05)
**MR0832715** Kąkol, J.: On countable codimensional subspaces in ultra-$(DF)$ spaces. Acta Math. Hung. **46**(3–4), 239–242 (1985) (46A15)
**MR0821575** Kąkol, J.: Nonlocally convex spaces and the Hahn-Banach extension property. Bull. Polish Acad. Sci. Math. **33**(7–8), 381–393 (1985) (46A15)
**MR0808330** Kąkol, J.: On subspaces of $b$-spaces in the sense of Noureddine. Math. Nachr. **120**, 43–51 (1985) (46A05) (46A22)
**MR0817345** Kąkol, J.: On dense subspaces of products of semibornological topological linear spaces. Funct. Approx. Comment. Math. **14**, 81–84 (1984) (46A09)
**MR0768811** Kąkol, J.: On subspaces of ultrabornological spaces. Comm. Math. Univ. Carolin. **25**(2), 247–256 (1984) (46A09) (46A07)
**MR0745337** Kąkol, J.: Countable codimensional subspaces of spaces with topologies determined by a family of balanced sets. Comment. Math. Prace Mat. **23**(2), 249–257 (1983) (46A09)
**MR0708597** Kąkol, J.: On subspaces and products of some topological linear spaces. Arch. Math. (Basel) **40**(4), 355–360 (1983) (46A09) (46A15)
**MR0721274** Kąkol, J.: On minimal inductive spaces. Math. Nachr. **110**, 195–203 (1983) (46A12) (46A30)
**MR0702324** Kąkol, J.: Some remarks on subspaces and products of ultrabornological spaces. Simon Stevin **57**(1–2), 83–97 (1983) (Reviewer: N. Adasch) (46A07) (46A09)

**MR0817328** Kąkol, J.: Topological linear spaces with some Baire-like properties. Funct. Approx. Comment. Math. **13**, 109–116 (1982) (46A15) (46A07)
**MR0679386** Kąkol, J.: On inductive limits of topological algebras. Colloq. Math. **47**(1), 71–78 (1982) (46H05) (46A12 46A15)
**MR0675763** Kąkol, J.: A note on functional representation of topological algebras equipped with a generalized inductive limit topology. Math. Nachr. **106**, 301–307 (1982) (46H15) (46A15)
**MR0664515** Kąkol, J.: A note on metrizable linear topologies strictly finer than a given metrizable complete linear topology. Math. Z. **180**(3), 287–290 (1982) (46A06)
**MR0646938** Kąkol, J.: Some remarks on countable codimensional subspaces of topological vector spaces. Rev. Roum. Math. Pures Appl. **26**(9), 1197–1202 (1981) (46A07)
**MR0641439** Kąkol, J.: Some remarks on Baire-like spaces. Comment. Math. Prace Mat. **22**(2), 257–262 (1980/81) (46A05)
**MR0616972** Kąkol, J.: Non-locally convex Baire-like, *b*-Baire-like spaces and spaces with generalized inductive limit topology. Rev. Roumaine Math. Pures Appl. **25**(10), 1523–1530 (1980) (46A15)
**MR0594128** Alexiewicz, A., Kąkol, J.: Garling algebras. Funct. Approx. Comment. Math. **8**, 111–118 (1980) (46H05) (46M40)
**MR0473773** Jurkiewicz, L., Kąkol, J.: Some remarks on Wiweger topology and generalization of inductive limit topology. Funct. Approx. Comment. Math. **5**, 179–182 (1977) (46A15) (46M10)

# Chapter 2
# Weak Barrelledness Versus P-spaces

**Stephen A. Saxon**

**Abstract** With Ferrando and Kąkol: *X is a P-space if and only if every bounded sequence in* $C_p(X)$ *is relatively compact, if and only if every pointwise eventually zero sequence in* $C_p(X)$ *is summable*. From these two extreme characterizations, already implicitly known, flow the better known ones. Weak barrelledness (with Sánchez Ruiz), briefly previewed here, affords clear motivation and prompted an open question herewith answered: *The closed absolutely convex hull of every bounded countable union of absolutely convex compact sets in* $C_p(X)$ *is compact if and only if every bounded set in* $C_p(X)$ *is relatively compact (if and only if X is discrete).*

**Keywords** Weak barrelledness · Relatively compact · P-spaces $X$ · $C_p(X)$

**MSC 2010 Classification:** 46A08, 54C35

Much of this paper synthesizes joint work with Professors Ferrando, Kąkol and Sánchez Ruiz [2, 3, 8]. Its main contribution answers the question of [2] by showing that, in the class of $L_m(X)$ spaces, the 16 distinct notions of weak barrelledness studied in [8] coalesce into just two.

We recall some definitions [8]. A *barrel* in a locally convex space (lcs) $E$ is an absorbing absolutely convex closed set; an $\aleph_0$-*barrel* is a barrel $U$ that is the intersection of a sequence $\{U_n\}_n$ of absolutely convex closed neighborhoods of the origin; if each member of $E$ is in all but finitely many of the $U_n$, then $U$ is also a $C$-*barrel*. An increasing sequence $\sigma = \{A_n\}_n$ of absolutely convex sets in a space $(E, \mathscr{T})$ is a ⟨*closed*⟩ *absorbing sequence* if $\bigcup_n A_n$ is absorbing ⟨and each $A_n$ is closed⟩; $\mathscr{T}_\sigma$ denotes the finest locally convex topology on $E$ that coincides with $\mathscr{T}$ on each $A_n$. A space $E$ is *semi-Baire-like* (sBL) if some member of each closed absorbing sequence is an absorbing set in $E$.

S. A. Saxon (✉)
Department of Mathematics, University of Florida, Gainesville, FL 32611-8105, USA
e-mail: stephen_saxon@yahoo.com

J. C. Ferrando and M. López-Pellicer (eds.), *Descriptive Topology and Functional Analysis*, Springer Proceedings in Mathematics & Statistics 80, DOI: 10.1007/978-3-319-05224-3_2,

© Springer International Publishing Switzerland 2014

A lcs $E$ is:

- *barrelled* if every barrel is a 0-neighborhood;
- $\langle\aleph_0\text{-}\rangle$ $(C\text{-})$*barrelled* if every $\langle\aleph_0\text{-}\rangle$ $(C\text{-})$barrel is a 0-neighborhood;
- $S_\sigma$ if there is an absorbing sequence of proper closed subspaces;
- *quasi-Baire* (QB) if it is barrelled and non-$S_\sigma$;
- *Baire-like* (BL) if it is barrelled and sBL;
- $\ell^\infty$*-barrelled* if every $\sigma\left(E', E\right)$-bounded sequence is equicontinuous;
- $c_0$*-barrelled* if every $\sigma\left(E', E\right)$-null sequence is equicontinuous;
- $mc_0$*-barrelled* if $\{a_n f_n\}_n$ is equicontinuous whenever $\{f_n\}_n$ is $\sigma\left(E', E\right)$-bounded and $\{a_n\}_n$ is a null scalar sequence;
- *dual locally complete* if $\left(E', \sigma\left(E', E\right)\right)$ is locally complete (cf. [6]);
- *inductive* if it is the inductive limit of every absorbing sequence of subspaces;
- $\varphi$*-complemental* if, for each closed $\aleph_0$-codimensional subspace $F$ of $E$, each algebraic complement of $F$ is a topological complement isomorphic to $\varphi$, an $\aleph_0$-dimensional space with its strongest locally convex topology;
- *primitive* if a linear form is continuous whenever it has continuous restrictions to each member of some absorbing sequence of subspaces.

A lcs $(E, \mathscr{T})$ has:

- *property (L)* if $\mathscr{T} = \mathscr{T}_\sigma$ for each absorbing sequence $\sigma$;
- *property (C)* if $\sigma\left(E', E\right)$-bounded sequences have $\sigma\left(E', E\right)$-accumulation points;
- *property (S)* if $\left(E', \sigma\left(E', E\right)\right)$ is sequentially complete.

All 16 notions are distinct, with interrelationships indicated in Fig. 2.1, taken from [8]. Our older picture had one awkward arrow pointing to the left, instead of down or to the right, and omitted $mc_0$*-barrelled* and $\varphi$*-complemental*, two previously unnamed but historically useful notions that nudge *dlc* and *primitive* one step to the right, while raising *property (S)* one level higher.

Within the class of metrizable lcs's, the notions in box (3) are equivalent, as are those in box (5). Thus metrizability reduces the 16 notions to just two distinct ones. Similarly, boxes (2) and (4) indicate notions that are equivalent for Mackey spaces; box (1), for separable lcs's. There can be no arrow from $c_0$*-barrelled* to *property (S)*: Every Mackey dlc space without property (S) [e.g., $\left(c, \mu\left(c, \ell^1\right)\right)$] is $c_0$-barrelled [7, Sect. 3.3], as indicated by box (2). (And, via Valdivia [10], has a *non* Mackey dense hyperplane; cf. [5].)

A completely regular Hausdorff space $X$ is a *P-space* if every countable intersection of open sets is open. $C_p(X)$ is the space $C(X)$ of continuous real-valued functions on $X$ endowed with the topology of pointwise convergence, and $L_m(X)$ denotes the dual $L(X)$ of $C_p(X)$ endowed with the Mackey topology $\mu(L(X), C(X))$. Routinely, $C_p(X)$ is a dense subspace of the product $\mathbb{R}^X$, $X$ is identified with a Hamel basis for $L(X)$, and $C_p(X)$ is the weak dual of $L_m(X)$. A sequence $\{f_n\}_n$ in $C(X)$ is *pointwise eventually zero* if for each point $x$ in $X$, the set $\{n \in \mathbb{N} : f_n(x) \neq 0\}$ is finite. Clearly, $\{f_n\}_n$ is eventually zero at the points of $X$ if and only if, when considered a sequence in the dual of $L_m(X)$, it is eventually zero at all points of $L(X)$. We saw in [8, Theorem 3.4] that a lcs $E$ is primitive if and only if each pointwise

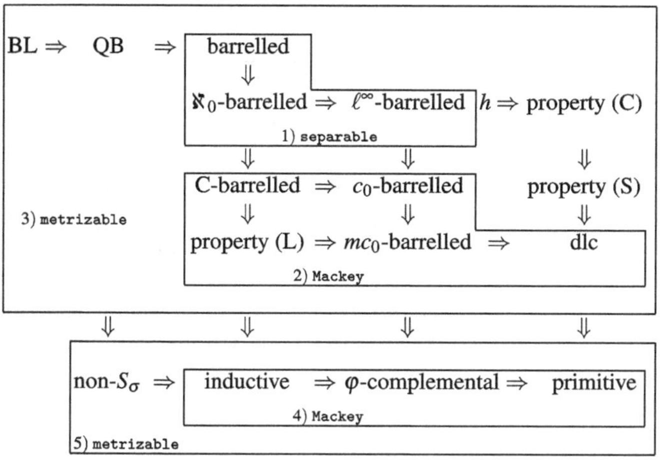


**Fig. 2.1** Relationships between 16 classes of lcs (from [8])

eventually zero sequence in $(E', \sigma(E', E))$ is summable. It follows that $L_m(X)$ is primitive if and only if every pointwise eventually zero sequence in $C_p(X)$ is summable. In [2, Theorem 6] we proved that $L_m(X)$ is primitive if and only if it is $\ell^\infty$-barrelled, if and only if $X$ is a P-space. Therefore, (i) *$X$ is a P-space if and only if every pointwise eventually zero sequence in $C_p(X)$ is summable.*

At the other extreme, a Mackey space $E$ is $\ell^\infty$-barrelled if and only if $S^{\circ\circ}$ is $\sigma(E', E)$-compact whenever $S$ is a $\sigma(E', E)$ -bounded sequence. This is equivalent to saying that every such sequence $S$ is, itself, relatively weakly compact, since $S^{\circ\circ}$ is the weak closure of a countable set of suitable rational linear combinations of points in $S$. Therefore, (ii) *$X$ is a P-space if and only if every bounded sequence in $C_p(X)$ is relatively compact.*

The beautifully simple statements (i) and (ii) do not require any knowledge of the weak barrelledness that motivated them. Now $\ell^\infty$-barrelled ⇒ property (C) ⇒ property (S) ⇒ dlc ⇒ primitive, as Fig. 2.1 indicates at a glance, and all these notions are defined in terms of the dual, so we have a corresponding scale of simple characterizations in terms of $C_p(X)$ which we reproduce from [3].

**Theorem 2.1** *The following six statements are equivalent:*

1. *$X$ is a P-space.*
2. *Every bounded sequence in $C_p(X)$ is relatively compact.*
3. *Every bounded set in $C_p(X)$ is relatively countably compact.*
4. *$C_p(X)$ is sequentially complete.*
5. *$C_p(X)$ is locally complete.*
6. *Every pointwise eventually zero sequence in $C_p(X)$ is summable.*

None of the experts we consulted could find characterizations (2) and (6) explicitly stated in the literature. However, as we detail in [3], characterizations (3, 4, 5) do appear explicitly, and a careful reading of the proof of [1, Sect. 2.2] suggests all of Theorem 2.1 could be credited to Buchwalter and Schmets. Nevertheless, our proof is more direct and simple.

Box (4) of Fig. 2.1 suggests possibly new characterizations associated with *inductive* which we give in [3].

Let us examine box (2), as well. We shall say that a sequence $\{A_n\}_n$ of non-empty subsets of a lcs $E$ is [*bounded*] ⟨*null*⟩ [if the union $\bigcup_n A_n$ is bounded in $E$] ⟨ if, given any neighborhood $U$ of the origin in $E$, all but finitely many $A_n$ are contained in $U$⟩. From [7, Sect. 3.1] we have:

(*) *If $\{A_n\}_n$ is a null sequence of sets in E and the absolutely convex closed hull $\overline{\text{acx}} A_n$ of each $A_n$ is compact, then $\overline{\text{acx}}\left(\bigcup_n A_n\right)$ is compact if and only if $\sum_n \lambda_n x_n$ converges in E whenever $x_n \in \overline{\text{acx}} A_n$ for each n and $(\lambda_n)_n$ is a point in the unit ball B of the Banach space $\ell^1$.*

**Theorem 2.2** *The following three statements are equivalent:*

($\alpha$) *X is a P-space.*
($\beta$) *$\overline{\text{acx}}\left(\bigcup_n A_n\right)$ is compact for every null sequence $\{A_n\}_n$ of absolutely convex compact sets in $C_p(X)$.*
($\gamma$) *$\overline{\text{acx}}\left(\bigcup_n a_n A_n\right)$ is compact when $\{A_n\}_n$ is a bounded sequence of absolutely convex compact sets in $C_p(X)$ and $\{a_n\}_n$ is a null scalar sequence.*

*Proof* $(\alpha) \Rightarrow (\beta)$. From Theorem 2.1(5), [6, Sect. 2.1], and (*).

$(\beta) \Rightarrow (\gamma)$. Trivially, since $\{a_n A_n\}_n$ is a null sequence of sets.

$(\gamma) \Rightarrow (\alpha)$. Let $\{f_n\}_n$ be a pointwise eventually zero sequence in $C_p(X)$ and set $A_n = \{a f_n : |a| \le n 2^n\}$ for each $n \in \mathbb{N}$. Since $\{n 2^n f_n\}_n$ is pointwise eventually zero, it is pointwise null, and $\{A_n\}_n$ is a null sequence of absolutely convex compact sets in $C_p(X)$. Hence $(\gamma)$ implies $\overline{\text{acx}}\left(\bigcup_n n^{-1} A_n\right)$ is compact, so (*) implies $\sum_n f_n = \sum_n 2^{-n} \cdot 2^n f_n$ converges in $C_p(X)$. Thus $X$ is a P-space by Theorem 2.1 (6). □

One may recognize $(\beta)$ and $(\gamma)$ as respective analogs of *C-barrelled* and *property (L)*. Analogs of $c_0$- and $mc_0$*-barrelled* produce already-familar characterizations of *locally complete*.

$L_m(X)$ is non-$S_\sigma$ if and only if $X$ is finite [2], and is barrelled if and only if $X$ is discrete [1, Sect. 2.1]. Remaining questions: When $X$ is a non-discrete P-space, can or must $L_m(X)$ be $\aleph_0$-barrelled? Our answer to the "must" part was *No*: *If $X$ is a non-discrete Lindelöf P-space, $L_m(X)$ is not $\aleph_0$-barrelled* [2, Theorem 7]. We shall now see that the answer to the "can" part is also negative.

**Theorem 2.3** *The following three statements are equivalent:*

*(A) X is discrete.*
*(B) $L_m(X)$ is barrelled.*
*(C) $L_m(X)$ is $\aleph_0$-barrelled.*

*Proof* Obviously, if $X$ is discrete then $L_m(X)$ has its strongest locally convex topology, hence is barrelled, and if barrelled, is $\aleph_0$-barrelled: (A) $\Rightarrow$ (B) $\Rightarrow$ (C).

Now assume $L_m(X)$ is $\aleph_0$-barrelled and $X$ is not discrete. To complete the proof, we argue to a contradiction. Since $\aleph_0$-*barrelled* $\Rightarrow$ $\ell^\infty$-*barrelled*, $X$ is a P-space [2, Theorem 6] and therefore has a neighborhood base of open-and-closed sets [4, 4K 8.]; i.e., has a base of neighborhoods $U$ whose characteristic functions $\mathbf{1}_U$ are continuous. Let $y$ be a non-isolated point in $X$. Zorn's lemma produces a maximal family $\mathcal{F}$ of non-empty pairwise disjoint open sets $U$ which miss $y$ and have continuous characteristic functions $\mathbf{1}_U$ $(U \in \mathcal{F})$, where $\mathbf{1}_U$ is defined to be identically 1 on $U$ and 0 on $X \backslash U$. Since $\mathcal{F}$ is maximal and $y$ is not isolated, every neighborhood of $y$ meets $\bigcup \mathcal{F}$. Thus the characteristic function $\chi$ of $\bigcup \mathcal{F}$ is not continuous at $y$.

We construct a special Hamel basis $H$ for $L(X)$. Let $A = X \backslash \bigcup \mathcal{F}$. For each $U \in \mathcal{F}$, fix $x_U \in U$ and set

$$H_U = \{x_U\} \bigcup \{x - x_U : x \in U \text{ and } x \neq x_U\}.$$

Each $H_U$ is a Hamel basis for the span of $U$ in $L(X)$. Clearly, the set

$$H = A \bigcup \left( \bigcup \{H_U : U \in \mathcal{F}\} \right)$$

is a Hamel basis for all of $L(X)$, and the coefficient functional for each $x_U$ is just the continuous $\mathbf{1}_U$. Let $E_C$ denote the span in $L(X)$ of the members of $H$ having continuous coefficient functionals, and $E_D$ the span of the complement in $H$. The splitting theorem in [9, Theorem 3.1] says that $L_m(X)$ is the direct sum of $E_C$ and $E_D$, and $E_C$ has its strongest locally convex topology. Consequently, every linear form on $L_m(X)$ which vanishes on $E_D$ is continuous. Now the characteristic function $\chi$ of $\bigcup \mathcal{F}$ vanishes at each point of $H$ except the $x_U$ $(U \in \mathcal{F})$, which are in $E_C$; hence $\chi$ vanishes on $E_D$ and is continuous. But this contradicts the fact that $\chi$ is discontinuous at $y$. □

One can relax the hypothesis of [2, Theorem 7]:

**Corollary 2.1** *For every non-discrete P-space $X$, the lcs $L_m(X)$ is $\ell^\infty$-barrelled but not $\aleph_0$-barrelled.*

**Corollary 2.2** *Let $X$ be a non-discrete P-space. For every bounded sequence $\{f_n\}_n$ of points in $C_p(X)$, we have $\overline{\mathrm{acx}}(\{f_n\}_n)$ is compact. But $\overline{\mathrm{acx}}(\bigcup_n A_n)$ is not compact for some bounded sequence $\{A_n\}_n$ of absolutely convex compact sets in $C_p(X)$.*

**Acknowledgments** Thanks to Profs. Ferrando and López Pellicer for editorial advice and encouragement.

## References


1. Buchwalter, H., Schmets, J.: Sur quelques propriétés de l'espace $C_s(T)$. J. Math. Pures Appl. **52**, 337–352 (1973)
2. Ferrando, J.C., Kąkol, J., Saxon, S.A.: The dual of the locally convex space $C_p(X)$. Functiones et Approximatio **50**(2), 1–11 (2014). doi:10.7169/facm/2014.50.2
3. Ferrando, J.C., Kąkol, J., Saxon, S.A.: Characterizing P-spaces $X$ in terms of $C_p(X)$ (preprint)
4. Gillman, L., Jerison, M.: Rings of Continuous Functions. GTM 43. Springer, New York (1976)
5. Saxon, S.A.: Mackey hyperplanes/enlargements for Tweddle's space. RACSAM (2013). doi:10.1004/s13398-013-0159-x
6. Saxon, S.A., Sánchez Ruiz, L.M.: Dual local completeness. Proc. Amer. Math. Soc. **125**, 1063–1070 (1997)
7. Saxon, S.A., Sánchez Ruiz, L.M.: Mackey weak barrelledness. Proc. Amer. Math. Soc. **126**, 3279–3282 (1998)
8. Saxon, S.A., Sánchez Ruiz, L.M.: Reinventing weak barrelledness (preprint)
9. Saxon, S.A., Tweddle, I.: Mackey $\aleph_0$-barrelled spaces. Adv. in Math. **145**, 230–238 (1999)
10. Valdivia, M.: On Mackey spaces. Duke Math J. **41**, 57–66 (1974)

# Chapter 3
# On the Topology of the Sets of the Real Projections of the Zeros of Exponential Polynomials

**Gaspar Mora**

**Abstract** It is analysed some topological properties of the set $P_{D(z)}$ of the real projections of the zeros of an exponential polynomial $D(z)$ of the form $1+\sum_{j=1}^{n} m_j \, e^{\omega_j z}$, where $n$ is a positive integer, $m_j \in \mathbb{C} \setminus \{0\}$ and $\omega_j > 0$, for all $1 \leq j \leq n$. It is pointed out the influence of the $\omega$'s, called frequencies, as well as the $m$'s, called coefficients of $D(z)$, on the topology of $P_{D(z)}$. Reciprocally, it is seen how the order of the zeros of $D(z)$ is also influenced by the topology of $P_{D(z)}$. Finally, has been proved that the set of the real projections of the zeros of every partial sum $\zeta_n(z) := \sum_{k=1}^{n} 1/k^z$, $n > 2$, of the Riemann zeta function situated in the critical strip of the Riemann zeta function is dense in itself. Therefore the closure of this set is a perfect set, for all $n > 2$.

**Keywords** Exponential polynomials · Zeros of the partial sums of the Riemann zeta function · Perfect sets

**MSC 2010 Classification:** 30AXX, 30D20

## 3.1 Introduction

The Riemann zeta function defined on $\Re z > 1$ as $\zeta(z) := \sum_{k=1}^{\infty} 1/k^z$ and extended by analytic continuation to the whole complex plane, except in $z = 1$, where it has a simple pole, has associated a strip, $(0, 1) \times \mathbb{R}$, called critical strip which contains the non-trivial zeros of $\zeta(z)$. It is well known that in 1859, B. Riemann, formulated his famous conjecture on the location of the zeros of $\zeta(z)$ under the form "*it is very likely that the complex zeros of $\zeta(z)$ all have real part equal to* $1/2$". This statement is known as the "Riemann hypothesis". In words of Edwards [8, p. 6], "*the experience of Riemann's successors with the Riemann hypothesis has been the same as Riemann's, they also consider its truth as 'very likely' and they also have been unable to prove it*".

G. Mora (✉)
Department of Mathematical Analysis, University of Alicante, 03080 Alicante, Spain
e-mail: gaspar.mora@ua.es

J. C. Ferrando and M. López-Pellicer (eds.), *Descriptive Topology and Functional Analysis*, Springer Proceedings in Mathematics & Statistics 80, DOI: 10.1007/978-3-319-05224-3_3,
© Springer International Publishing Switzerland 2014

Borwein says us in [6, p. 5] "*in one of the largest calculations done to date, it was checked that the first ten trillion of these zeros lie on the correct line*". In spite of the numerical evidence concerning the exact location on the line $x = 1/2$ of the first ten trillion of zeros of $\zeta(z)$, in the present paper we show that the zeros of its partial sums $\zeta_n(z) := \sum_{k=1}^{n} 1/k^z$, $n > 2$, situated in the critical strip are placed in such a way that their real projections define a set dense in itself. Therefore, for any $n > 2$, no vertical strip $[a, b] \times \mathbb{R}$ with $0 \leq a < b \leq \min\left\{1, b_{\zeta_n(z)}\right\}$ is zero-free, where

$$b_{\zeta_n(z)} := \sup\left\{\Re z : \zeta_n(z) = 0\right\} = 1 + \left(\frac{4}{\pi} - 1 + o(1)\right)\frac{\log\log n}{\log n}, \qquad n \to \infty$$

(the last equality is justified by means of the inequalities (3.1.6) in this paper). A similar result for the truncations $\sum_{k=1}^{n}(-1)^{k-1}1/k^z$ of the alternating Dirichlet series $\sum_{k=1}^{\infty}(-1)^{k-1}1/k^z$ was given in [9, Theorem 1].

The position and distribution of the zeros of exponential polynomials has been a subject studied, mostly in the first third of twentieth century, due to its relation with the theory of differential equations (see [2, 12, 14, 34, 35, 39]). However, as far as we know, the first work on topological properties of density of the set of the real projections of the zeros of an exponential polynomial was made by Moreno [29] in 1973. He proved the following result:

**Theorem 3.1** (Moreno [29]) *Consider the exponential polynomial*

$$E(z) := \sum_{j=1}^{n} c_j\, e^{\omega_j z}, \;\; n \geq 3, \;\;\; c_j \in \mathbb{C} \setminus \{0\},$$

*where the set of real numbers $\{1, \omega_1, \omega_2, \ldots, \omega_n\}$ is linearly independent over the rationals, and let $S := (\sigma_0, \sigma_1) \times \mathbb{R}$ be an open vertical strip. Thus $E(z)$ has zeros near any line inside $S$ if and only if the $n$ inequalities*

$$\left|c_j\right| e^{\omega_j x} \leq \sum_{k=1,\, k\neq j}^{n} \left|c_k\right| e^{\omega_k x} \;\; \textit{for all } x \textit{ with } x + iy \in S, \;\; j = 1, \ldots, n, \qquad (3.1.1)$$

*are satisfied.*

It is worthwhile, firstly, to remark that in the preceding result the expression "to have zeros near any line inside a strip $S := (\sigma_0, \sigma_1) \times \mathbb{R}$" means that the set $P_{E(z)} := \{\Re z : E(z) = 0\}$ is dense in the interval $(\sigma_0, \sigma_1)$. Then Moreno's theorem states that the condition (3.1.1) is necessary and sufficient for the set $P_{E(z)}$ to be dense in $(\sigma_0, \sigma_1)$, provided that the set $\{1, \omega_1, \omega_2, \ldots, \omega_n\}$ is linearly independent (briefly l.i.) over the rationals. Secondly, it is easy to prove that Moreno's result also follows if we assume the linear independence over the rationals of the set $\{\omega_1, \omega_2, \ldots, \omega_n\}$.

Any exponential polynomial $E(z)$ with at least two terms and real exponents (if necessary, divide $E(z)$ by the monomial of smallest exponent), for the purpose of studying its zeros, can be substituted by a complex function of the form

$$D(z) = 1 + \sum_{j=1}^{n} m_j \, e^{\omega_j z}, \quad z = x + iy, \tag{3.1.2}$$

where, $n \geq 1$ is a positive integer, $m_j \in \mathbb{C} \setminus \{0\}$ for all $j = 1, 2, \ldots, n$ and $0 < \omega_1 < \cdots < \omega_n$. The topological properties of the set of the real projections of the zeros of $D(z)$,

$$P_{D(z)} := \{\Re z : D(z) = 0\},$$

will be studied by means of its closure set

$$R_{D(z)} := \overline{P_{D(z)}}.$$

For $n = 1$, since $D(z) = 1 + m_1 \, e^{\omega_1 z}$ is a periodic function, all the zeros of $D(z)$ lie on a vertical line. In this case, the sets $P_{D(z)}$ and $R_{D(z)}$ are equal and both are reduced to only one point. Thus, $n = 1$ is the trivial case and hence, from now on, we will assume that $n > 1$.

By following the proof of Moreno's result, we have obtained the following easy pointwise version of such theorem for exponential polynomials of the form (3.1.2).

**Theorem 3.2** *Consider an exponential polynomial of the form* (3.1.2) *with the set* $\{\omega_1, \ldots, \omega_n\}$ *l.i. over the rationals. Then* $\sigma_0 \in R_{D(z)} := \overline{\{\Re z : D(z) = 0\}}$ *if and only if the* $n + 1$ *inequalies*

$$1 \leq \sum_{j=1}^{n} |m_j| \, e^{\omega_j \sigma_0}, \quad |m_k| \, e^{\omega_k \sigma_0} \leq 1 + \sum_{j=1, \, j \neq k}^{n} |m_j| \, e^{\omega_j \sigma_0}, \quad k = 1, \ldots, n,$$

*are satisfied.*

In 1980 Avellar and Hale [2, Theorem 3.1] introduced a new result on the density of the set of real projections of the zeros of an exponential polynomial $E(z)$ under the weaker hypothesis of the existence of a basis in $\{\omega_1, \omega_2, \ldots, \omega_n\}$. Under this hypothesis, Avellar and Hale demonstrated:

**Theorem 3.3** (Avellar and Hale [2]) *Let* $N$, $M$ *be positive integers, let us denote* $\mathbb{R}^+ = [0, \infty)$, $a_j \in \mathbb{R}$ *with* $a_j \neq 0$, $j = 1, 2, \ldots, N$, $r = (r_1, r_2, \ldots, r_M) \in (\mathbb{R}^+)^M$, $\gamma_j = (\gamma_{j1}, \gamma_{j2}, \ldots, \gamma_{jM})$, $\gamma_{jk}$ *non-negative integers,* $k = 1, 2, \ldots, M$, $\gamma_j.r = \sum_{k=1}^{M} \gamma_{jk} r_k$. *Consider the functions*

$$h(z, r) = 1 + \sum_{j=1}^{N} a_j e^{-z\gamma_j . r}, \quad z = x + iy,$$

$$H(\sigma, \theta, r) = 1 + \sum_{j=1}^{N} a_j \mathrm{e}^{-\sigma \gamma_j . r} \mathrm{e}^{i\gamma_j . \theta}, \quad \sigma \in \mathbb{R},$$

*and the set*

$$Z_r := \{\Re z : h(z, r) = 0\},$$

*where all the components of r are rationally independent and* $\theta = (\theta_1, \theta_2, \ldots, \theta_M)$ *such that* $0 \leq \theta_k \leq 2\pi$, *for all* $k = 1, 2, \ldots, M$. *Then* $\sigma \in \overline{Z_r}$ *if and only if there is a* $\theta$ *such that* $H(\sigma, \theta, r) = 0$.

Thus, for exponential polynomials of the form $h(z, r) = 1 + \sum_{j=1}^{N} a_j \mathrm{e}^{-z\gamma_j . r}$, the theorem of Avellar and Hale supplies a pointwise characterization of the set $\overline{Z_r := \{\Re z : h(z, r) = 0\}}$ in terms of $H(\sigma, \theta, r)$ provided that there exists a basis of frequencies (the set defined by the components of $r$). Note that $H(\sigma, \theta, r)$ is a rather complicated complex function depending on a number of real variables equal to the cardinal of the basis of frequencies plus 1. There, the authors admitted that such a theorem is essentially in Henry [12] (stated without proof) and Moreno [29].

Observe that Avellar and Hale's theorem applies to the partial sums of the Riemann zeta function $\zeta_n(z) := \sum_{k=1}^{n} 1/k^z$ for every $n > 1$. Indeed, by defining $h(z, r) := \zeta_n(z)$, it is enough to take $N = n - 1$ , $M = k_n$ ($k_n$ denotes the number of primes not exceeding $n$), $a_j = 1$ for all $j$. The vector $r = \left(\log 2, \log 3, \log 5, \ldots, \log p_{k_n}\right)$ is defined by the logarithm of each prime number, from 2 to $p_{k_n}$, where $p_{k_n}$ denotes the last prime not exceeding $n$. The components $\gamma_{jk}$, $k = 1, 2, \ldots, M$, of the vector $\gamma_j$ are the unique nonnegative integers such that the number $j + 1$, for $j = 1, \ldots, n - 1$, is expressed of unique form as $2^{\gamma_{j1}} \cdot 3^{\gamma_{j2}} \cdots p_{k_n}^{\gamma_{jM}}$ by virtue of the Fundamental Theorem of the Arithmetic.

In 1948 the interest on the distribution of the zeros of the partial sums $\zeta_n(z) := \sum_{k=1}^{n} \frac{1}{k^z}$ of the Riemann zeta function increased as consequence of a famous result of Turán, [37, Theorem 1], where he pointed out an intriguing connection betweeen the position of the zeros of $\zeta_n(z)$ near the line $x = 1$ and the Riemann Hypothesis (briefly R.H.). Namely, Turán [37, Theorem I] proved that if

$$\zeta_n(z) \neq 0, \text{ for } \Re z \geq 1 + n^{-\frac{1}{2}+\varepsilon}, \quad n \geq N_0(\varepsilon),$$

then the Riemann Hypothesis is true.

Neverthelees, Turán's dream of proving R.H. via [37, Theorem I] was seriously impeded by Haselgrove [11] few years later. Indeed, by means of an added note at the end of [11], Haselgrove showed that the suggestion of Turán [37, (25.1)] on the non-negativity for all $x \geq 1$ of the function $T(x) := \sum_{k \leq x} \lambda(k)/k$ was false, where $\lambda(k) := (-1)^{p(k)}$ and $p(k)$ is the number of prime factors of $k$, counting multiplicities. It means that there exists some $n$ such that

$$\sum_{k=1}^{n} \frac{\lambda(k)}{k} < 0, \tag{3.1.3}$$

which allows to define the Dirichlet polynomial

$$D_n(z) := \sum_{k=1}^{n} \frac{\lambda(k)}{k^z},$$

and then the corresponding partial sum

$$\zeta_n(z) := \sum_{k=1}^{n} \frac{1}{k^z},$$

because the equivalence theorem of Bohr (see, for instance, [1] or [4]), attains the same set of values in any half-plane as $D_n(z)$ does. That is, for arbitrary real $a$,

$$\{D_n(z) : \Re z > a\} = \{\zeta_n(z) : \Re z > a\}$$

holds. Since, for sufficiently large real $z$, $D_n(z)$ is close to 1 and, by (3.1.3), $D_n(1) < 0$, there exists a real zero, say $x$, of $D_n(z)$ such that $x > 1$. Consequently, there is a zero of $\zeta_n(z)$ in the half-plane $\Re z > 1$ .

In 1968 Spira [32, 33] demonstrated that the same happens for $n = 19$, that is, he proved that $\zeta_{19}(z)$ has a zero in the half-plane $\Re z > 1$. Levinson [19, Theorem 1] in 1973 found an asymptotic formula, for large $n$, for the location of the zeros of $\zeta_n(z)$ near the point $z = 1$. In particular, he proved that those zeros have real part less than 1. Voronin [38] in 1974 showed that $\zeta_n(z)$ has zeros in $\Re z > 1$, for infinitely many $n$.

A sharp result on the upper bound

$$b_{\zeta_n(z)} := \sup\{\Re z : \zeta_n(z) = 0\}$$

was given in 2001 by Montgomery and Vaughan in [21]. Namely, they proved that there exists $N_0$ such that if $n > N_0$ then $\zeta_n(z) \neq 0$ whenever

$$\Re z \geq 1 + \left(\frac{4}{\pi} - 1\right) \frac{\log \log n}{\log n}. \tag{3.1.4}$$

In the opposite direction, in 1983, Montgomery [20] had already shown that for every $0 < c < \dfrac{4}{\pi} - 1$ there is an $N_0(c)$ such that $\zeta_n(z)$ has zeros in the half-plane

$$\Re z > 1 + c\frac{\log \log n}{\log n}, \quad n > N_0(c). \tag{3.1.5}$$

Since for any positive real number $c$, $n$ large and any $\varepsilon$ such that $0 < \varepsilon < \dfrac{1}{2}$, one has

$$c\frac{\log\log n}{\log n} > n^{-\frac{1}{2}+\varepsilon},$$

then inequality (3.1.5) implies the existence of zeros of $\zeta_n(z)$ with real part much larger than Turán's bound [37, Theorem I]. Hence, the old theorem of Turán of 1948 became vacuous and the hope of proving R.H. via such hypothetical localization of the zeros of the partial sums $\zeta_n(z)$ died.

The results of Montgomery and Vaughan, (3.1.4) and (3.1.5), imply that, fixed $0 < c < \frac{4}{\pi} - 1$, for every $n > N_1 := \max\{N_0, N_0(c)\}$, the upper bound $b_{\zeta_n(z)}$ satisfies

$$1 + c\frac{\log\log n}{\log n} < b_{\zeta_n(z)} \leq 1 + \left(\frac{4}{\pi} - 1\right)\frac{\log\log n}{\log n}. \tag{3.1.6}$$

Thus, since $\lim_{n\to\infty}\frac{\log\log n}{\log n} = 0$, it follows that

$$\lim_{n\to\infty} b_{\zeta_n(z)} = 1.$$

On the other hand, Balazard and Velásquez-Castañón [3] demonstrated that the lower bound

$$a_{\zeta_n(z)} := \inf\{\Re z : \zeta_n(z) = 0\}$$

is so that

$$\lim_{n\to\infty}\frac{a_{\zeta_n(z)}}{n} = -\log 2. \tag{3.1.7}$$

Therefore

$$\lim_{n\to\infty} a_{\zeta_n(z)} = -\infty.$$

In the present paper our goal is, firstly, to survey our main results, contained in [7, 22–28], on the distribution of the real parts of the zeros of exponential polynomials having linearly independent frequencies over the rationals. To do it is necessary previously to recall some results (see, for instance, [7, 27, 28]) that complete the basic and crucial theorems of Moreno [29] and Avellar and Hale [2]. Secondly, we would like to present our results, mostly contained in [7, 22–28], on the zeros of the partial sums of the Riemann zeta function. Finally, we provide in this paper new results which complete the main theorem [27, Theorem 12], of asymptotic validity, on the distribution of the zeros of the partial sums. This is done by proving the validity of the aforementioned main theorem, for all $n > 2$, on the set of the real projections of the zeros of $\zeta_n(z)$ situated on the critical strip of the Riemann zeta function (see our Theorem 3.22 of the present paper).

### 3.1.1 *The Critical Interval of $D(z)$ and the Set $R_{D(z)}$*

Given an exponential polynomial $D(z)$ of the form (3.1.2), the topological properties of $P_{D(z)}$, the set of the real projections of the zeros of $D(z)$, obviously depend on the coefficients $m_j$ and frequencies $\omega_j$. But, as we will soon see, the linear independence, or dependence, of the frequencies over the rationals has a great influence on the topological structure of $P_{D(z)}$. In the first case, i.e., when the $\omega$'s are linearly independent (briefly l.i.) over the rationals, we have proved that the sets $P_{D(z)}$ are dense in themselves [28]. That is, the set $P_{D(z)}$ is a subset of the set of its accumulation points, denoted by $AcP_{D(z)}$, which is equivalent to say that $P_{D(z)}$ has no isolated point. In consequence, the sets $R_{D(z)} := \overline{P_{D(z)}}$ are perfect, i.e., satisfy the property $R_{D(z)} = AcR_{D(z)}$ (for details on topological properties of these sets see, for instance, the classical book of Kuratowski [13, p. 75–80]).

An exponential polynomial $D(z)$ of the form (3.1.2) satisfies

$$\lim_{x\to-\infty} D(z) = 1, \quad \lim_{x\to+\infty} \frac{D(z)}{m_n\, e^{\omega_n z}} = 1, \quad \text{for any real } y,$$

which means that $D(z)$ has no zero on the half-planes $\Re z < A$ and $\Re z > B$, for some real numbers $A$, $B$. Thus, the numbers

$$a_{D(z)} := \inf\{\Re z : D(z) = 0\}, \; b_{D(z)} := \sup\{\Re z : D(z) = 0\}$$

exist, and, therefore, the interval $\left[a_{D(z)}, b_{D(z)}\right]$, called *critical interval* of $D(z)$, contains the real projections of all the zeros of $D(z)$. Consequently, the strip $\left[a_{D(z)}, b_{D(z)}\right] \times \mathbb{R}$, called *critical strip* of $D(z)$, contains all the zeros of $D(z)$.

Any exponential polynomial $D(z)$ of the form (3.1.2) is a non constant entire function of order 1. So it has at most countably many zeros. By Hadamard's factorization theorem $D(z)$ has an infinite quantity of zeros (see [22, Proposition 1]). Hence $P_{D(z)}$ is a nonvoid countable (finite or infinite) set contained in the critical interval $\left[a_{D(z)}, b_{D(z)}\right]$ of $D(z)$. Therefore the optimal relation between the set $P_{D(z)}$ and $\left[a_{D(z)}, b_{D(z)}\right]$ is reached when $R_{D(z)}$, the closure of $P_{D(z)}$, coincides with the whole interval $\left[a_{D(z)}, b_{D(z)}\right]$.

## 3.2 The Sets $P_{D(z)}$ When the Frequencies of $D(z)$ Are l.i.

Consider an exponential polynomial $D(z)$ of the form (3.1.2) with positive frequencies

$$\omega_1 < \omega_2 < \cdots < \omega_n$$

linearly independent over the rationals, and let

$$a_{D(z)} := \inf \{\Re z : D(z) = 0\}, \quad b_{D(z)} := \sup \{\Re z : D(z) = 0\}.$$

Then in [28, Theorem 3] we have proved that the set $R_{D(z)}$ contains at least two nondegenerated closed intervals, where $R_{D(z)}$ denotes the closure of the set $P_{D(z)}$ of the real projecions of the zeros of $D(z)$.

**Theorem 3.4** (Mora et al. [28]) *Let $D(z)$ be an exponential polynomial of the form with frequencies linearly independent over the rationals. Then the extreme points of the critical interval $[a_{D(z)}, b_{D(z)}]$ are the solutions of the equations*

$$1 = \sum_{j=1}^{n} |m_j| e^{\omega_j x} \tag{3.2.1}$$

*and*

$$|m_n| e^{\omega_n x} = 1 + \sum_{j=1}^{n-1} |m_j| e^{\omega_j x}, \tag{3.2.2}$$

*respectively. Furthermore, there exist $x_1, x_2$ with $a_{D(z)} < x_1 \leq x_2 < b_{D(z)}$ such that the intervals $[a_{D(z)}, x_1]$ and $[x_2, b_{D(z)}]$ are both contained in $R_{D(z)}$.*

An immediate conclusion is derived from the above theorem.

**Corollary 3.1** *The set of real projections of the zeros of an exponential polynomial of the form* (3.1.2) *with frequencies linearly independent over the rationals is not nowhere dense.*

The Eq. (3.2.1) can be written as an equation with increasing exponents under the form,

$$1 - \sum_{j=1}^{n} |m_j| e^{\omega_j x} = 0,$$

and then $W$, the number of changes of sign of its coefficients, is 1. On the other hand, if $N$ denotes the number of real solutions (counting multiplicities) of the above equation, by Pólya and Szëgo's formula [30, p. 46], $W - N$ is an even nonnegative integer. Thus, $N$ is necessarily 1. An analogous analysis can be made on the equation (3.2.2). Therefore, the Eqs. (3.2.1) and (3.2.2) each have only one solution.

Another important conclusion from Theorem 3.4 is that any exponential polynomial $D(z)$ of the form (3.1.2) with frequencies linearly independent over the rationals, independently of its coefficients, has a set $R_{D(z)}$ which always contains two nondegenerated extreme intervals $[a_{D(z)}, x_1]$ and $[x_2, b_{D(z)}]$.

In [28, Theorem 9] we have given a complete description of $R_{D(z)}$, namely, the set $R_{D(z)}$ is the union of at most $n$ disjoint nondegenerated closed intervals.

**Theorem 3.5** (Mora et al. [28]) *Given an exponential polynomial $D(z)$ of the form* (3.1.2) *with frequencies linearly independent over the rationals, the set $R_{D(z)}$ is either* *the critical interval* $\left[a_{D(z)}, b_{D(z)}\right]$ *or is the union of at most n disjoint nondegenerated closed intervals. In the last case, the gaps are produced by those equations*

$$|m_k| \, e^{\omega_k x} = 1 + \sum_{j=1, j\neq k}^{n} |m_j| \, e^{\omega_j x}, \quad k = 1,2,\ldots,n-1, \tag{3.2.3}$$

*having two distinct real solutions.*

It is worthwhile to remark that in the above theorem we have given an explanation on the appearance of disjoint intervals in the set $R_{D(z)}$ and, in consequence, the existence of gaps between such as intervals. Some easy consequences are derived from the above theorem.

**Theorem 3.6** (Mora et al. [28]) *Let* $D(z)$ *be an exponential polynomial of the form* (3.1.2) *with frequencies linearly independent over the rationals. Then the set* $P_{D(z)}$ *is dense in itself and* $R_{D(z)}$ *is a perfect set.*

**Theorem 3.7** (Mora et al. [28]) *Let* $D(z)$ *be an exponential polynomial of the form* (3.1.2) *with frequencies linearly independent over the rationals and coefficients* $m_j$ *such that* $|m_j| = 1$ *for all* $j = 1, \ldots, n$*. Then* $R_{D(z)} = \left[a_{D(z)}, b_{D(z)}\right]$.

*Proof* Whenever $|m_j| = 1$ for all $j = 1, \ldots, n$, it is immediate to check that each equation in (3.2.3), for any $k = 1, \ldots, n-1$, has no real solution, so there is no gap. □

**Theorem 3.8** (Mora et al. [28]) *Let* $D(z) = 1 + \sum_{j=1}^{n} m_j e^{\omega_j z}$ *and* $E(z) = 1 + \sum_{j=1}^{n} n_j e^{\omega_j z}$ *be two exponential polynomials of the form* (3.1.2) *with frequencies linearly independent over the rationals such that* $|m_j| = |n_j|$ *for all* $j = 1, \ldots, n$*. Then* $R_{D(z)} = R_{E(z)}$.

*Proof* By Theorem 3.4, the extreme points of the critical intervals of $D(z)$ and $E(z)$ depend on the Eqs. (3.2.1) and (3.2.2). By Theorem 3.6, the extreme points of the gaps, if there are, depend on Eq. (3.2.3). Then the sets $R_{D(z)}$ and $R_{E(z)}$ depend on (3.2.1), (3.2.2) and (3.2.3), and these equations are written in terms of the moduli of the coefficients of $D(z)$ and $E(z)$, and their common frequencies. Thus, since by hypothesis, the moduli are equal, the Eqs. (3.2.1), (3.2.2) and (3.2.3) are the same for $D(z)$ and $E(z)$, so $R_{D(z)} = R_{E(z)}$. □

An easy consequence of the above theorem is that the set $R_{D(z)}$ associated to an exponential polynomial $D(z)$ of the form (3.1.2) with frequencies linearly independent over the rationals is invariant with respect to the moduli of its coefficients. Hence, in particular, all them can be taken positive.

The next result is an immediate consequence of Theorem 3.6.

**Theorem 3.9** (Mora et al. [28]) *Let* $D(z)$ *be an exponential polynomial of the form* (3.1.2) *with frequencies linearly independent over the rationals. Then the boundary of* $R_{D(z)}$*, denoted by* $\partial R_{D(z)}$*, is a finite set. Furthermore, if we denote by p the number of points of* $\partial R_{D(z)}$*, then p is even and satisfies* $2 \leq p \leq 2n$.

The next result [28, Theorem 11] points out how the order of the zeros of $D(z)$ situated on the vertical lines passing through the boundary points of $R_{D(z)}$ is influenced by the topology of the set $R_{D(z)}$.

**Theorem 3.10** (Mora et al. [28]) *Let $D(z)$ be an exponential polynomial of the form* (3.1.2) *with frequencies linearly independent over the rationals. If $D(z)$ has a zero situated on a vertical line passing through a boundary point of $R_{D(z)}$, then the zero is simple.*

A surprising property [28, Theorem 12] is again obtained from Theorem 3.6, namely, the zeros of $D(z)$ situated on the two vertical lines passing through the extreme points of a gap have the same imaginary part.

**Theorem 3.11** (Mora et al. [28]) *Let $D(z)$ be an exponential polynomial of the form* (3.1.2) *with frequencies linearly independent over the rationals. Assume that $z_0$ is zero of $D(z)$ situated on a vertical line that passes through a point of $\partial R_{D(z)} \setminus \{a_{D(z)}, b_{D(z)}\}$, then there exists another zero, say $z_1$, situated on the parallel line that determines the corresponding gap, such that $\Im z_1 = \Im z_0$.*

Finally, from our Theorem 3.6, we have also obtained applications to the theory of fractal strings (for details on this subject see [15–18]) by proving in [28] that the three conjectures [16, Conjecture 8.3], [17, Conjecture 4.9] and [18, Conjecture 3.55] of Lapidus and Van Frankenhuysen on non-lattice fractal strings are false.

## 3.3 The Sets $P_{D(z)}$ When the Frequencies of $D(z)$ Are l.d.

Whenever the frequencies of an exponential $D(z)$ are linearly dependent over the rationals, the set $P_{D(z)}$ of the real projections of its zeros could be chaotic and we do not know any general result on a topological description of such a set. However, it can be interesting to study such polynomials under certain additional conditions.

If an exponential polynomial

$$D(z) = 1 + \sum_{j=1}^{n} m_j e^{\omega_j z}, \; n > 1, \; m_j \in \mathbb{C} \setminus \{0\}, \; 0 < \omega_1 < \cdots < \omega_n, \tag{3.3.1}$$

is such that its frequencies are commensurable, i.e.,

$$\frac{\omega_j}{\omega_1} \in \mathbb{Q}, \text{ for all } \; j = 1, \ldots, n,$$

$D(z)$ becomes an algebraic polynomial in a variable, say $Y$, defined as $Y := X^{\frac{1}{m}}$, for some positive integer $m$, where the variable $X := e^{\omega_1 z}$. Then $P_{D(z)}$ is a finite set, so it coincides with $R_{D(z)}$, the closure of $P_{D(z)}$, and consequently the study of the distribution of the zeros of $D(z)$ lacks of interest, in general. Therefore, we will assume that there exists some $j = 1, \ldots, n$ such that

$$\frac{\omega_j}{\omega_1} \notin \mathbb{Q}. \tag{3.3.2}$$

An exponential polynomial of the form (3.3.1) satisfying the condition (3.3.2) is called *non-lattice* in the language of the theory of fractal strings introduced by Lapidus and Pomerance in 1993, [15]. Since, for $n = 2$, the condition (3.3.2) means that $D(z)$ is of the form (3.1.2) with two frequencies linearly independent over the rationals, and this case has been already treated in the previous section, we will suppose that $n > 2$ . Hence we will deal with exponential polynomials of the form (3.3.1), with $n > 2$, having frequencies l.d. over the rationals and satisfying the condition (3.3.2). Such a class of exponential polynomials will be denoted by $\mathscr{D}_{ld}$.

The topological properties of the set of real projections of an exponential polynomial $D(z)$ of the class $\mathscr{D}_{ld}$ are unknown in general. As we said in the Introduction, we only have the theorem of characterization of Avellar and Hale of the set $R_{D(z)}$ in terms of an auxiliary function, provided that there exists a basis of frequencies of $D(z)$. Even in such a case, the topological properties on the density of the set of real projections of its zeros, $P_{D(z)}$, or the closure set, $R_{D(z)}$, are quite distinct from the other case, as we can see for instance by means of the exponential polynomial

$$D(z) := 1 + \frac{1}{3}e^{z\log\left(\frac{27}{8}\right)} + e^{z\log 8} - \frac{1}{3}e^{z\log 216}.$$

Indeed, the above polynomial is of the form (3.3.1), it has three frequencies $\omega_1 = \log(27/8)$, $\omega_2 = \log 8$ and $\omega_3 = \log 216$ linearly dependent over the rationals (is immediate to see that $\omega_3 = \omega_1 + 2\omega_2$) and the quotient $\omega_2/\omega_1 \notin \mathbb{Q}$, as we can easily check. Thus, $D(z)$ belongs to class $\mathscr{D}_{ld}$ and it has a basis of frequencies $\{\omega_1, \omega_2\}$ linearly independent over the rationals. Nevertheless, the complex values of $z$ satisfying the equation $8^z = -1$ are zeros of $D(z)$ and then is immediate to prove that 0 is an isolated point of $P_{D(z)}$. It means that Theorem 3.6 fails on the class $\mathscr{D}_{ld}$. Moreover, the preceding example proves that Theorem 3.6 does not even hold on the subclass of polynomials of $\mathscr{D}_{ld}$ having a basis of frequencies.

### *3.3.1 The Partial Sums of the Riemann Zeta Function*

The partial sums of the Riemann zeta function,

$$\zeta_n(z) := \sum_{k=1}^{n} \frac{1}{k^z}, \quad n > 1,$$

for the purpose of studying the location of their zeros, can be substituted by the functions

$$G_n(z) := \sum_{k=1}^{n} k^z = \sum_{k=1}^{n} e^{z \log k}, \quad n > 1 .$$

Indeed, noting that $\zeta_n(z) = G_n(-z)$, for all $z \in \mathbb{C}$, the sets of the zeros of both functions, denoted by $Z_{\zeta_n(z)}$ and $Z_{G_n(z)}$, are related by

$$Z_{\zeta_n(z)} = -Z_{G_n(z)}. \tag{3.3.3}$$

Then the bounds of the critical interval of $G_n(z)$,

$$a_{G_n(z)} := \inf \{\Re z : G_n(z) = 0\}, \; b_{G_n(z)} := \sup \{\Re z : G_n(z) = 0\},$$

satisfy

$$a_{G_n(z)} = -b_{\zeta_n(z)}, \;\; b_{G_n(z)} = -a_{\zeta_n(z)},$$

where, as usual, the bounds corresponding to the partial sums are defined as

$$a_{\zeta_n(z)} := \inf \{\Re z : \zeta_n(z) = 0\}, \;\; b_{\zeta_n(z)} := \sup \{\Re z : \zeta_n(z) = 0\}.$$

Therefore we have

$$a_{G_n(z)} = -1 - \left(\frac{4}{\pi} - 1 + o(1)\right) \frac{\log \log n}{\log n}, \quad b_{G_n(z)} = n \log 2 + o(n), \tag{3.3.4}$$

by virtue of the estimation of the bounds $a_{\zeta_n(z)}$, $b_{\zeta_n(z)}$ by Balazard and Velásquez-Castañón [3], and by Montgomery [20] and Montgomery and Vaughan [21], respectively (see (3.1.7) and (3.1.6) in this paper).

On the other hand, for any $n > 3$, $G_n(z)$ is an exponential polynomial of the class $\mathscr{D}_{ld}$ because its frequencies,

$$\{\log(k + 1) : k = 1, \dots, n - 1\},$$

are linearly dependent over the rationals, and for $k = 2$ one has

$$\frac{\omega_2}{\omega_1} = \frac{\log 3}{\log 2} \notin \mathbb{Q}.$$

Furthermore, for every $n > 3$, $G_n(z)$ has a basis of frequencies defined by the set

$$\left\{\log p_j : j = 1, \dots, k_n\right\},$$

where $\left\{p_j : j = 1, \dots, k_n\right\}$ is the set of all prime numbers not exceeding $n$. The *ad hoc* version of Avellar and Hale's theorem, [2, Theorem 3.1], applied to every $G_n(z)$, on the pointwise characterization of the set $R_{G_n(z)} := \overline{P_{G_n(z)}}$, where

$$P_{G_n(z)} := \{\Re z : G_n(z) = 0\},$$

was given in [7, Theorem 1] as follows.

**Theorem 3.12** (Dubon et al. [7]) *Given an integer $n > 2$, let $\{p_1, p_2, \ldots, p_{k_n}\}$ be the set of prime numbers not exceeding n. For each $1 \leq m \leq n$, let us denote by $c_m$ the vector with nonnegative integers components such that* $\log m = c_m.p$, *where* (.) *denotes the usual inner product in $\mathbb{R}^{k_n}$ and* $p = (\log p_1, \log p_2, \ldots, \log p_{k_n})$. *We define the function $F_n(\sigma, x) := \sum_{m=1}^{n} m^{\sigma} e^{ic_m.x}$, $\sigma \in \mathbb{R}$, $x \in \mathbb{R}^{k_n}$. Then a real number $x_0 \in R_{G_n(z)} := \overline{\{\Re z : G_n(z) = 0\}}$ if and only if there exists a vector $x \in \mathbb{R}^{k_n}$ such that $F_n(x_0, x) = 0$.*

Because the auxiliary function $F_n(\sigma, x)$ is hard to handle, our purpose is to find a new characterization of the set $R_{G_n(z)}$ in terms of more simple functions. Indeed, the old notion of level curve will allow us to do it (see, for instance, the classical book of Titchmarsh [36, p. 121].

By recalling that given an analytic function $f(z)$ on a region $\Omega$ of the complex plane and a positive constant $A$, the level curve is defined by the equation $|f(z)| = A$, in [7, Theorem 2] we have proved the following result:

**Theorem 3.13** (Dubon et al. [7]) *Given an integer $n > 2$, a real number $x_0$ belongs to $R_{G_n(z)} := \overline{\{\Re z : G_n(z) = 0\}}$ if and only if the vertical line $x = x_0$ intersects the level curve $\left|G_n^*(z)\right| = p_{k_n}^{x_0}$, where $G_n^*(z) := G_n(z) - p_{k_n}^{z}$ and $p_{k_n}$ is the last prime not exceeding n.*

By defining, for every integer $n > 2$, the real function

$$A_n(x, y) := \left|G_n^*(x + iy)\right| - p_{k_n}^{x}, \quad x, y \in \mathbb{R},$$

the above result was completed in [27, Theorem 2] under the form:

**Theorem 3.14** (Mora [27]) *For every integer $n > 2$, a real number $x$ belongs to $R_{G_n(z)}$ if and only if there exists some $y \in \mathbb{R}$ such that $A_n(x, y) = 0$. Furthermore, $A_n(x, 0) \geq 0$ for all $x \in \left[a_{G_n(z)}, b_{G_n(z)}\right]$.*

We recently have proved [27, Theorem 11] that, asymptotically, the functions $G_n(z)$ have a behavior, with respect to the real projection of their zeros, exactly equal to the exponential polynomials having frequencies linearly independent over the rationals and coefficients of modulus 1 (see Theorem 3.7 of the present paper). Therefore, by virtue of (3.3.3), the partial sums of the Riemann zeta function, $\zeta_n(z)$, have the same property. That is, in [27, Theorem 12], we have demonstrated the existence of a positive integer $N$ such that the set

$$P_{\zeta_n(z)} := \{\Re z : \zeta_n(z) = 0\}$$

is dense in itself, so $R_{\zeta_n(z)} := \overline{P_{\zeta_n(z)}}$ is a perfect set, for all $n > N$. In fact, it has been shown in [27, Theorem 12] that $R_{\zeta_n(z)}$ has only one connected component, so there is no gap, and, finally, the set $R_{\zeta_n(z)}$ coincides with the critical interval $\left[a_{\zeta_n(z)}, b_{\zeta_n(z)}\right]$, for all $n > N$. Consequently, is attained the best possible relation between the critical interval of $\zeta_n(z)$ and the set $R_{\zeta_n(z)}$, provided that $n > N$. That means that, given an arbitrary vertical line in $S_{\zeta_n(z)} := \left[a_{\zeta_n(z)}, b_{\zeta_n(z)}\right] \times \mathbb{R}$, the critical strip of $\zeta_n(z)$, every $\zeta_n(z)$ possesses infinitely many zeros arbitrarily close to that line, for all $n > N$.

In order to prove the mentioned Theorem 12 in [27], it was necessary to show that the boundary of $R_{G_n(z)}$, denoted by $\partial\left(R_{G_n(z)}\right)$, has only two points, namely, the extreme points of the critical interval. Previously were proved the following lemmas [27, Lemmas 6–8]:

**Lemma 3.1** (Mora [27]) *Given an integer $n > 2$, let us assume that there exists $x_0 \in \partial\left(R_{G_n(z)}\right) \setminus \left\{a_{G_n(z)}, b_{G_n(z)}\right\}$. Then there is at least a $y_0 > 0$, satisfying $A_n(x_0, y_0) = 0$, such that*

$$\min\left\{\left|G_n^*(z)\right| : \Re z = x_0\right\}$$

*exists and it is attained at the point $z_0 = x_0 + iy_0$.*

The next lemma says us that The Prime Number Theorem [10, p. 371] provides primes relatively near every integer sufficiently large. This lemma can be considered as a strong version of a property deduced from Bertrand's Postulate. Indeed, the postulate of Bertrand (see, for instance, Hardy and Wright's book [10, Theorem 418]) says that always there is a prime between $n$ and $2n$, for any positive integer $n \geq 2$. It is easy to prove that this postulate implies the property consisting in to affirm that, for every integer $n \geq 2$, the last prime not exceeding $n$, denoted by $p_{k_n}$, satisfies $2p_{k_n} > n$. Thus, following this property of the numbers we have demonstrated [27, Lemma 7]:

**Lemma 3.2** (Mora [27]) *Given an integer $n \geq 7$, let $p_{k_n-3} < p_{k_n-2} < p_{k_n-1} < p_{k_n}$ be the last four prime numbers not exceeding $n$. Then there exists an integer $n_0 \geq 7$ such that for all $n > n_0$ one has*

$$2p_{k_n-3} > n.$$

**Lemma 3.3** (Mora [27]) *Let $n$ be an integer such that $n > n_0$, where $n_0$ has been determined in* Lemma 3.2. *Then, if there exists $x_0 \in \partial\left(R_{G_n(z)}\right) \setminus \left\{a_{G_n(z)}, b_{G_n(z)}\right\}$, the complex number $z_0 = x_0 + iy_0$ determined in* Lemma 3.1 *is so that the three numbers*

$$\frac{G_n^*(z_0)}{p_{k_{n-l}}^{z_0}}, \quad l = 1, 2, 3,$$

*are real and distinct from* 0, *where $\left\{p_{k_{n-l}}\right\}_{l=1,2,3}$ are the three consecutive primes preceding $p_{k_n}$ (the last prime not exceeding $n$).*

Thus, from the previous lemmas we have [27, Theorem 9] whose statement is as follows.

**Theorem 3.15** (Mora [27]) *Let $n_0$ be the integer determined in* Lemma 3.2. *Then* $\partial\left(R_{G_n(z)}\right) = \left\{a_{G_n(z)}, b_{G_n(z)}\right\}$, *for any integer $n > n_0$.*

At this point we need to prove the existence of zeros of $G_n(z)$ having real part in the interior of the critical interval $\left[a_{G_n(z)}, b_{G_n(z)}\right]$. The existence of such zeros for sufficiently large $n$ is settled in the next result by using, on one hand, the asymptotic estimations (3.3.4) of the bounds corresponding to the functions $G_n(z)$, and, on the other hand, by applying a Ritt's formula [31, formula (9)] to estimate the sum of the real parts of a special class of exponential polynomials which contains in particular to the functions $G_n(z)$, for all $n \geq 2$. The statement of the result on the existence of zeros of $G_n(z)$ in the interior of its critical strip [27, Theorem 10] is as follows.

**Theorem 3.16** (Mora [27]) *There exists a positive integer $n_1$ such that $G_n(z)$ possesses infinitely many zeros whose real part is in the interior of its critical interval, for all $n > n_1$.*

Now we obtain the main theorems [27, Theorems 11 and 12] on the sets $R_{G_n(z)}$ and $R_{\zeta_n(z)}$.

**Theorem 3.17** (Mora [27]) *There exists a positive integer $N$ such that $R_{G_n(z)} = \left[a_{G_n(z)}, b_{G_n(z)}\right]$, for all $n > N$.*

Finally, from Theorem 3.17 and noting the relation (3.3.3) on the sets of zeros of $\zeta_n(z)$ and $G_n(z)$, it follows the main result on the distribution of the real parts of the zeros of the partial sums of the Riemann zeta function.

**Theorem 3.18** (Mora [27]) *There exists a positive integer $N$ such that $R_{\zeta_n(z)} = \left[a_{\zeta_n(z)}, b_{\zeta_n(z)}\right]$, for all $n > N$.*

Consequently, for all $n > N$, the sets of the real projections of the zeros of the partial sums of the Riemann zeta function, $\zeta_n(z)$, are dense in themselves and their closure sets are perfect. Furthermore, the best possible relation between the closure sets of the real projections of the zeros of the partial sums and their critical intervals is attained, namely, every closure set $R_{\zeta_n(z)}$ coincides with its corresponding critical interval $\left[a_{\zeta_n(z)}, b_{\zeta_n(z)}\right]$. That means that given any line contained in the critical strip $\left[a_{\zeta_n(z)}, b_{\zeta_n(z)}\right] \times \mathbb{R}$, the corresponding partial sum $\zeta_n(z)$ possesses infinitely many zeros arbitrarily near that line, so the distribution of real parts of the zeros of $\zeta_n(z)$ is completely uniform, for all $n > N$.

## 3.4 The Critical Strip of the Riemann Zeta Function and the Real Projection of the Zeros of Its Partial Sums

For a given integer $n > 2$, we consider the function

$$G_n^*(z) := G_n(z) - p_{k_n}^z = 1 + 2^z + \cdots + n^z - p_{k_n}^z,$$

$p_{k_n}$ being the last prime not exceeding $n$. Then, given a real number $x_0$, by squaring the equation of the level curve $\left|G_n^*(z)\right| = p_{k_n}^{x_0}$ and using elementary formulas of trigonometry, we obtain its cartesian equation,

$$\sum_{m=1}^{n} m^{2x} + 2.1^x \sum_{m=2}^{n} m^x \cos\left(y \log\left(\frac{m}{1}\right)\right) + 2.2^x \sum_{m=3}^{n} m^x \cos\left(y \log\left(\frac{m}{2}\right)\right)$$

$$+ \cdots + 2(n-1)^x \sum_{m=n}^{n} m^x \cos\left(y \log\left(\frac{m}{n-1}\right)\right) = p_{k_n}^{2x_0}, \qquad (3.4.1)$$

where in the left side of (3.4.1) the integer variable $m$ does not take the value $p_{k_n}$.

Since, for any value of $y$, the limit of the left side of (3.4.1) is $+\infty$ when $x \to +\infty$ and the rigth side of (3.4.1) is fixed, the range of $x$ is always upper bounded for every $x_0$. However, as the limit of the left side of (3.4.1) is 1 when $x \to -\infty$, if $x_0 \neq 0$, the range of the variable $x$ must be lower bounded. Finally, from equation (3.4.1), it is immediate that the level curve neither contains a vertical segment nor has a vertical asymptote. Thus, fixed $n > 2$ and $x_0 \in \mathbb{R}$, the variable $x$ in the Eq. (3.4.1) either runs on a finite interval

$$\left[b_{n,x_0}^-, b_{n,x_0}\right], \quad \text{if } x_0 \neq 0,$$

or on the infinite interval

$$\left(-\infty, b_{n,x_0}\right], \quad \text{if } x_0 = 0,$$

where the extremes $b_{n,x_0}^-$, $b_{n,x_0}$ are defined as follows.

**Definition 3.1** Fixed an integer $n > 2$ and a real number $x_0$, we define $b_{n,x_0}$, called upper extreme associated with $x_0$, as the maximum value that we can assign to the variable $x$ to have, for at least a value of $y$, a point $\left(b_{n,x_0}, y\right)$ of the level curve $\left|G_n^*(z)\right| = p_{k_n}^{x_0}$. When $x_0 \neq 0$, we define $b_{n,x_0}^-$, called lower extreme associated with $x_0$, as the minimum value that we can assign to the variable $x$ to have, for at least a value of $y$, a point $\left(b_{n,x_0}^-, y\right)$ of the level curve $\left|G_n^*(z)\right| = p_{k_n}^{x_0}$.

To illustrate the above definition, in the next example, we determine the level curve for $n = 3$ and the extreme values corresponding to a real number $x_0$.

*Example 3.1* The level curve $\left|G_3^*(z)\right| = p_{k_3}^{x_0}$, $x_0 \in \mathbb{R}$.

The last prime not exceeding 3 is 3. Then $p_{k_3} = 3$ and the function $G_3^*(z) := G_3(z) - p_{k_3}^z = 1 + 2^z$. Hence the level curve is $|1 + 2^z| = 3^{x_0}$. By squaring we obtain its cartesian equation,

$$1 + 2^{2x} + 2^{x+1} \cos(y \log 2) = 3^{2x_0}. \qquad (3.4.2)$$

The lower extreme associated with $x_0 \neq 0$ is

$$b_{3,x_0}^{-} = \begin{cases} \dfrac{\log(1-3^{x_0})}{\log 2} & \text{if } x_0 < 0 \\ \dfrac{\log(3^{x_0}-1)}{\log 2} & \text{if } x_0 > 0 \end{cases}$$

and the upper extreme associated with any $x_0$ is

$$b_{3,x_0} = \frac{\log(1+3^{x_0})}{\log 2}.$$

If $x_0 = 0$, by dividing (3.4.2) by $2^{x+1}$, the equation of the level curve is

$$2^{x-1} + \cos(y \log 2) = 0$$

and the variable $x$ runs on $(-\infty, 1]$.

Thus, for $x_0 < 0$, it is immediate that (3.4.2) represents an infinite family of closed curves containing each of them one zero of $G_3^*(z)$, so the level curve has infinitely many arc-connected components. For $x_0 = 0$, the above equation represents an infinite family of open curves having horizontal asymptotes oriented to $-\infty$ of equations $y = (2k+1)\dfrac{\pi}{2\log 2}$, $k \in \mathbb{Z}$, so the level curve has infinitely many arc-connected components. Finally, for $x_0 > 0$, (3.4.2) is the equation of a unique open curve, that is, the level curve has only one arc-connected component. The curve meets to the real axis at the point $b_{3,x_0}^{-} = \log(3^{x_0}-1)/\log 2$, and the variable $y$ takes all real numbers.

We have just seen that $b_{3,x_0} = \log(1+3^{x_0})/\log 2$, so $b_{3,x_0}$ is a positive number for any real number $x_0$. Next, we will prove that this property is also true for all $n > 2$. Previously we need to demonstrate the existence of zeros of $G_n^*(z)$ on the half-plane $\Re z \geq 0$.

**Theorem 3.19** *For every integer $n > 2$, the function $G_n^*(z)$ has at least one zero on the half-plane $\Re z \geq 0$.*

*Proof* Since $G_n^*(z) = G_n(z) - p_{k_n}^z$ is an exponential polynomial, by Bochner's theorem [5, p. 105], $G_n^*(z)$ is an almost periodic function on any closed strip $[\alpha, \beta] \times \mathbb{R}$ of the complex plane. Following the definition of Bohr (see [5, p. 101]), it means that given any open substrip $S_{a,b} := (a, b) \times \mathbb{R}$, for every $\varepsilon > 0$ there exists a length $l = l(\varepsilon)$ such that every interval of length $l$ on the imaginary axis contains at least one translation number $\tau = \tau(\varepsilon)$ associated with $\varepsilon$, i.e., a number $\tau$ satisfying the inequality

$$\left|G_n^*(z+i\tau) - G_n^*(z)\right| \leq \varepsilon,$$

for all $z \in (a, b) \times \mathbb{R}$.

On the other hand, $G_n^*(z)$ is an exponential polynomial of the form (3.1.2). Then as we have just seen at the end of Sect. 3.1.1, $G_n^*(z)$ has infinitely many zeros in a vertical strip of the complex plane. If all the zeros of $G_n^*(z)$ were situated on $\Re z < 0$, let $w_0$ be one of them. Now we claim that there exist positive numbers $\delta$ and $l$ such that on any segment of length $l$ on the line $x = \Re w_0$ there is a point $\Re w_0 + iT$ verifying

$$\left|G_n^*(\Re w_0 + iT)\right| \geq \delta. \tag{3.4.3}$$

Indeed, we fix a closed strip $[\alpha, \beta] \times \mathbb{R}$ such that $\alpha < \Re w_0 < \beta$ and let us denote by $S_{a,b} := (a, b) \times \mathbb{R}$ a substrip whose extreme points satisfy $\alpha < a < \Re w_0 < b < \beta$. By taking $\delta = \dfrac{G_n^*(\Re w_0)}{2} > 0$, the almost periodicity of $G_n^*(z)$ involves the existence of a real number $l = l(\delta)$ such that every interval of length $l$ on the imaginary axis contains at least a translation number $\tau_1 = \tau_1(\delta)$ satisfying

$$\left|G_n^*(z + i\tau_1) - G_n^*(z)\right| \leq \delta, \quad \text{for all } z \in S_{a,b}. \tag{3.4.4}$$

By substituting $z = \Re w_0$ in (3.4.4) and according to the choice of $\delta$, the inequality (3.4.3) follows for $T = \tau_1$, as claimed. By reiterating this process for every $\delta_m := \dfrac{\delta}{2^{m-1}}$, $m \geq 1$, the almost periodicity of $G_n^*(z)$ assures the existence of a real number $l_m = l_m(\delta_m)$ such that every interval of length $l_m$ contains at least a translation number $\tau_m = \tau_m(\delta_m)$ which, without loss of generality, can be chosen in such a way that $(\tau_m)_{m=1,2,\ldots}$ be an increasing sequence of positive numbers with $\tau_{m+1} - \tau_m > 1$, for all $m$, satisfying

$$\left|G_n^*(z + i\tau_m) - G_n^*(z)\right| \leq \delta_m, \quad \text{for all } \ z \in S_{a,b}. \tag{3.4.5}$$

Then, by just making $z = w_0$ in (3.4.5), we have

$$\lim_{m\to\infty} G_n^*(w_0 + i\tau_m) = G_n^*(w_0) = 0. \tag{3.4.6}$$

We now define the sequence of analytic functions on $S_{a,b}$

$$H_m(z) := G_n^*(z + i\tau_m), \quad z \in S_{a,b}, \ m = 1, 2, \ldots, \tag{3.4.7}$$

which are uniformly bounded on the compacts of $S_{a,b}$ by virtue of the boundedness of $G_n^*(z)$ on any vertical strip of the complex plane (this is an immediate consequence of the property

$$\max\left\{\left|G_n^*(z)\right| : \Re z \leq \sigma\right\} = G_n^*(\sigma),$$

for an arbitrary real number $\sigma$). Hence, by Montel's theorem, there exists a uniformly convergent subsequence $\left\{H_{m_k}(z) : k = 1, 2, \right\}$ to an analytic function, say $H(z)$, on $S_{a,b}$. The function $H(z)$, by (3.4.3), is not identically 0 on $S_{a,b}$ and, because of (3.4.7) and (3.4.6), has a zero at the point $w_0$ which, by analyticity, is not isolated. Given

$\varepsilon > 0$, we determine a positive number

$$r < \min\left\{\Re w_0 - a, b - \Re w_0, \varepsilon, \frac{1}{2}\right\}$$

such that $|H(z)| \neq 0$ for all $z$ on the circle $|z - w_0| = r$, so the number

$$M := \min\{|H(z)| : |z - w_0| = r\}$$

is positive. Noticing $|z - w_0| = r$ is a compact set in the strip $S_{a,b}$ and $\lim_{k\to\infty} H_{m_k}(z) = H(z)$ is uniform on the compacts of $S_{a,b}$, there exists a positive integer $k_0$ such that

$$\left|H(z) - H_{m_k}(z)\right| < M \leq |H(z)|, \quad \text{for all } |z - w_0| = r \text{ and for all } \ k > k_0.$$

By Rouché's theorem, this means that every function $H_{m_k}(z)$, for all $k > k_0$, has in the disk $|z - w_0| < r$ as many zeros as $H(z)$ does, so at least one, say $w_{m_k}$. Then, from (3.4.7), one has

$$G_n^*(w_{m_k} + i\tau_{m_k}) = 0, \quad \text{for all } \ k > k_0.$$

Finally, since $\left|w_{m_k} - w_0\right| < r < \frac{1}{2}$ for all $k > k_0$ and $\tau_{m_{k+1}} - \tau_{m_k} > 1$ for all $k$, the points $w_{m_k} + i\tau_{m_k}$ are distinct for all $k > k_0$. Hence, the function $G_n^*(z)$ has a zero at each point $w_{m_k} + i\tau_{m_k}$ for any $k > k_0$. Consequently, $G_n^*(z)$ has infinitely many zeros in the strip $S_\varepsilon := \{z : \Re w_0 - \varepsilon < \Re z < \Re w_0 + \varepsilon\}$, for arbitrary $\varepsilon > 0$. Now, by taking $\varepsilon = -\frac{\Re w_0}{2}$, $G_n^*(z)$ has infinitely many zeros having real part less than $\frac{\Re w_0}{2}$. Therefore, as we are assuming that all the zeros of $G_n^*(z)$ are situated on $\Re z < 0$, it means that

$$\sum_{l=1}^{\infty} \Re z_{n,l} = -\infty, \tag{3.4.8}$$

where $\left(z_{n,l}\right)_{l=1,2,\dots}$ denotes the sequence of all the zeros of $G_n^*(z)$. Now, we write

$$G_n^*(z) = 1 + e^{z\log 2} + \cdots + e^{z\log n}, \tag{3.4.9}$$

where the right side of (3.4.9) has not the term $e^{z\log p_{k_n}}$, $p_{k_n}$ being the last prime not exceeding $n$. Then, by Ritt's formula [31, formula (9)], we have

$$\sum_{l=1}^{\infty} \Re z_{n,l} = O(1), \tag{3.4.10}$$

which contradicts (3.4.8). Thus, $G_n^*(z)$ has at least a zero on the half-plane $\Re z \geq 0$, so the theorem follows. □

Recalling the meaning of the upper extreme associated with a real number $x_0$ given in Definition 3.1, in the next lemma we prove that any level curve $\left|G_n^*(z)\right| = p_{k_n}^{x_0}$, for any integer $n > 2$ and an arbitrary real number $x_0$, has points inside the half-plane $\Re z > 0$.

**Lemma 3.4** *Given any integer $n > 2$ and any real number $x_0$, its upper extreme associated $b_{n,x_0}$ is a positive number.*

*Proof* The level curve $\left|G_n^*(z)\right| = p_{k_n}^{x_0}$ divides to the complex plane in the two open regions $\Omega_1 := \left\{z : \left|G_n^*(z)\right| < p_{k_n}^{x_0}\right\}$ and $\Omega_2 := \left\{z : \left|G_n^*(z)\right| > p_{k_n}^{x_0}\right\}$. By using the preceding theorem, let $w$ be a zero of $G_n^*(z)$ such that $\Re w \geq 0$. Since $G_n^*(w) = 0$, one has $\left|G_n^*(w)\right| < p_{k_n}^{x_0}$. Then $w$ is either an interior point to a closed arc-connected component of the level curve $\left|G_n^*(z)\right| = p_{k_n}^{x_0}$ or is a point situated on the left to an open arc-connected component (the first case is obvious, the second case holds by noting that $\max\left\{\left|G_n^*(z)\right| : \Re z \leq \sigma\right\} = G_n^*(\sigma)$, for any real number $\sigma$). Therefore, in both cases, there exists a point $z_1 = (x_1, y_1)$ of the level curve $\left|G_n^*(z)\right| = p_{k_n}^{x_0}$ such that $\Re w < x_1$. Now, by Definition 3.1, it follows that $x_1 \leq b_{n,x_0}$ and then

$$0 \leq \Re w < x_1 \leq b_{n,x_0},$$

which proves the lemma. □

Since the zeros of $G_2(z) = 1 + \frac{1}{2^z}$, or those of the partial sum $\zeta_2(z) := 1 + 2^z$, are of the form

$$\frac{\pi(2k+1)i}{\log 2}, \quad k \in \mathbb{Z},$$

the sets $R_{G_2(z)} = R_{\zeta_2(z)} = \{0\}$. Then $n = 2$ is the trivial case and so we will assume that $n > 2$.

Let us observe that we have already demonstrated, by means of Theorems 3.17 and 3.18, that $R_{G_n(z)}$ and $R_{\zeta_n(z)}$ are perfect sets (actually they are closed intervals) provided that $n > N$, where $N$ is a positive integer whose existence has been proved in Lemma 3.2 depending on the Prime Number Theorem. Now we are going to prove that the set of the real projections of the zeros of every $G_n(z)$ situated on $\Re z \leq 0$ is dense in itself, for all $n > 2$.

**Theorem 3.20** *For every integer $n > 2$, the set of the real projections of the zeros of $G_n(z)$ has no isolated point on the interval $(-\infty, 0]$.*

*Proof* Firstly we claim that the bound $b_{G_n(z)} := \sup\{\Re z : G_n(z) = 0\}$ is a positive number for all $n > 2$. Indeed, by [22, Proposition 3], $G_n(z)$ has at least a zero, say $w_0$, outside the imaginary axis. By assuming that $w_0$ satisfies $\Re w_0 < 0$ and by making an analogous proof to that of Theorem 3.19, we would be lead to the same conclusion, namely, that there are infinitely many zeros of $G_n(z)$ situated on the strip $S_\varepsilon := \{z : \Re w_0 - \varepsilon < \Re z < \Re w_0 + \varepsilon\}$, for arbitrary $\varepsilon > 0$. Hence $G_n(z)$ would have infinitely many zeros on the half-plane $\Re z < 0$. Thus, taking into account formula (3.4.10), i.e. , the formula of Ritt, $G_n(z)$ must have infinitely many zeros on the half-plane $\Re z > 0$. If we suppose that $\Re w_0 > 0$, then directly one has $b_{G_n(z)} > 0$. Therefore, in both cases, the upper bound $b_{G_n(z)}$ is a positive number, as claimed.

Let denote by $P_{G_n(z)}$ the set of the real projections of the zeros of $G_n(z)$ and by $R_{G_n(z)} := \overline{P_{G_n(z)}}$. Let assume that $P_{G_n(z)}$ has an isolated point, say $x_0$, on $(-\infty, 0]$. Then, $x_0 \in \left[a_{G_n(z)}, b_{G_n(z)}\right]$, where as usual $a_{G_n(z)} := \inf\{\Re z : G_n(z) = 0\}$. From above, $b_{G_n(z)} > 0$. Then we determine a $\delta > 0$ with $x_0 + \delta < b_{G_n(z)}$ such that

$$[x_0 - \delta, x_0 + \delta] \cap R_{G_n(z)} = \{x_0\}, \tag{3.4.11}$$

which means that $G_n(z)$ has no zero on the strip $[x_0 - \delta, x_0 + \delta] \times \mathbb{R}$, except on the line $x = x_0$ where there is at least one. Since $x_0 \in R_{G_n(z)}$, by Theorem 3.13, the line $x = x_0$ meets to the level curve $\left|G_n^*(z)\right| = p_{k_n}^{x_0}$ at a point, say $z_0 = (x_0, y_0)$. Now we claim that the level curve does not traverse the line $x = x_0$ (from the left to the right). Indeed, if this is so, we can find a point $z_1 = (x_1, y_1)$ of the level curve $\left|G_n^*(z)\right| = p_{k_n}^{x_0}$ such that $x_0 < x_1 < x_0 + \delta$. Then, since $x_0 \in \left[a_{G_n(z)}, b_{G_n(z)}\right]$ and for the choice of $\delta$ is $x_0 + \delta < b_{G_n(z)}$, it follows that

$$x_1 \in \left[a_{G_n(z)}, b_{G_n(z)}\right]. \tag{3.4.12}$$

On the other hand, as $z_1$ belongs to the level curve, one has $\left|G_n^*(z_1)\right| = p_{k_n}^{x_0} < p_{k_n}^{x_1}$. Hence the function $A_n(x, y) := \left|G_n^*(x + iy)\right| - p_{k_n}^{x}$, on the point $z_1$, satisfies

$$A_n(x_1, y_1) = \left|G_n^*(z_1)\right| - p_{k_n}^{x_1} = p_{k_n}^{x_0} - p_{k_n}^{x_1} < 0.$$

Now, noting (3.4.12), from Theorem 3.14, we obtain $A_n(x_1, 0) \geq 0$. Thus, the continuity of the function $A_n(x_1, y)$ implies the existence of a point $(x_1, y_2)$, with $0 \leq y_2 < y_1$, such that $A_n(x_1, y_2) = 0$. By using again the Theorem 3.14, it means that $x_1 \in R_{G_n(z)}$ and, since $x_0 < x_1 < x_0 + \delta$, this contradicts (3.4.11). Thus, the claim follows. Therefore, $x_0$ is the maximum value that we can assign to the variable $x$ to have, for at least a value of $y$, a point of the level curve $\left|G_n^*(z)\right| = p_{k_n}^{x_0}$. Consequently, taking into account the Definition 3.1, $x_0$ coincides with $b_{n,x_0}$, the upper extreme associated with $x_0$. This is a contradiction because, by Lemma 3.4, $b_{n,x_0}$ is a positive number and $x_0 \leq 0$. This completes the proof. □

From the preceding theorem an immediate consequence follows.

**Corollary 3.2** *For any integer $n > 2$, the closure of set of the real projections of the zeros of $G_n(z)$ on $(-\infty, 0]$ is a perfect set.*

Now we obtain the main results of the present paper.

**Theorem 3.21** *For every integer $n > 2$, the set of real projections of the zeros of the nth partial sum of the Riemann zeta function, $\zeta_n(z) := \sum_{k=1}^{n} 1/k^z$, has no isolated point on $[0, \infty)$.*

*Proof* By Theorem 3.20 and taking into account the relation (3.3.3), the theorem follows. □

The corresponding conclusion from this theorem is the following.

**Corollary 3.3** *For every integer $n > 2$, the closure of the set of the real projections of the zeros of $\zeta_n(z)$ on $[0, \infty)$ is a perfect set.*

In particular, the above corollary follows for the interval $(0, 1)$. That is, if we consider the critical strip of the Riemann zeta function $(0, 1) \times \mathbb{R}$, according to inequalities (3.1.6), we have the following result.

**Theorem 3.22** *The zeros of any partial sum $\zeta_n(z) := \sum_{k=1}^{n} 1/k^z$, $n > 2$, situated in the critical strip of the Riemann zeta function are distributed in such a way that their real projections define a set dense in itself. Therefore, its closure is a perfect set. Consequently, for any $n > 2$, no vertical strip $[a, b] \times \mathbb{R}$ with $0 \leq a < b \leq \min\left\{1, b_{\zeta_n(z)}\right\}$ is zero-free, where*

$$b_{\zeta_n(z)} := \sup\left\{\Re z : \zeta_n(z) = 0\right\} = 1 + \left(\frac{4}{\pi} - 1 + o(1)\right)\frac{\log\log n}{\log n}, \quad n \to \infty.$$

## References


1. Apostol, T.M.: Modular Functions and Dirichlet Series in Number Theory. Springer, New York (1990)
2. Avellar, C.E., Hale, J.K.: On the zeros of exponential polynomials. J. Math. Anal. Appl. **73**, 434–452 (1980)
3. Balazard, M., Velásquez-Castañón, O.: Sur l'infimum des parties réelles des zéros des sommes partielles de la fonction zêta de Riemann. C. R. Acad. Sci. Paris Ser. I **347** 343–346 (2009)
4. Bohr, H.: Zur theorie der allgemeinen dirichletschen reihen. Math. Ann. **79**, 136–156 (1919)
5. Bohr, H.: Almost Periodic Functions. Chelsea Publishing Company, New York (1947)
6. Borwein, P., Choi, S., Rooney, B., Weirathmueller, A.: The Riemann Hypothesis. Springer, Canada (2008)
7. Dubon, E., Mora, G., Sepulcre, J.M., Ubeda, J.I., Vidal, T.: A note on the real projection of the zeros of partial sums of the Riemann zeta function. RACSAM. doi:10.1007/s13398-012-0094-2
8. Edwards, H.M.: Riemann's Zeta Function. Academic Press, New York (1974)
9. Farag, H.M.: Dirichlet truncations of the riemann zeta function in the critical strip possess zeros near every vertical line. Inter. J. Number Theory **4**(4), 653–662 (2008)
10. Hardy, G.H., Wright, E.M.: An Introduction to the Theory of Numbers. Clarendon Press, Oxford (1954)
11. Haselgrove, C.B.: A disproof of a conjecture of Pólya. Mathematika **5**, 141–145 (1958)
12. Henry, D.: Linear autonomous neutral functional differential equations. J. Differ. Eqn. **15**, 106–128 (1974)
13. Kuratowski, K.: Topology. Academic Press, New York (1966)
14. Langer, R.E.: On the zeros of exponential sums and integrals. Bull. Amer. Math. Soc. **37**, 213–239 (1931)
15. Lapidus, M.L., Pomerance, C.: The riemann zeta-function and the one-dimensional Weyl-Berry conjecture for fractal drums. Proc. London Math. Soc. **66**(3), 41–69 (1993)

16. Lapidus, M.L., Van Frankenhuysen, M.: Complex dimensions of self-similar fractal strings and diophantine approximation. Exp. Math. **12**, 41–69 (2003)
17. Lapidus, M.L., van Frankenhuysen, M.: Fractality, self-similarity and complex dimensions. In: Proceeding of Symposia in Pure Mathematics 72, Part 1. American Mathematical Society, Providence (2004)
18. Lapidus, M.L., Van Frankenhuysen, M.: Fractal geometry, complex dimensions and zeta functions: geometry and spectra of fractal strings. Springer, New York (2006)
19. Levinson, N.: Asymptotic formula for the coordinates of the zeros of sections of the zeta function, $\zeta_N(s)$, near $s = 1$. Proc. Nat. Acad. Sci. USA **70**, 985–987 (1973)
20. Montgomery, H.L.: Zeros of approximations to the zeta function. In: Studies in Pure Mathematics: To the Memory of Paul Turán, pp. 497–506. Birkhäuser, Basel (1983)
21. Montgomery, H.L., Vaughan, R.C.: Mean values of multiplicative functions. Period. Math. Hungar. **43**, 199–214 (2001)
22. Mora, G.: A note on the functional equation $F(z) + F(2z) + \cdots + F(nz) = 0$. J. Math. Anal. Appl. **340**, 466–475 (2008)
23. Mora, G., Sepulcre, J.M.: On the distribution of zeros of a sequence of entire functions approaching the riemann zeta function. J. Math. Anal. Appl. **350**, 409–415 (2009)
24. Mora, G., Sepulcre, J.M.: The critical strips of the sums $1 + 2^z + \cdots + n^z$. Abstr. Appl. Anal. **909674**, 15 (2011). doi:10.1155/2011/909674
25. Mora, G., Sepulcre, J.M.: The zeros of riemann zeta partial sums yield solutions to $f(x) + f(2x) + \cdots + f(nx) = 0$. Mediterr. J. Math. **10**, 1221–1233 (2013)
26. Mora, G., Sepulcre, J.M.: Privileged regions in critical strips of non-lattice dirichlet polynomials. Complex. Anal. Oper. Theory **7**, 1417–1426 (2013)
27. Mora, G.: On the asymptotically uniform distribution of the zeros of the partial sums of the riemann zeta function. J. Math. Anal. Appl. **403**, 120–128 (2013)
28. Mora, G., Sepulcre, J.M., Vidal, T.: On the existence of exponential polynomials with prefixed gaps. Bull. London Math. Soc. (2013). doi:10.1112/blms/bdt043
29. Moreno, J.C.: The zeros of exponential polynomials (I). Compos. Math. **26**(1), 69–78 (1973)
30. Pólya, G., Szëgo, G.: Problems and theorems in analysis, vol. II. Springer, New York (1976)
31. Ritt, J.F.: On the zeros of exponential polynomials. Trans. Amer. Math. Soc. **31**, 680–686 (1929)
32. Spira, R.: Zeros of sections of the zeta function. I. Math. Comp. **20**, 542–550 (1966)
33. Spira, R.: Zeros of sections of the zeta function. II. Math. Comp. **22**, 168–173 (1968)
34. Tamarkin, J.D.: The zeros of certain integral functions. J. London Math. Soc. **2**, 66–69 (1927)
35. Tamarkin, J.D.: Some general problems of the theory of ordinary linear differential equations and expansions of an arbitrary function in series of fundamental functions. Math. Z. **27**, 1–54 (1928)
36. Titchmarsh, E.C.: The Theory of Functions, 2nd edn. Oxford University Press, London (1939)
37. Turán, P.: On some approximate dirichlet polynomials in the theory of the zeta-function of riemann. Dansk. Vid. Selsk. Mat.-Fys Medd. **24**(17), 3–36 (1948)
38. Voronin, S.M.: On the zeros of partial sums of the dirichlet series for the riemann zeta-function. Dolk. Akad. Nauk. SSSR. **216**, 964–967 (1974) (Trans. Soviet. Math. Doklady **15**, 900–903, 1974)
39. Wilder, C.E.: Expansion problems of ordinary linear differential equations with auxiliary conditions at more than two points. Trans. Amer. Math. Soc. **18**, 415–442 (1917)

# Chapter 4
# The Density Character of the Space $C_p(X)$

**María Muñoz Guillermo, J. C. Ferrando and M. López-Pellicer**

**Abstract** The main purpose of this survey is to introduce to the reader the adequate framework and motivation for the recent results obtained relating the density character and the space of the continuous functions, [16]. The interest in this cardinal function has been continuous over the years. We will offer a vision of the process along the time and we will point out different general results. Specially, we are interested in those in which the space of the continuous functions appears as well as those in which duality plays an important role. Of course, precise classes of spaces are considered in each case to apply the results, which will take us forward to expose a parallel development and description of a specific class, in fact it will be the development of a different cardinal function, the number of Nagami, which measures the specific property of the space what makes things work well.

**Keywords** Density character · Lindelöf $\Sigma$-space · Locally convex space · Space of the continuous functions · Weakly compact generated

**MSC 2010 Classification:** Primary 54A25, Secondary 54C35

---

M. Muñoz Guillermo (✉)
Departamento de Matemática Aplicada y Estadística, Universidad Politécnica de Cartagena, 30202 Cartagena, Murcia, Spain
e-mail: maria.mg@upct.es

J. C. Ferrando
Centro de Investigación Operativa, Universidad Miguel Hernández, 03202 Elche, Alicante, Spain
e-mail: jc.ferrando@umh.es

M. López-Pellicer
Departmento de Matemática Aplicada and IUMPA, Universitat Politècnica de València, 46002 Valencia, Spain
e-mail: mlopezpe@mat.upv.es

J. C. Ferrando and M. López-Pellicer (eds.), *Descriptive Topology and Functional Analysis*, Springer Proceedings in Mathematics & Statistics 80, DOI: 10.1007/978-3-319-05224-3_4, 
© Springer International Publishing Switzerland 2014

## 4.1 Introduction

Following Hödel [21, p. 10] a *cardinal function* $\Psi$ is a function from the class of all topological spaces (or some precisely defined subclass) into the class of all cardinals such that $\Psi(X) = \Psi(Y)$ whenever $X$ and $Y$ are homeomorphic. Cardinal functions have been used to establish general topological properties. In some sense cardinal functions measure the cardinality of "good" topological properties.

The *density character* is a cardinal function whose definition can be found in early papers, see for example [18, p. 329].

**Definition 4.1** Let $X$ be a topological space. The *density character of* $X$, denoted by $d(X)$, is the least infinite cardinal number of a dense subset of $X$.

When the density character is countable we say that $X$ is a *separable space*.

The interest in this property has been continuous over the years. Recently, in [16] a general result involving the space of the continuous functions with the pointwise convergence topology and the density character has been proved. Throughout this survey we will offer a general framework for the density character along the time. We will point out different general results, specially we are interested in those in which the space of the continuous functions appears as well as those in which duality plays an important role. Of course, precise classes of spaces are considered in each case to apply that results, which will take us forward to expose a parallel development of a description of a specific class, in fact it will be the development of a different cardinal function which measures the specific property of the space what makes things work well.

Proofs are included, in other case we indicate the reference where they can be found.

As we have already mentioned, classic results refer to the density character. B. Pospíšil pointed out in [33] that for a topological space $X$ the cardinality of $X$, $|X|$, is bounded by $2^{2^{d(X)}}$.

About the stability properties we must mentioned that the density character is not stable respect to subspaces. A cardinal function $\phi$ is *monotone* if $\phi(Y) \leq \phi(X)$ for every subspace $Y$ of $X$. After this definition we have that the density character is not monotone. An easy example is the following one, take $X = \beta\mathbb{N}$ (the Stone-Čech compactification of $\mathbb{N}$) which is separable while $X \backslash \mathbb{N}$ is not. Respect to the product we have the following theorem:

**Theorem 4.1** ([19, 32]) *Let* $\mathfrak{m}$ *be an infinite cardinal number. Let* $X = \prod_{i \in I} X_i$ *where* $X_i$ *is a topological space with* $d(X_i) \leq \mathfrak{m}$. *If* $|I| \leq 2^{\mathfrak{m}}$, *then* $d(X) \leq \mathfrak{m}$.

Throughout this chapter *topological space* means completely regular Hausdorff topological space. We use $X, Y$ to denote topological spaces. Notation and terminology are standard, our basic references are [13, 15, 23–25, 39] and it will be provided as needed. The set of the continuous functions from $X$ to $Y$ is denoted by $C(X, Y)$, when $Y = \mathbb{R}$, the set is simply denoted by $C(X)$. $C_c(X)$ and $C_p(X)$ denote the space of the continuous functions endowed with the compact-open topology and pointwise convergence topology, respectively. The *weight* of $X$, $w(X)$, is the least cardinality of an open base of $X$. The *weak weight* of $X$, $iw(X)$, is the smallest infinite cardinal number $\mathfrak{m}$ such that there exists a continuous bijection from $X$ onto a space $Y$with $w(Y) \leq \mathfrak{m}$. The density character of the space of the continuous functions was considered by N. Noble in [30]. In the following theorem the density character of the space of the continuous functions is related with the weak weight of $X$.

**Theorem 4.2** ([30, Theorem 1]) *Let $X$ be a topological space and let $C(X)$ be the set of the continuous functions on $X$ endowed with any topology between the pointwise convergence topology and the compact-open topology. Then*

$$d(C(X)) = iw(X). \tag{4.1}$$

This result can also be found in [27, Theorem 4.2.1]. The proof that follows is obtained from this last reference.

*Proof* ([27, p. 53]) To prove $d(C(X)) = iw(X)$, it is enough to prove that

$$d(C_c(X)) \leq iw(X) \leq d(C_p(X)).$$

Let $\phi : X \to Y$ be a continuous bijection where $w(Y) = iw(X)$. Then the induced function $\phi^* : C_c(Y) \to C_c(X)$ defined as $\phi^*(f) := f \circ \phi$ gives us the following chain of inequalities

$$d(C_c(X)) \leq d(C_c(Y)) \leq nw(C_c(Y)) \leq w(Y) = iw(X),$$

where has been used that the density character is less than or equal to the network weight, see Definition 4.8.3. To see that $iw(X) \leq d(C_p(X))$ consider an infinite dense subset $D$ of $C_p(X)$ having cardinality $d(C_p(X))$. Define $\psi : X \to \mathbb{R}^D$ by $\pi_f \circ \psi(x) = f(x)$ for each $x \in X$ and $f \in D$. Now $\psi$ is a bijective continuous map since $D$ separates points of $X$. Thus, $iw(X) \leq w(\psi(X)) \leq |D| = d(C_p(X))$. □

The next corollary follows immediately from Theorem 4.2.

**Corollary 4.1** ([27, p. 54]) *The following statements are equivalent:*

1. *$C_c(X)$ is separable.*
2. *$C_p(X)$ is separable.*
3. *$X$ has a coarser separable metrizable topology.*

## 4.2 The *Nice* Lindelöf-$\Sigma$ Class

In order to get the adequate framework, in this section we point out the importance of a "nice" class of spaces that will give us the key to develop a general cardinal function that will be used in the results that involve, on one hand, the density character and, on the other hand, the space of the continuous functions.

Let us recall the definition of Lindelöf-$\Sigma$-space. Following Arkhangelskiĭ, [1, p. 6] the class of Lindelöf $\Sigma$-spaces is defined as

> *the smallest class of spaces containing all compacta, all spaces with a countable base, and closed under the following three operations: taking the product of two spaces, transition to a closed subspace, and transition to a continuous image.*

The class of Lindelöf-$\Sigma$-spaces has become to be very popular not only in Topology but also in Functional Analysis in which it is called the class of countably $K$-determined spaces. The class of $\Sigma$-spaces was introduced by Nagami in 1969, [29]. Later, the class of $\Sigma$-spaces with the Lindelöf property was considered. In the frame of Functional Analysis the class of $K$-analytic spaces was developed, see [10, 11, 17]. In fact, the class of $K$-analytic spaces was proved as an important class since M. Talagrand proved that each weakly compactly generated Banach space is $K$-analytic in the weak topology. M. Talagrand introduced $K$-analyticity in Banach spaces, see [35, 36], in order to find a *good* class of Banach spaces with the property of being Lindelöf for the weak topology. The more general class of countably $K$-determined spaces was developed, [40]. The related literature is full of references to the Lindelöf $\Sigma$-class (or equivalently, countably $K$-determined class), mainly because their good properties. The title of the survey of V. Tkachuck: *Lindelöf $\Sigma$-spaces: an omnipresent class* [37] is quite representative of this fact. Other reason for the presence of the Lindelöf $\Sigma$-class is that there are tools to use it in different ways. This means that the definition with which we began this section is descriptive but not too useful to work with it. We will show different characterizations of the notion of Lindelöf $\Sigma$-class.

In [10] a topological space is called *K-analytic* if it is the continuous image of a $K_{\sigma\delta}$ of a compact space. In fact, the notion of countably $K$-determined space was described as follows:

**Definition 4.2** ([40, Definition 1]) Let $Y$ be a topological space, $X \subset Y$ is *countably determined in* $Y$ if and only there are a countable family of compact sets $\{A_n : n \in \mathbb{N}\}$ in $Y$ such that for any $x \in X$ there exists $J \subset \mathbb{N}$ such that

$$x \in \bigcap_{n \in J} A_n \subset X.$$

Nevertheless, M. Talagrand showed up the importance of using the notion of upper semicontinuous maps, [35].

**Definition 4.3** Let $X$ and $Y$ be topological spaces. A multivalued map $\phi : X \to 2^Y$ is said to be *upper semicontinuous* in $x_0 \in X$ if $\phi(x_0)$ is not empty and for each open set $V$ in $Y$ with $\phi(x_0) \subset V$ there exists an open set $U$ of $x_0$ in $X$ such that $\phi(U) \subset V$. A multivalued map $\phi$ is said to be upper semicontinuous if it is upper semicontinuous for each point in $X$. We will say that a multivalued map $\phi : X \to 2^Y$ is *usco* if $\phi$ is upper semicontinuous and the set $\phi(x)$ is compact for each $x \in X$.

More about usco maps can be found in [7] and the references therein.

Now we can introduce the definition of $K$-analyticity in different terms.

**Definition 4.4** A topological space is *$K$-analytic* if there exists an usco map from $\mathbb{N}^{\mathbb{N}}$ into $2^X$ covering $X$.

Using the notion of usco map, an equivalent definition for the class of countably $K$-determined can also be obtained.

**Definition 4.5** ([35, Proposition 1.1]) A topological space is called *countably $K$-determined* is there exists a subspace $\Sigma \subset \mathbb{N}^{\mathbb{N}}$ and an usco map $\phi : \Sigma \to 2^X$ covering $X$.

And the notion of Lindelöf $\Sigma$-space can also be given in terms of usco maps.

**Definition 4.6** The following conditions are equivalent for any topological space $X$:

1. $X$ is a Lindelöf $\Sigma$-space.
2. There exist a second countable space $M$ and an usco map $\phi : M \to 2^X$ such that $X = \bigcup\{\phi(x) : x \in M\}$.
   The previous definitions are equivalent to the following one.
3. There exist a compact cover $\mathscr{C}$ of the space $X$ and a countable family $\mathscr{N}$ of subsets of $X$ such that for any $C \in \mathscr{C}$ and any open set $U$ such that $C \subset U$ there exists $N \in \mathscr{N}$ such that $C \subset N \subset U$.

In fact, there are many more equivalent definitions to the notion of Lindelöf $\Sigma$-space. In [37, Theorem 1] a summary of the different equivalences can be found.

The proposition that follows give us the key to see that the class of countably $K$-determined spaces and the class of Lindelöf $\Sigma$-spaces are the same class.

**Proposition 4.1** *Let $M$ be a metric space and $\mathfrak{m}$ an infinite cardinal number such that the weight of $M$ is $\mathfrak{m}$, $w(M) = \mathfrak{m}$. Then there exist a set $I$ with $|I| = \mathfrak{m}$, a subspace $\Sigma \subset I^{\mathbb{N}}$, where $I$ is endowed with the discrete topology and a continuous onto map $f : \Sigma \to (M, d)$.*

Now, since the different equivalent definitions, the following inclusions are obvious.

$K$-analytic space $\Rightarrow$ Lindelöf $\Sigma$-space $\Rightarrow$ Lindelöf space.

We close this section with a well-known result of M. Talagrand that relates the density character and the weight (of a topology coarser than the original one) for a Lindelöf $\Sigma$-space, see [35, Theorem 2.4]. This result will be of great interest for their applications, as we will see later.

**Theorem 4.3** ([35]) *Let $(X, \tau)$ be a Lindelöf $\Sigma$-space and $\tau'$ a topology coarser than $\tau$, then*

$$d(X, \tau) \leq w(X, \tau').$$

## 4.3 Generalizing the Notion of Lindelöf $\Sigma$-Spaces

In the light of the results obtained with the class of Lindelöf $\Sigma$-spaces, the following step is trying to generalize this notion. In this sense, there are some cardinal functions which have been introduced with the idea of generalizing the notion of Lindelöf $\Sigma$-space. Some of them are the following:

**Definition 4.7** Let $X$ be a topological space.

1. The strong $\Sigma$-degree, $\Sigma(X)$, [20], is the smallest infinite cardinal number $\mathfrak{m}$ such that $X$ has a strong $\Sigma$-net $\{\mathscr{F}_\alpha : \alpha \in A\}$ with $|A| = \mathfrak{m}$. A family $\{\mathscr{F}_\alpha : \alpha \in A\}$ of locally finite closed covers of a space $X$ is a strong $\Sigma$-net for $X$ if the following hold for each $x \in X$:

    a. $C(x) = \bigcap\{C(x, \mathscr{F}_\alpha) : \alpha \in A\}$ is compact;
    b. $\{C(x, \mathscr{F}_\alpha) : \alpha \in A\}$ is a base for $C(x)$ in the sense that given any open set $U$ containing $C(x)$, there exists $\alpha$ in $A$ such that $C(x, \mathscr{F}_\alpha) \subset U$,

    where $C(x, \mathscr{F}_\alpha) := \bigcap\{F \in \mathscr{F}_\alpha : x \in F\}$.
2. The $\ell\Sigma$-number of $X$, $\ell\Sigma(X)$, [6], is the smallest infinite cardinal number such that there exist a metric space $M$ and an usco map $\phi : M \to 2^X$ covering $X$.
3. The Nagami number of $X$, $Nag(X)$, [6, 29], is the smallest infinite cardinal number $\mathfrak{m}$ such that there exist a topological space $Y$ of weight $\mathfrak{m}$ and an usco map $\phi : Y \to 2^X$ covering $X$.
4. The index of $\mathscr{K}$-analyticity, $\ell K(X)$, [3] is the least infinite cardinal $\mathfrak{m}$ for which there exists a complete metric space $M$ of weight $\mathfrak{m}$ and an usco map $\phi : M \to 2^X$ covering $X$.

5. The index of compact generation of $X$, $CG(X)$, [3] is the least infinite cardinal number $\mathfrak{m}$ such that there exists a family $\{K_\lambda : \lambda < \mathfrak{m}\}$ of compact subsets of $X$ whose union is a dense subset of $X$.

Analyzing the previous definitions it is obvious that all of them are related in some sense. The following results show the relationships between them. Recall that for a topological space $X$, the *Lindelöf number*, $\ell(X)$, is the smallest infinite cardinal $\mathfrak{m}$ such that for any open cover of $Y$ there exists a subcover of cardinality less than or equal to $\mathfrak{m}$.

**Proposition 4.2** ([3, 6]) *Let X be a topological space. Then, the following inequalities hold*

$$\ell(X) \le Nag(X) \le \ell\Sigma(X) \le \ell K(X) \le CG(X).$$

**Proposition 4.3** ([6]) *Let X be a topological space. Then,*

$$Nag(X) = \max\{\ell(X), \Sigma(X)\}.$$

## 4.4 The Cardinal Function $Nag(X)$

The cardinal functions $Nag(X)$ and $\ell\Sigma(X)$ generalize the notion of Lindelöf $\Sigma$-space, in the sense that if $Nag(X)$ or $\ell\Sigma(X)$ are countable then $X$ is a Lindelöf $\Sigma$-space. Indeed, both cardinal functions (with some differences) measure the minimal number of compact sets we need to get a continuous cover of a topological space $X$. The cardinal function $Nag(X)$ is less than or equal to $\ell\Sigma(X)$, although they are in general different. In [6] an example in which both cardinal functions are not the same is showed. We know that the Lindelöf $\Sigma$-class has a good behavior with respect to the stability properties. The following natural question is if the generalization of this notion, in our case, the number of Nagami, verifies the same stability properties. We have that the answer is positive and the same occurs for the general case. The following proposition, that can be found in [6], give us the mentioned properties.

**Proposition 4.4** 1. *For any family* $(X_i)_{i\in I}$ *we have that*

$$Nag\left(\prod_{i\in I} X_i\right) \le \max\left\{|I|, \sup_{i\in I} Nag(X_i)\right\}.$$

2. *Let* $X$ *be a topological space and* $Z \subset X$ *a closed subspace. Then* $Nag(Z) \le Nag(X)$.
3. *Let* $X$ *and* $Y$ *be topological spaces and* $\phi : X \to 2^Y$ *an usco map such that* $Y = \bigcup_{x\in X} \phi(x)$. *Then*
$$Nag(Y) \le Nag(X).$$
*In particular, the statement is true for onto continuous functions.*
4. *Let* $(X_i)_{i\in I}$ *a family of subspaces of* $X$, *then*

$$Nag\left(\bigcup_{i\in I} X_i\right) \le \max\left\{|I|, \sup_{i\in I} Nag(X_i)\right\}.$$

We must recall that we are interested in applying these cardinal functions to state some topological cardinal inequalities involving $C_p(X)$, that is, the space of the real-valued continuous functions endowed with the pointwise convergence topology. For that we need also other cardinal functions. For a topological space $X$, we consider the following ones.

**Definition 4.8** Let $X$ be a topological space.

1. The *tightness* of $X$, $t(X)$, is the smallest infinite cardinal $\mathfrak{m}$ such that for any set $A \subset X$ and any point $y \in \overline{A}$ there is a set $B \subset A$ for which $|B| \le \mathfrak{m}$ and $y \in \overline{B}$.
2. The *hereditarily density* of $X$, $hd(X)$, is the maximal cardinality of $\{d(Y) : Y \subset X\}$.
3. The *network weight* of $X$, $nw(X)$ is the minimal cardinality of a network in $X$ (a network in a space $X$ is a family $S$ of subsets of the set $X$ such that for any point $x \in X$ and any open set $O_x$ such that $x \in O_x$ there is $P \in S$ such that $x \in P \subset O_x$.
4. The *Hewitt-Nachbin number* of $X$, $q(X)$, is the minimal cardinality $\mathfrak{m}$ such that $X$ is $\mathfrak{m}$-placed in $\beta X$, that is for each $x \in \beta X\backslash X$ there is a set $P$ of type $G_{\mathfrak{m}}$ (intersection of a family of open sets of cardinality less than or equal to $\mathfrak{m}$) such that $x \in P \subset \beta X \setminus X$. A Hewitt $\mathfrak{m}$-extension of $X$ is the subspace $\upsilon_{\mathfrak{m}}X$ of $\beta X$ consisting of all $x \in \beta X$ for which any set of type $G_{\mathfrak{m}}$ in $\beta X$ containing $x$ intersects $X$. If $\mathfrak{m} = \omega_0$ the set $\upsilon_{\mathfrak{m}}X$ is just the (Hewitt) realcompactification of $X$. When $q(X)$ is countable, $X$ is called a realcompact space.

The following proposition generalizes the Theorem 4.3 of M. Talagrand.

**Proposition 4.5** ([6]) *Let* $(X, \tau)$ *be a topological space and* $\tau'$ *a topology in* $Y$ *coarser than* $\tau$. *Then*

$$d(X, \tau) \le \max\{Nag(X, \tau), nw(X, \tau')\}. \tag{4.2}$$

The following result establishes the relationship between the network and the weak weight of a topological space through the number of Nagami.

**Proposition 4.6** ([28]) *Let $X$ and $Y$ be topological spaces and $f : X \to Y$ a continuous bijection, then*

$$nw(X) \leq \max\{Nag(X), w(Y)\}. \tag{4.3}$$

Respect to the relationship between the number of Nagami and the number of Hewitt-Nachbin we have that the inclusions $X \subset \upsilon_\gamma X \subset \upsilon_\lambda X$ for $\lambda \leq \gamma$ are obvious, in fact, $\upsilon_\gamma X$ is $C$ -embedded in $\upsilon_\lambda X$.

In particular, $X$ is $C$-embedded in $\upsilon_\lambda X$ for every infinite cardinal number $\lambda$. We recall that a subspace $Y$ of a space $X$ is *$C$-embedded in $X$* if every real-valued continuous function on $Y$ can be extended to a real-valued continuous function on $X$.

Observe also that if $Y$ is a subspace of $X$ which is $C$-embedded in $X$, then the restriction map, $\pi : C_p(X) \to C_p(Y)$, is a continuous onto map, hence

$$Nag(C_p(Y)) \leq Nag(C_p(X)).$$

The following proposition establishes different bounds which will be useful later, see [38, Theorem 1].

**Proposition 4.7** *Let $X$ be a topological space. The following inequalities hold.*

1. $q(X) \leq d(C_p(X)) \leq hd(C_p(X))$.
2. $q(X) \leq Nag(X)$.

## 4.5 The Density Character and the Duality

In previous sections we have shown some general results about the density character and its relationship with other cardinal functions. In this section all our efforts are directed in attempting to motivate the recent results for the space of the continuous functions given in [16]. For that, we will show some results that involve not only the density character but also the duality. In particular, we are interested in those results in which we can obtain information about the density character of a space and the density character of the dual.

Additional terminology is required. All topological vector spaces and all locally convex spaces are assumed over the real field. If $E$ is a topological vector space, the space $E'$ denotes the set of all continuous linear functionals on $E$. Let $F$ be a subset of $E'$, we write $\sigma(E, F)$ to denote the locally convex topology on $X$ of pointwise convergence on $F$. The topology $\sigma(E, E')$ is the weak topology of $E$ and $\sigma(E', E)$ is the weak* topology of $E'$. The locally convex topology on $E$ of uniform convergence of all $\sigma(E', E)$-compact convex balanced sets in $E'$ is called the Mackey topology on $E$ and denoted by $\mu(E, E')$.

### 4.5.1 The Density Character for WCG

D. Amir and J. Lindenstrauss [2, 26] studied the structure of the class of weakly compact sets in Banach spaces. In this section we consider a well-known result for the class of weakly compactly generated spaces.

**Definition 4.9** A locally convex space $E$ is *weakly compactly generated* (WCG) if there exists an absolutely convex weakly compact total subset of $E$.

Recall that a subset of a locally convex space $E$ is *total* if its closed linear span is equal to $E$. Subspaces of WCG spaces behave badly. Rosenthal [34], has given an example of a closed linear subspace of a WCG Banach space which is not WCG.

The following result can be found in [26, Proposition 2.2] in a slightly different terminology. The proof follows the steps described in the book [15, Theorem 11.3].

**Proposition 4.8** ([26, pp. 239–240]) *Let E be a weakly compactly generated (WCG) Banach space then,*

$$d(E) \leq d(E', \sigma(E', E)).$$

*Proof* [15] Let $K$ be a weakly compact convex subset of $E$ that generates $E$, and let $D$ be a set of cardinality $d(E', \sigma(E', E))$ that is $\sigma(E', E)$-dense in $E'$.

Define an operator $T : E' \to C(K)$ by $T(f) : K \to \mathbb{R}$ given by $T(f)(k) = f(k)$ for each $k \in K$.

Since $T(D)$ separates points of $K$, the algebra generated by $T(D)$ and the constant functions contains a dense set $N$ of cardinality $d(E', \sigma(E', E)) = |T(D)|$ that is, by the Stone-Weierstrass Theorem, dense in $C(K)$.

For every $g \in N$, choose $k_g \in K$ such that $g(k_g) = \sup_{k \in K} g(k)$. We claim that $S = \{k_g : g \in N\}$ is $\sigma(E, E')$ dense in $K$. Indeed, we have that $\sup_K(\tilde{g}) = \sup_{g \in N}(\tilde{g}(k_g))$ for every $\tilde{g} \in C(K)$ since otherwise for $g \in N$ sufficiently close in $C(K)$ to $\tilde{g}$ we can not have that $\sup_{k \in K}(g) = g(k_g)$.

Therefore, $S$ is weakly dense in $K$ by Urysohn's Theorem, so *span*$(S)$ is $\sigma(E, E')$-dense in $E$. Since *span*$(S)$ is weakly and thus norm dense in $E$, we have that $d(E) \leq d(E', \sigma(E', E))$. □

The inequality $d(E) \geq d(E', \sigma(E', E))$ also holds. Let $K$ be a weakly compact set generating $E$, and let $D$ be a set of cardinality $d(E', \sigma(E', E))$ that is $\sigma(E', E)$-dense in $E'$. The topology $\tau$ of pointwise convergence on $D$ coincides with the weak topology of $K$ because $K$ is $\sigma(E, E')$-compact. The topology $\tau$ has a base of cardinality $|D|$ and thus there is a set $S$ of cardinality $|D|$ which is $\sigma(E, E')$ dense in $K$. Therefore, according to [12, Theorem 13, p. 422], $d(E, \sigma(E, E')) \leq |D|$.

### 4.5.2 The Density Character for Weakly Compactly Generated Locally Convex Spaces

R. J. Hunter and J. Lloyd considered in [22] the class of *weakly compactly locally convex spaces*, denoted cWCG. Let's see the definition.

**Definition 4.10** A locally convex space $E$ is *countably weakly compactly generated* if there exists a sequence of absolutely convex weakly compact subsets of $E$ whose union is total (or equivalently, dense) in $E$.

Every separable locally convex space is cWCG. If $E$ is a Fréchet space, then $(E', \sigma(E', E))$ is cWCG. In general, WCG and cWCG spaces are different. Each cWCG Fréchet space is a WCG space, [22, Proposition 1.1], but $\mathbb{R}^{(\mathbb{N})}$, which denotes the linear subspace of $\mathbb{R}^{\mathbb{N}}$ consisting of all maps taking all but finitely many elements of $\mathbb{N}$ equal to zero is cWCG but not WCG, see [22].

The following proposition [22, Proposition 1.2], give us the key to prove a result in which the density character for this class is involved.

**Proposition 4.9** *Let $E$ be a locally convex space. Then $E$ is cWCG if and only if there exists a metrizable locally convex topology $\tau$ on $E'$, which is coarser than $\mu(E', E)$.*

Now, a theorem for this class involving the density character and duality follows, extending the result of J. Lindenstrauss to the class of cWCG locally convex spaces.

**Theorem 4.4** *Let $E$ be a cWCG locally convex space. Then*

$$d(E) \leq d(E', \sigma(E', E)).$$

*Proof* Let $E$ be a cWCG locally convex space. By Proposition 4.9, there exists a metrizable locally convex topology $\tau$ on $E'$ coarser than $\mu(E', E)$. The topological dual $G$ of $(E', \tau)$ is contained in $E$ and $\sigma(E', G)$ is a Hausdorff topology on $E'$ which is coarser than $\sigma(E', E)$. Let $D$ be a $\sigma(E', E)$-dense subset of $E'$ of cardinality equal to $\mathfrak{m}$. Since $(E', \tau)$ is metrizable, $G$ has a $\sigma(G, E')$-dense subset of cardinality equal to $\mathfrak{m}$. But $G$ is $\sigma(E, E')$-dense in $E$ because $G$ separates points of $E'$. Thus $E$ has a $\sigma(E, E')$-dense subset of cardinality $\mathfrak{m}$ and the proof is over. □

### 4.5.3 The Density Character for Weakly Countably K-Determined Spaces

A Banach space $E$ is said to be *weakly countably $K$-determined* if $E$ endowed with the weak topology is countably $K$-determined (or equivalently a Lindelöf $\Sigma$-space). We have used in the title of this subsection the notation *countably $K$-determined* instead of *Lindelöf $\Sigma$-class* in order to be coherent with the original notation since the following result was established in these terms by L. Vašák [40, Corollary 2].

**Theorem 4.5** *Let $E$ be a weakly countably $K$-determined space, then*

$$d(E) = d(E', \sigma(E', E)).$$

The same theorem was proved by M. Talagrand [35, Théorème 6.1] in a different way, as an application of Theorem 4.3, with $\tau$ the weak topology and $\tau'$ the poinwise convergence topology on a weak*-dense set of the dual.

### *4.5.4 The Density Character for K-Analytic Locally Convex Spaces*

As we have mentioned, M. Talagrand introduced $K$-analyticity in Banach spaces, see [35, 36], in order to find a *good* class of Banach spaces with the property of being Lindelöf for the weak topology and countably $K$-determined.

M. Talagrand has proved that a WCG Banach space is weakly $K$-analytic and has constructed an example of type $C(K)$ to show that both cases are not the same (this is a consequence of the fact that closed subspaces of WCG Banach spaces can fail to be WCG, which was proved previously by Rosenthal [34]).

Now we can introduce a theorem that involves the density character for this class of spaces.

**Theorem 4.6** *Let $E$ be a weakly $K$-analytic locally convex space. Then the following statements follow:*

1. $d(E) \leq d(E', \sigma(E', E))$.
2. *If $E$ is metrizable,* $d(E) = d(E', \sigma(E', E))$.

*Proof* 1. Let $D$ be an infinite $\sigma(E', E)$-dense subset of $E'$ of cardinality $\mathfrak{m}$. Now the topology $\sigma(E, D)$ is a Hausdorff topology on $E$ coarser than $\sigma(E, E')$, which admits a fundamental system of zero neighborhoods of cardinality $\mathfrak{m}$. Since $(E, \sigma(E', E))$ is $K$-analytic, $(E, \sigma(E, E'))$ is Lindelöf, hence $(E, \sigma(E, D))$ is Lindelöf too and it admits a base of open sets of cardinality $\mathfrak{m}$. Now using Theorem 4.3, we have that $d(E, \sigma(E, E')) \leq \mathfrak{m}$.

2. Let $E$ be a dense subset of cardinality $\mathfrak{m}$ and $K$ be a $\sigma(E', E)$-compact subset of $E'$, then $K$ admits a base of uniformity of cardinality $\mathfrak{m}$ for the $\sigma(E', E)$ topology, and therefore a base of open sets of cardinality $\mathfrak{m}$. Then, $K$ has a dense subset of cardinality $\mathfrak{m}$. Now the result follows because $E'$ is a countable union of $\sigma(E', E)$-compact subsets. □

### *4.5.5 The Density Character for the Class $\mathfrak{G}$*

B. Cascales and J. Orihuela have introduced in [8] a large class of locally convex spaces that it is called the class $\mathfrak{G}$. The definition of this class is as follows.

**Definition 4.11** A locally convex space $E$ is said to be in the *class* $\mathfrak{G}$ if there exists a family $\{A_\alpha : \alpha \in \mathbb{N}^\mathbb{N}\}$ of subsets of $E'$ (called a $\mathfrak{G}$-representation of $E$) such that

1. $E' = \bigcup\{A_\alpha : \alpha \in \mathbb{N}^\mathbb{N}$.
2. $A_\alpha \subset A_\beta$ if $\alpha \le \beta$.
3. In each $A_\alpha$ all sequences are equicontinuous.

Condition 3 implies that every $A_\alpha$ is $\sigma(E', E)$-relatively countably compact. Therefore, if $E$ is in the class $\mathfrak{G}$, the space $(E', \sigma(E', E))$ has a relatively countably compact resolution. The class $\mathfrak{G}$ contains the metrizable and dual metric space and it is stable by taking subspaces, separated quotients, completions, countable direct sums and countable products. A good reference in which a completed description and more results about the class $\mathfrak{G}$ can be found is [24].

**Theorem 4.7** ([8, Theorem 13]) *Let $E$ be a locally convex space of the class $\mathfrak{G}$ which is weakly $K$-countably determined. Then,*

$$d(E, \sigma(E, E')) = d(E', \sigma(E', E)).$$

*Proof* Let $D$ be a dense subset in $(E, \sigma(E, E'))$. By [31, Theorem 6] the space $(E', \sigma(E', E))$ is angelic. Now, the space $(E', \sigma(E', E))$ is $K$-analytic using the $\mathfrak{G}$-representation of $E$ in $E'$. Let $\sigma(E', D)$ be the topology on $E'$ of pointwise convergence on $D$. Applying Theorem 4.3, we obtain that $d(E', \sigma(E', E))$ is less than or equal to $|D|$.

Conversely, if $D$ is a dense subset of $(E', \sigma(E', E))$, using the topology $\sigma(E, D)$ on $E$ and since $(E, \sigma(E, E'))$ is countably $K$-determined, again using Theorem 4.3, we obtain that $d(E, \sigma(E, E'))$ is less than or equal to the cardinality of $D$ and we are done. □

## 4.6 The Density Character for $C_p(X)$

A Banach space $C(K)$ of continuous functions on a compact space $K$ is WCG in and only if $K$ is an *Eberlein compact*, (i.e. if and only if $K$ is homeomorphic to a weakly compact set in some $c_0(\Gamma)$), [2]. On the other hand, every closed linear subspace $Y$ of a WCG Banach space is a Lindelöf $\Sigma$-space in the weak topology of $Y$, [14, Proposition 7.1.6]. Recall that a compact space $X$ is said to be a *Gul'ko compact* if $C_p(X)$ is a Lindelöf $\Sigma$-space. According to Theorem [14, 7.1.8] a compact space is Gul'ko compact if and only if the Banach space $C(X)$ is weakly Lindelöf $\Sigma$-space.

**Theorem 4.8** *If $X$ and $C_p(X)$ are Lindelöf $\Sigma$-spaces, then $d(X) = d(C_p(X))$.*

*Proof* If $L_p(X)$ denotes the weak* dual of $C_p(X)$, Theorem 4.3 asserts that

$$d(C_p(X)) \le d(L_p(X)).$$

If $D$ is a dense subspace of $X$, the rational span $H$ of $D$ is dense in $L_p(X)$. Since $|H| = |D|$, it follows that $d(L_p(X)) \leq d(X)$. On the other hand, if $\mathscr{F}$ is a family of real-valued functions dense in $C_p(X)$, let $\eta$ be the weakest topology on $X$ that makes continuous each $f \in \mathscr{F}$. Since $\eta$ is coarser than the original topology of $X$, Theorem 4.3 ensures that $d(X) \leq w(X, \eta) = |\mathscr{F}|$. So $d(X) \leq d(C_p(X))$ and we are done.

In particular, if $K$ is a Gul'ko compact space, then

$$d(K) = d(C_p(K)) = iw(C_p(K)),$$

by Theorem 4.2. This implies in particular that there exists on $C(K)$ a weaker completely regular topology $\tau'$ such that $w(C(K), \tau') = d(C_p(K)) = d(K)$.

All these results give us reasons to think about establishing a general result involving the density character and the space of the continuous functions. Using the number of Nagami, a version of the previous result (relating duality and density character) is provided in [16] for spaces $C_p(X)$. The main theorem is as follows.

**Theorem 4.9** ([16]) *Let $X$ be a topological space and $L \subset C_p(X)$. Then there exist a space $Y$ and completely regular Hausdorff topologies $\tau' \leq \tau$ for $Y$ such that:*

1. $Nag(Y, \tau) \leq Nag(X)$, $\ell\Sigma(Y, \tau) \leq \ell\Sigma(X)$, $d(Y, \tau) \leq d(X)$.
2. $w(Y, \tau') \leq d(L)$.
3. $nw(C_p(Y, \tau)) = nw(Y, \tau) \leq \max\{Nag(X), d(L)\}$, *hence*

$$d(Y, \tau) \leq \max\{Nag(X), d(L)\},$$

$$d(C_p(Y, \tau)) \leq \max\{Nag(X), d(L)\}.$$

4. *$L$ is embedded into $C_p(Y, \tau)$.*
5. $d(L) \leq \max\{Nag(L), d(Y, \tau)\}$.

*Proof* Let $D$ be a dense subset of $L$, such that $|D| = d(L)$. Let $\mathscr{T}_D$ and $\mathscr{T}_L$ be the weakest topologies on $X$ that make continuous all those real-valued functions that belong to $D$ or $L$, respectively. By density, $f(x) = f(y)$ for each $f \in D$ implies $f(x) = f(y)$ for each $f \in L$. Let $(\widehat{X}, \widehat{\mathscr{T}_D})$ and $(\widehat{X}, \widehat{\mathscr{T}_L})$ be the topological quotients of $(X, \mathscr{T}_D)$ and $(X, \mathscr{T}_L)$ with respect to the relation $x \sim y$ if and only if $f(x) = f(y)$ for all $f$ of $D$ and $L$, respectively. If we define the map $F : (X, \mathscr{T}_D) \to \mathbb{R}^D$ by $F(z) = \rho_z$, where $\rho_z(f) = f(z)$ for all $f \in D$, then clearly $F$ is continuous and $x \sim y$ if and only if $F(x) = F(y)$. Hence, $(\widehat{X}, \widehat{\mathscr{T}_D})$ is homeomorphic to a subspace of $\mathbb{R}^D$ and consequently

$$w(\widehat{X}, \widehat{\mathscr{T}_D}) \leq w(\mathbb{R}^D) \leq |D| = d(L).$$

On the other hand, since $(\widehat{X}, \widehat{\mathscr{T}_L})$ is a continuous image of $X$, we note that

$$Nag(\widehat{X}, \widehat{\mathscr{T}_L}) \leq Nag(X), \ \ell\Sigma(\widehat{X}, \widehat{\mathscr{T}_L}) \leq \ell\Sigma(X), \ d(\widehat{X}, \widehat{\mathscr{T}_L}) \leq d(X),$$

see [6, Proposition 7 (iv), Remark 8] and [13, Theorem 1.4.10]. Now, applying Proposition 4.6 we have that

$$nw(\widehat{X}, \widehat{\mathscr{T}_L}) \leq \max\{Nag(\widehat{X}, \widehat{\mathscr{T}_L}), w(\widehat{X}, \widehat{\mathscr{T}_D})\} \leq \max\{Nag(X), d(L)\}.$$

On the other hand, by [1, Theorem I.1.3] we have

$$nw(C_p(\widehat{X}, \widehat{\mathscr{T}_L})) = nw(\widehat{X}, \widehat{\mathscr{T}_L}).$$

In particular, as the density is less than or equal to the network weight we have that

$$d(\widehat{X}, \widehat{\mathscr{T}_L}) \leq \max\{Nag(X), d(L)\},$$
$$d(C_p(\widehat{X}, \widehat{\mathscr{T}_L})) \leq \max\{Nag(X), d(L)\}.$$

Therefore $(Y, \tau) := (\widehat{X}, \widehat{\mathscr{T}_L})$ is a completely regular Hausdorff space such that

$$d(Y, \tau) \leq \max\{Nag(X), d(L)\}$$

and

$$d(C_p(Y, \tau)) \leq \max\{Nag(X), d(L)\}.$$

Now set $(Y, \tau') := (\widehat{X}, \widehat{\mathscr{T}_D})$. Clearly $\tau' \leq \tau$ on $Y$.

Recall that

$$\widehat{x} := \{y \in X : f(y) = f(x) \text{ for all } f \in L\},$$

define $T : L \to C_p(Y, \tau)$ by $T(f) = \widehat{f}$, where $\widehat{f}(\widehat{x}) := f(x)$. Note that if $y \in \widehat{x}$, then

$$\widehat{f}(\widehat{y}) = f(y) = f(x) = \widehat{f}(\widehat{x}),$$

since $f \in L$, so that $\widehat{f}$ is well-defined. On the other hand, $\widehat{f} \in C_p(Y, \tau)$. Indeed, if $\widehat{x_d} \to \widehat{x}$ in $(Y, \tau)$, then

$$\widehat{f}(\widehat{x_d}) = f(x_d) \to f(x) = \widehat{f}(\widehat{x}).$$

If $\widehat{f}(\widehat{x}) = \widehat{g}(\widehat{x})$ for all $x \in X$ then $f(x) = g(x)$ for all $x \in X$, which implies that $f = g$ and consequently that $\widehat{f} = \widehat{g}$, so that $T$ is injective. Moreover, if $f_p \to f$ in $L$ then $f_p(x) \to f(x)$ for every $x \in X$, which implies that $\widehat{f_p}(\widehat{x}) \to \widehat{f}(\widehat{x})$. Hence, $T$ is continuous. Finally, if $\widehat{f_p} \to \widehat{f}$ in $T(L)$ under the pointwise convergence topology, then $\widehat{f_p}(\widehat{x}) \to \widehat{f}(\widehat{x})$ and hence $f_p(x) \to f(x)$ for all $x \in X$. This shows that $T$ is a homeomorphism from $L$ into $C_p(Y, \tau)$.

Lastly, let $B$ be a dense subset of $(Y, \tau)$ of cardinality $d(Y, \tau)$. The restriction map

$$j : C_p(Y, \tau) \to C_p(B)$$

is bijective and continuous. Since $w(C_p(B)) = |B|$, we deduce that $C_p(Y, \tau)$ admits a weaker Hausdorff topology $\xi$ such that $w(C(Y, \tau), \xi) \leq |B|$. Hence, $L$ admits a weaker topology $\xi_L$ such that $w(L, \xi_L) \leq |B|$. By Proposition 4.5 we get that

$$d(L) \leq \max\{Nag(L), w(L, \xi_L)\} \leq \max\{Nag(L), |B|\}. \qquad \square$$

From statement 5 of Theorem 4.9, if $L$ is a Lindelöf $\Sigma$-space and $Nag(X) \leq d(L)$ (this holds for example if $X$ is a Lindelöf $\Sigma$-space), then

$$d(L) = d(Y, \tau).$$

Consequently, if $X$ is a Lindelöf $\Sigma$-space and $Nag(C_p(X)) \leq d(X)$, since $X$ is embedded into $C_p(C_p(X))$ there exists a space $Y$ admitting a weaker completely regular Hausdorff topology $\tau'$ such that $X$ embeds into $C_p(Y)$ and

$$d(X) = d(Y) \geq w(Y, \tau').$$

The following corollary, [16, Corollary 1], provides bounds considering extensions of the space $X$.

**Corollary 4.2** ([16]) *Let $X$ be a topological space and $\mathfrak{m}$ an infinite cardinal number with $\aleph_0 \leq \mathfrak{m} \leq q(X)$. Let $L \subset C_p(X)$ be a subspace of $C_p(X)$ such that $d(L) = \mathfrak{m}$. Then there exist a space $Y$ and completely regular Hausdorff topologies $\tau$ and $\tau'$ on $Y$ such that $\tau' \leq \tau$ and:*

1. $Nag(Y, \tau) \leq \min\{Nag(\upsilon_\lambda X) : \lambda \geq \mathfrak{m}\} \leq Nag(\upsilon_{\mathfrak{m}} X)$,
   $\ell\Sigma(Y, \tau) \leq \min\{\ell\Sigma(\upsilon_\lambda X) : \lambda \geq \mathfrak{m}\} \leq \ell\Sigma(\upsilon_{\mathfrak{m}} X)$.
2. $w(Y, \tau') \leq d(L)$.
3. $d(Y, \tau) \leq \min\{Nag(\upsilon_\lambda X) : \lambda \geq \mathfrak{m}\} \leq Nag(\upsilon_{\mathfrak{m}} X)$.
4. *$L$ is embedded into $C_p(Y, \tau)$.*

Particular results are obtained simply applying the results to the countable case, see [16]. In order to establish a clear relationship with the previous results we point out some applications to the case of locally convex spaces. The following applications can be found in [16].

**Corollary 4.3** ([9]) *A $\sigma(E, E')$-compact set $Y$ in a lcs $E$ from the class $\mathfrak{G}$ is $\sigma(E, E')$-metrizable if and only if $Y$ is contained in a $\sigma(E, E')$-separable subset of $E$.*

As we have been noted in [16], Theorem 4.9 applies to show Theorem 4.7.

**Corollary 4.4** *Let $E$ be a lcs from the class $\mathfrak{G}$ such that $(E, \sigma(E, E'))$ is a Lindelöf $\Sigma$-space, then $d(E, \sigma(E, E')) = d(E', \sigma(E', E))$.*

*Proof* Since $E$ belongs to $\mathfrak{G}$ and $L := (E, \sigma(E, E'))$ is a Lindelöf $\Sigma$-space, the space $X := (E', \sigma(E', E))$ is $K$-analytic by applying [31, Theorem 21] and [5, Corollary 1.1]. Clearly, $L \subset C_p(X)$, so by Theorem 4.9 (5,1) there exists $Y$ such that $d(L) = d(Y) \leq d(X)$. Note that $d(X) \leq d(L)$ also holds. Indeed, since $X \subset C_p(C_p(X))$, by Theorem 4.9 (5,1) there exists $Z$ such that

$$d(X) \leq \max\{Nag(X), d(Z)\} = d(Z) \text{ and } d(Z) \leq d(C_p(X)).$$

On the other hand, $d(C_p(X)) = iw(X)$, Theorem 4.2. If $B$ is a dense subset of $L$, then $\sigma(E', B) \leq \sigma(E', E)$, so $iw(X) \leq d(L)$. Hence, $d(X) \leq d(L)$. □

## References


1. Arkhangel'skiĭ, A.V.: Topological Function Spaces, Mathematics and Its Applications (Soviet Series), vol. 78. Kluwer, Dordrecht (1992)
2. Amir, D., Lindenstrauss, J.: The structure of weakly compact sets in Banach spaces. Ann. Math. **88**, 35–46 (1968)
3. Avilés, A.: The number of weakly compact sets which generate a Banach space. Israel J. Math. **159**, 189–204 (2007)
4. Canela, M.A.: $K$-analytic locally convex spaces. Port. Math. **41**, 105–117 (1982)
5. Cascales, B.: On $K$-analytic locally convex spaces. Arch. Math. **49**, 232–244 (1987)
6. Cascales, C., Muñoz, M., Orihuela, J.: The number of $K$-determination of topological spaces. RACSAM **106**, 341–357 (2012)
7. Cascales, C., Oncina, L.: Compactoid filters and USCO maps. J. Math. Anal. Appl. **282**, 826–845 (2003)
8. Cascales, B., Orihuela, J.: On compactness in locally convex spaces. Math. Z. **195**, 365–381 (1987)
9. Cascales, B., Orihuela, J.: On pointwise and weak compactness in spaces of continuous functions. Bull. Soc. Math. Belg. Ser. B **40**, 331–352 (1988)
10. Choquet, G.: Theory of capacities. Ann. Inst. Fourier **5**, 131–297 (1953)
11. Choquet, G.: Ensembles $K$-analytiques et $K$-sousliniens. Cas général et cas métrique. Ann. Inst. Fourier **9**, 75–89 (1959)
12. Dunford, N., Schwartz, J.T.: Linear Operators. Part I. Interscience Pub, New York (1958)
13. Engelking, R.: General Topology. PWN-Polish Scientific Publishers, Warsaw (1977)
14. Fabian, M.: Gâteaux Differentiability of Convex Functions and Topology. Weak Asplund Spaces. Canadian Mathematical Society Series of Monographs and Advanced Texts. John Wiley and Sons, New York (1997)
15. Fabian, M., Habala, P., Hájek, P., Montesinos, V., Pelant, J., Zizler, V.: Functional Analysis and Infinite Dimensional Geometry. CMS Books in Mathematics, Springer, New York (2001)
16. Ferrando, J.C., Kąkol, J., López-Pellicer, M., Muñoz Guillermo, M.: Some cardinal inequalities for spaces $C_p(X)$. Topology Appl. **160**, 1102–1107 (2013)
17. Fremlin, D.H.: $K$-analytic spaces with metrizable compacts. Mathematika **24**, 257–261 (1977)
18. Hewitt, E.: A problem of set-theoretic topology. Duke Math. J. **10**, 309–333 (1943)
19. Hewitt, E.: A remark on density characters. Bull. Amer. Math. Soc. **52**, 641–643 (1946)

20. Hödel, R.: On a theorem of Arkhangel'skiĭ concerning Lindelöf $p$-spaces. Can. J. Math. **27**, 459–468 (1975)
21. Hödel, R.: Cardinal Functions I. In: Kunen, K., Vaughan, J.E. (eds.) Handbook of Set-Theoretic Topology, pp. 1–61. Elsevier, Amsterdam (1984)
22. Hunter, R.J., Lloyd, J.: Weakly compactly generated locally convex spaces. Proc. Math. Proc. Cambridge Philos. Soc. **82**, 85–98 (1977)
23. Jarchow, H.: Locally Convex Spaces. B.G. Teubner, Stuttgart (1981)
24. Kąkol, J., Kubiś, W., López-Pellicer, M.: Descriptive Topology in Selected Topics of Functional Analysis. Springer, New York (2010)
25. Kelley, J.L.: General Topology. Springer, New York (1975)
26. Lindenstrauss, J.: Weakly compact sets: their topological properties and the Banach spaces they generate. Proc. Int. Symp. Topology Ann. Math. Stud. **69**, 235–273 (1972)
27. McCoy, R.A., Ntantu, I.: Topological Properties of Spaces of Continuous Functions. Springer, Berlin (1980)
28. Muñoz, M.: Bounds using the index of Nagami. Rocky Mountain J. Math. **41**, 1977–1986 (2011)
29. Nagami, K.: $\Sigma$-spaces. Fund. Math. **61**, 169–192 (1969)
30. Noble, N.: The density character of function spaces. Proc. Amer. Math. Soc. **42**, 228–233 (1974)
31. Orihuela, J.: Pointwise compactness in spaces of continuous functions. J. London Math. Soc. **s2–36**, 143–152 (1987)
32. Pondiczery, E.S.: Power problems in abstract spaces. Duke Math. J. **11**, 835–837 (1944)
33. Pospíšil, B.: Sur la puissance d'un espace contenant une partie dense de puissance donnée. Časopis pro pěstování Matematiky a Fysiky **67**, 89–96 (1937–1938)
34. Rosenthal, H.P.: The heredity problem for weakly compactly generated Banach spaces. Comp. Math. **124**, 83–111 (1970)
35. Talagrand, M.: Espaces de Banach faiblement $K$-analytiques. Ann. Math. **110**, 407–438 (1979)
36. Talagrand, M.: Sur la $K$-analyticite de certains espaces d'operateurs. Israel J. Math. **32**, 124–130 (1979)
37. Tkachuk, V.: Lindelöf $\Sigma$-spaces: an omnipresent class. RACSAM **104**, 221–244 (2010)
38. Uspenskiĭ, V.V.: A characterization of realcompactness in terms of the topology of pointwise convergence on the function space. Comm. Math. Univ. **24**, 121–126 (1983)
39. Valdivia, M.: Topics in Locally Convex Spaces. Mathematics Studies, vol. 67. North-Holland Publishing, Amsterdam (1982)
40. Vašák, L.: On a generalization of weakly compactly generated Banach spaces. Studia Math. **70**, 11–19 (1981)

# Chapter 5
# Compactness and Distances to Spaces of Continuous Functions and Fréchet Spaces

**Carlos Angosto and Manuel López-Pellicer**

**Abstract** In recent years, several *quantitative* counterparts for several classical such as Krein-Šmulyan, Eberlein-Šmulyan, Grothendieck, etc. have been proved by different authors. These new versions strengthen the original theorems and lead to new problems and applications in topology and analysis. In this survey, we present several of these quantitative versions of theorems about compactness in Banach spaces with the weak topology, Fréchet spaces with the weak topology and spaces of continuous functions with the pointwise convergence topology. For example if $H$ is a subset of a Banach space $E$, and $w^*$ is the weak* topology in $E''$, the index $k(H) := \sup\{d(x^{**}, E), x^{**} \in \overline{H}^{w^*}\}$ is zero if and only if $H$ is relatively compact in $(E, w)$. Then $k(H)$ measures how far is $H$ from being relatively compact in $(E, w)$. The following inequalty $k(co(H)) \leq 2k(H)$ is a quantitative version of the Krein-Šmulian theorem about the $w$-relative compactness of the convex hull of a weakly compact set.

**Keywords** Compactness · Fréchet space · Space of continuous functions · Webcompact · Distances

**MSC 2010 Classification:** 46A50, 54C35

---

C. Angosto (✉)
Departamento de Matemática Aplicada y Estadistica, Universidad Politécnica de Cartagena, 30203 Cartagena, Spain
e-mail: carlos.angosto@upct.es

M. López-Pellicer
Departamento de Matemática Aplicada and IMPA, Universitat Politècnica de València, 46022 Valencia, Spain
e-mail: mlopezpe@mat.upv.es

J. C. Ferrando and M. López-Pellicer (eds.), *Descriptive Topology and Functional Analysis*, Springer Proceedings in Mathematics & Statistics 80, DOI: 10.1007/978-3-319-05224-3_5, 
© Springer International Publishing Switzerland 2014

## 5.1 Preliminary Results

Recall that a subset $H$ of a topological space $X$ is countably compact (resp. relatively countably compact) if every sequence in $H$ has a cluster point in $H$ (resp. in $X$) and $H$ is sequentially compact (resp. relatively sequentially compact) if every sequence in $H$ has a convergent subsequence with limit in $H$ (resp. in $X$). Clearly a (relatively) compact set is (relatively) countably compact and a (relatively) sequentially compact set is (relatively) countably compact, but in general other implications do not hold.

Šmulian showed in 1940 [18] that if $H$ is a relatively compact set in the weak topology of a Banach space $E$, then $H$ is also relatively sequentially compact and he also proved that the converse implication is also true if the weak* dual is separable. Eberlein [8] proved in 1947 that the converse to Šmulian's theorem is true, i.e., for the weak topology of a Banach space a subset is relatively compact if and only if it is relatively countably compact. Grothendieck proved in 1952 [12] that this result also holds in any locally convex space that is quasicomplete for its Mackey topology. Grothendieck's result is based upon a similar result on spaces $C_p(K)$ of continuous functions on a compact set with the pointwise convergence topology. Kelley and Namioka [14] pointed out that a cluster point of a relatively countably compact subset of $C_p(K)$ is the limit of a sequence in $H$. J. D. Pryce [17] extended the Eberlein-Šmulian theorem to spaces $C_p(X)$ where $X$ has a dense subset that is the countable union of compact sets. A very useful tool that was used in the previous results was the following notion that was introduced by Grothendieck.

**Definition 5.1** Let $Z$ be a topological space, let $X$ be a set and $H \subset Z^X$. It is said that $H$ *interchanges limits with* $X$ if for each sequences $(f_n) \subset H$ and $(x_m) \subset X$

$$\lim_n \lim_m f_n(x_m) = \lim_m \lim_n f_n(x_m)$$

whenever all the limits involved exist.

If $H \subset C(X, Z)$ where $X$ is a countably compact space and $Z$ a compact metric space, then $H$ is relatively compact in the pointwise convergence topology if and only if $H$ interchanges limits with $X$ ($X$ can be replaced by a dense subset of $X$). The interchange of limits property is also very useful in locally convex spaces. For example a bounded subset of a Banach space is relatively compact in the weak topology of a Banach space if it interchanges limits with the unit dual ball.

Fremlin's notion of angelic space and some of its consequences can be used for proving the above results in a clever and clear way, see the book by Floret [10].

**Definition 5.2** A topological space is called *angelic* if for every relatively countably compact set $H \subset X$ the following holds:

- $H$ is relatively compact;
- for each $x \in \overline{H}$ there is a sequence in $H$ which converges to $x$.

In angelic spaces, a subset is (relatively) compact if and only if it is (relatively) countably compact if and only if it is (relatively) sequentially compact. Orihuela [15] introduced the following class of topological spaces:

**Definition 5.3** A topological space $X$ is a *web-compact space* if there is a subset $\Sigma$ of $\mathbb{N}^{\mathbb{N}}$, a family $\{A_\alpha : \alpha \in \Sigma\}$ of subsets of $X$ whose union $D$ is dense in $X$ and if we denote

$$C_{n_1,\ldots,n_k} := \bigcup\{A_\beta : \beta = (m_k) \in \Sigma, m_j = n_j, j = 1, \ldots, k\}$$

for $\alpha = (n_k) \in \Sigma$, then if $\alpha = (n_k) \in \Sigma$ and $x_k \in C_{n_1,n_2,\ldots,n_k}$, the sequence $(x_k)_k$ has a cluster point in $X$. If $X$ is web-compact with $D = X$, we call $X$ *strongly web-compact.*

This definition is very technical but includes a big class of spaces, for example spaces with a dense union of compact sets (in particular separable spaces), countably determined topological spaces or quasi-Suslin spaces are web-compact spaces.

**Theorem 5.1** (Orihulela [15]) *Let $X$ be a web-compact space. The space $C_p(X)$ is angelic.*

The main tool that Orihule used to prove the previous theorem was the following result.

**Theorem 5.2** (Orihulela [15]) *Let $(Z, d)$ be a compact metric space, $X$ a set and $H$ a subset of the space $(Z^X, \tau_p)$. We assume that:*

*(i) There is $\Sigma \subset \mathbb{N}^{\mathbb{N}}$ and a family $\{A_\alpha : \alpha \in \Sigma\}$ of non-void subsets of the set $X$ such that $X = \bigcup\{A_\alpha : \alpha \in \Sigma\}$.*
*(ii) For every $\alpha = (a_1, a_2, \ldots) \in \Sigma$ the set $H$ interchanges limits in $Z$ with every sequence $(x_n)_n$ in $X$ that is eventually in each set*

$$C_{\alpha|m} = \bigcup\{A_\beta : \beta \in \Sigma, b = (b_m) \text{ and } b_j = a_j \text{ for } j = 1, 2, \ldots, m\},$$

*for $m \in \mathbb{N}$.*

*Then for any $f \in \overline{H}^{Z^X}$ there exists a sequence $(f_n)_{n\in\mathbb{N}}$ in $H$ that converges pointwise to $f$ on $X$.*

*Proof* It follows from Lemma 5.4 □

In recent years, several *quantitative* counterparts for the previous and other classical results have been proved by different authors. These new versions strengthen the original theorems and lead to new problems and applications in topology and analysis. In this survey, we present several of that quantitative versions.

## 5.2 Distance to Banach Spaces

If $H$ is a bounded subset of a Banach space $E$, then $H$ is relatively compact in the weak topology if and only if when we consider $H$ as a subset of the bidual space $E''$ with the weak* topology, the weak* closure $\overline{H}^{w*}$ is a subset of $E$. Then we can think that the index

$$k(H) := \hat{d}(\overline{H}^{w^*}, E) = \sup\{d(x^{**}, E), x^{**} \in \overline{H}^{w^*}\}$$

measures how far is $H$ from being $w$-relatively compact. Fabian, Hajek, Montesinos and Zizler used this notion and proved the following quantitative version of the Krein theorem. The second part of the theorem was obtained by Granero.

**Theorem 5.3** [9, 11] *Let $H$ be a bounded subset of a Banach space $E$, then*

$$k(co(H)) \leq 2k(H).$$

*If $H$ is a bounded subset of the bidual space $E''$ then*

$$k(co(H)) \leq 5k(H).$$

Observe that if $H$ is $w$-relatively compact then $k(H) = 0$ and then by the previous theorem $k(co(H)) = 0$ which means that the convex hull of $H$ is also relatively compact in $(E, w)$. One tool that have been used in the proof of previous theorem is the *$\varepsilon$-interchange of limits* (quantitative version of Grothendieck notion when $\varepsilon = 0$). But this notion was previous used in [5] with the following index:

$$\gamma(H) := \sup\{|\lim_m \lim_n f_m(x_n) - \lim_n \lim_m f_m(x_n)| : (f_m)_m \subset B_{E'}, (x_n)_n \subset H\}$$

assuming that the iterated limit exist. Observe that $\gamma(H) = 0$ if and only if $H$ interchanges limits with the unit ball. In the definition of the index used in [5], the supremum is taked for $(x_n)_n \subset co(H)$ instead of $H$. Fabian et al. [9] also proved that the definition that we are using here is the same that the definition that appears in [5], i.e. $\gamma(H) = \gamma(co(H))$. This last result combined with the following inequalities

$$k(H) \leq \gamma(H) \leq 2k(H).$$

implies first part of Theorem 5.3.

We shall also consider the following index

$$ck(H) := \sup\{d(clust_{E''}(\phi), E), \phi \in H^{\mathbb{N}}\}$$

where $clust_{E''}(\phi)$ denotes the set of cluseter points of $\phi$ in $E''$ and $d(A, B) = \inf\{d(a, b) : a \in A, b \in B\}$. It is clear that if $H$ is relatively weakly countably compact then $ck(H) = 0$. The converse implication also holds by the following result.

**Theorem 5.4** [2] *If $H$ is a bounded subset of a Banach space $E$ then*

$$ck(H) \leq k(H) \leq \gamma(H) \leq 2ck(H).$$

*For any $x^{**} \in \overline{H}^{w*}$ there is a sequence $(x_n)$ in $H$ such that*

$$\|x^{**} - y^{**}\| \leq \gamma(H)$$

*for any cluster point $y^{**}$ of $(x_n)$ in $E''$.*

**Corollary 5.1** *If $E$ is a Banach space then $(E, w)$ is angelic.*

*Proof* Let $H$ be a $w$-relatively countably compact subset of $E$. By the very definition every sequence in $H$ has a $w$-cluster point in $E$ and therefore $ck(H) = 0$. Then by Theorem 5.4 we have $H$ is $w$-relatively compact in $E$. On the other hand, let us pick $x \in \overline{H}^{w}$. Note that then $\gamma(H) = 0$ and thus if we use Theorem 5.4 we obtain the existence of a sequence $(x_n)$ in $H$ such that every $w$-cluster point $y \in E$ of $(x_n)$ satisfies that $0 \leq \|y - x\| \leq \gamma(H) = 0$. Since $H$ is $w$-relatively compact and $(x_n)$ in $H$ and $x$ is the unique $w$-cluster point of $(x_n)$ we conclude that $(x_n)$ weakly converges to $x$ and the proof is over. □

**Corollary 5.2** [2] *If $H$ is a bounded subset of a Banach space $E$, the following conditions are equivalent:*

1. *$ck(H) = 0$.*
2. *$k(H) = 0$.*
3. *$H$ is weakly relatively countably compact.*
4. *$H$ is weakly relatively compact.*

In Theorem 5.4 the involved constants are sharp but sometimes we can obtain some equalities. Recall that a Banach space $E$ is said to have Corson property $\mathscr{C}$ if each collection of closed convex subsets of $E$ with empty intersection has a countable subcollection with empty intersection. If $(E, w)$ is Lindelöf, then $E$ has property $\mathscr{C}$. There are Banach spaces with Corson property $\mathscr{C}$ which are not weakly Lindelöf, [16, p. 146]. It is shown in [16] that a Banach space $E$ has the property $\mathscr{C}$ if and only if, whenever $A \subset E'$ and $x^* \in \overline{A}^{w^*}$, there is a countable subset $C$ of $A$ such that $x^* \in \overline{convC}^{w^*}$. In particular Banach spaces with $w^*$ angelic dual unit balls have Corson property $\mathscr{C}$.

**Proposition 5.1** [2] *If $E$ is a Banach space with Corson property $\mathscr{C}$, then for every bounded set $H \subset E$ we have $ck(H) = k(H)$.*

Several of the results of this section can be deduced from similar results in spaces of continuous functions, see Sect. 5.4. Next theorem gives the tool to export results obtained in the context of distances to spaces of continuous functions on a compact set to the context of Banach spaces.

**Theorem 5.5** [7] *Let $E$ be a Banach space and let $B_{E'}$ be the closed unit ball in the dual space $E'$ endowed with the $w^*$-topology. Let $i : E \to E''$ and $j : E'' \to \ell_\infty(B_E)$ the canonical embeddings. Then for every $z \in E''$ we have*

$$d(z, i(E)) = d(j(z), C(B_{E'})).$$

If we consider $\ell_\infty(B_{E'})$ as a subspace of $(\mathbb{R}^{B_{E'}}, \tau_p)$, then the natural embedding $j : (E'', w^*) \to (\ell_\infty(B_{E'}), \tau_p)$ is continuous. For a bounded set $H \subset E''$, the closure $\overline{H}^{w^*}$ is $w^*$-compact and therefore, the continuity of $j$ gives that $\overline{j(H)}^{\tau_p} = j(\overline{H}^{\tau_p})$. Then

$$d(\overline{j(H)}^{\tau_p}, C(B_{E'}), w^*)) = d((\overline{H}^{w^*}, i(E))$$

and

$$\hat{d}(\overline{j(H)}^{\tau_p}, C(B_{E'}), w^*)) = \hat{d}((\overline{H}^{w^*}, i(E)).$$

## 5.3 Distance to Fréchet Spaces

In this section we present generalizations of the results given in Sect. 5.2 when we deal con Fréchet spaces. These results can be found in [3, 4]. First introduce some notation. Let $E$ be a Fréchet space and let $(U_n)_n$ be a decreasing basis of absolutely convex neighborhoods of zero. By $(E', \beta(E', E))$ and $(E'', \beta(E'', E'))$ we mean the strong dual of $E$ and $(E', \beta(E', E))$, respectively. In $(E'', \beta(E'', E'))$ the sequence of bipolars $(U_n^{00})_n$ is a decreasing basis of absolutely convex neighborhoods of zero. By $\|h\|_n = \sup\{|h(u)| : u \in U_n^0\}$ we denote the seminorm in $E''$ associated with $U_n^0$ and $d_n$ means the pseudometric defined by $\|.\|_n$. The restriction of $\|.\|_n$ to $E$, also denoted by $\|.\|_n$, is the seminorm defined by $U_n$. The topology of $E$ can be defined by the $F$-norm $d(x, y) := \sum_n 2^{-n}\|x - y\|_n(1 + \|x - y\|_n)^{-1}$ for $x, y \in E$. The topology of the space $(E'', \beta(E'', E'))$ is defined by the $F$-norm $d(x^{**}, y^{**}) := \sum_n 2^{-n}\|x^{**} - y^{**}\|_n(1 + \|x^{**} - y^{**}\|_n)^{-1}$ for all $x^{**}, y^{**} \in E''$. Additionally without loss of generality, we assume in this paper that $2U_{n+1} \subset U_n$ for $n \in \mathbb{N}$; and this clearly implies that $2\|x^{**}\|_n \le \|x^{**}\|_{n+1}$ for $n \in \mathbb{N}$ and each $x^{**} \in E''$.

If $H$ is a bounded subset of $E$ then $H^0$ is a neighborhood of zero in $(E', \beta(E', E))$ and the bypolar $H^{00}$ is a compact subset of $(E'', \sigma(E'', E'))$. Therefore an $E$-bounded subset $H$ is weakly relatively compact if and only if $\overline{H}^{\sigma(E'', E')}$ is contained in $E$.

Next concepts are the natural extensions of the given above:

$$\gamma_n(H) := \sup\left\{\left|\lim_p \lim_m u_p(h_m) - \lim_m \lim_p u_p(h_m)\right| : (u_p) \subset U_n^0, (h_m) \subset H\right\}$$

assuming the involved limits exist. Let

$$ck_n(H) := \sup \left\{ d_n \left( clust_{E''}(\varphi), E \right) : \varphi \in H^{\mathbb{N}} \right\}$$

and

$$ck(H) := \sup \left\{ d \left( clust_{E''}(\varphi), E \right) : \varphi \in H^{\mathbb{N}} \right\}$$

where $clust_{E''}(\varphi) := \bigcap_p \overline{\{\varphi(m) : m > p\}}^{\sigma(E'',E')}$ is the set of all cluster points in $E''$ of the sequence $\varphi \in H^{\mathbb{N}}$ and $d_n(A, B) = \inf\{d_n(a, b) : a \in A, b \in B\}$. Also define

$$k_n(H) := \sup \left\{ d_n(h, E) : h \in \overline{H}^{\sigma(E'',E')} \right\},$$

and

$$k(H) := \sup \left\{ d(h, E) : h \in \overline{H}^{\sigma(E'',E')} \right\}.$$

We say that $H$ $\varepsilon$-interchanges limits with a subset $B$ of $E'$ if

$$\sup \left\{ \left| \lim_p \lim_m u_p(h_m) - \lim_m \lim_p u_p(h_m) \right| : (u_p) \subset B, (h_m) \subset H \right\} \leq \varepsilon$$

where $\varepsilon \geq 0$ and the involved limits exist. $\gamma_n(H) \leq \varepsilon$ ($\gamma_n(H) = 0$) means: $H$ $\varepsilon$-interchanges (interchanges) limits with $U_n^0$. Note that

$$2\gamma_n(H) \leq \gamma_{n+1}(H), \quad 2ck_n(H) \leq ck_{n+1}(H), \quad 2k_n(H) \leq k_{n+1}(H).$$

Hence $\sup_n \gamma_n(H) < \infty$, $\sup_n ck_n(H) < \infty$, $\sup_n k_n(H) < \infty$ iff $\gamma_n(H) = 0$, $ck_n(H) = 0$, $k_n(H) = 0$, $n \in \mathbb{N}$, respectively.

**Lemma 5.1** [3] *Let $H$ be a bounded subset of a Fréchet space $E$ and let $h \in \overline{H}^{\sigma(E'',E')}$. Then for each $n \in \mathbb{N}$ there exists a net $(u_\beta)_\beta$ in $U_n^0$ that $\sigma(E', E)$-converges to 0 and such that for each net $(h_\alpha)_\alpha$ in $H$ that $\sigma(E'', E')$-converges to $h$ we have $d_n(h, E) = \lim_\beta \lim_\alpha u_\beta(h_\alpha)$. Consequently, there exist sequences $(h_m)_m$ in $H$ and $(u_p)_p$ in $U_n^0$ such that $d_n(h, E) = \lim_p \lim_m u_p(h_m)$ and $\lim_m \lim_p u_p(h_m) = 0$. Hence $k_n(H) \leq \gamma_n(H)$.*

*Proof* The linear functional $u$ defined on the linear hull of $E$ and $h$ by $u(e + \lambda h) = \lambda d_n(h, E)$ for $e \in E$ verifies $|u(e + \lambda h)| = |\lambda| \, d_n(h, E) = d_n(\lambda h, E) = d_n(e + \lambda h, E) \leq \|e + \lambda h\|_n$. By the Hahn-Banach theorem $u$ admits a linear extension to $E''$, also named $u$, such that

$$\left| u(x^{**}) \right| \leq \left\| x^{**} \right\|_n$$

for each $x^{**} \in E''$. Clearly $u \in \left(U_n^{00}\right)^0 = (U_n^0)^{00}$ and we obtain a net $(u_\beta)_\beta$ in $U_n^0$ such that

$$u(x^{**}) = \lim_\beta u_\beta(x^{**})$$

for each $x^{**} \in E''$. In particular

$$d_n(h, E) = u(h) = \lim_{\beta} u_\beta(h), \ \ 0 = d_n(e, E) = u(e) = \lim_{\beta} u_\beta(e)$$

for each $e \in E$ so $(u_\beta)_\beta$ $\sigma(E', E)$-converges to 0. If $(h_\alpha)_\alpha$ is a net in $H$ that $\sigma(E'', E')$-converges to $h$, then each $u_\beta(h)$ is the limit of the net $\left(u_\beta(h_\alpha)\right)_\alpha$ and

$$d_n(h, E) = u(h) = \lim_{\beta} \lim_{\alpha} u_\beta(h_\alpha), \ \ 0 = \lim_{\alpha} u(h_\alpha) = \lim_{\alpha} \lim_{\beta} u_\beta(h_\alpha).$$

By Lemma 2.1 of [8] there exist sequences $(h_m)_m$ in $H$ and $\left(u_p\right)_p$ in $U_n^0$ such that

$$d_n(h, E) = \lim_{p} \lim_{m} u_p(h_m), \ \ 0 = \lim_{m} \lim_{p} u_p(h_m).$$

By $d_n(h, E) = \lim_p \lim_m u_p(h_m) - \lim_m \lim_p u_p(h_m)$ we have $k_n(H) \leq \gamma_n(H)$. □

**Lemma 5.2** [3] *Let $(h_\alpha)_\alpha$ be a net in a bounded subset $H$ of a Fréchet space $E$. Let $h$ be a $\sigma(E'', E')$-cluster point of $(h_\alpha)_\alpha$. If $(v_\beta)_\beta$ is a net in $U_n^0$ such that the involved limits* $\lim_\beta \lim_\alpha v_\beta(h_\alpha)$ *and* $\lim_\alpha \lim_\beta v_\beta(h_\alpha)$ *exist, then*

$$\left| \lim_{\beta} \lim_{\alpha} v_\beta(h_\alpha) - \lim_{\alpha} \lim_{\beta} v_\beta(h_\alpha) \right| \leq 2d_n(h, E).$$

*Hence $\gamma_n(H) \leq 2ck_n(H)$ for each $n \in \mathbb{N}$.*

*Proof* If $(u_\beta)_\beta$ is a net in $U_n^0$ that $\sigma(E', E)$-converges to 0 and the involved limits in $\lim_\beta \lim_\alpha u_\beta(h_\alpha)$ exist, then

$$\left| \lim_{\beta} \lim_{\alpha} u_\beta(h_\alpha) \right| \leq d_n(h, E). \tag{5.1}$$

Indeed, for each $\varepsilon > 0$ let $h_\varepsilon \in E$ be such that

$$d_n(h, h_\varepsilon) < d_n(h, E) + \varepsilon.$$

By the hypothesis $\lim_\alpha u_\beta(h_\alpha) = u_\beta(h)$ and $\lim_\beta u_\beta(h_\varepsilon) = 0$. Then

$$\left| \lim_{\beta} \lim_{\alpha} u_\beta(h_\alpha) \right| = \left| \lim_{\beta} u_\beta(h) \right| = \left| \lim_{\beta} u_\beta(h - h_\varepsilon) \right| \leq d_n(h, h_\varepsilon) < d_n(h, E) + \varepsilon.$$

The inequality $\left|\lim_\beta \lim_\alpha u_\beta(h_\alpha)\right| < d_n(h, E) + \varepsilon$ is true for each positive number $\varepsilon$, so we have

$$\left| \lim_{\beta} \lim_{\alpha} u_\beta(h_\alpha) \right| \leq d_n(h, E).$$

To prove the main inequality pick $v$ a $\sigma(E', E)$-cluster point of $(v_\beta)_\beta$. By hypothesis, $\lim_\alpha \lim_\beta v_\beta(h_\alpha)$ exists and $v(h_\alpha)$ is a cluster point of $(v_\beta(h_\alpha))_\beta$ so $\lim_\beta v_\beta(h_\alpha) = v(h_\alpha)$. Then $\lim_\alpha v(h_\alpha)$ exists and $v(h)$ is a cluster point of $(v(h_\alpha))_\alpha$ so $\lim_\alpha v(h_\alpha) = v(h)$. Therefore $u_\beta := 2^{-1}(v_\beta - v)$ is a net in $U_n^0$ such that 0 is a $\sigma(E', E)$-cluster point. Choosing a subnet we can suppose that $(u_\beta)_\beta$ $\sigma(E', E)$ converges to 0. The involved limits in $\lim_\beta \lim_\alpha u_\beta(h_\alpha)$ exist, because by hypothesis the limits in $\lim_\beta \lim_\alpha v_\beta(h_\alpha)$ exist and $\lim_\beta \lim_\alpha v(h_\alpha) = \lim_\beta v(h) = v(h)$. Then

$$\left|\lim_\beta \lim_\alpha v_\beta(h_\alpha) - \lim_\alpha \lim_\beta v_\beta(h_\alpha)\right| = \left|\lim_\beta \lim_\alpha v_\beta(h_\alpha) - \lim_\alpha v(h_\alpha)\right|$$
$$= \left|\lim_\beta \lim_\alpha v_\beta(h_\alpha) - \lim_\beta \lim_\alpha v(h_\alpha)\right| = 2\left|\lim_\beta \lim_\alpha 2^{-1}\left(v_\beta - v\right)(h_\alpha)\right| \le 2d_n(h, E),$$

where the last inequality follows from (5.1). Hence $\gamma_n(H) \le 2ck_n(H)$ for each $n \in \mathbb{N}$. □

Next result generalizes Theorem 5.4.

**Theorem 5.6** [3] *$ck_n(H) \le k_n(H) \le \gamma_n(H) \le 2ck_n(H)$ for a bounded subset $H$ of a Fréchet space $E$ and each $n \in \mathbb{N}$. Then $ck(H) = 0$ iff $k(H) = 0$.*

*Proof* The second and third inequalities follow from previous lemmas. The first inequality is obvious. □

**Corollary 5.3** [3] *If $H$ is a bounded subset of a Fréchet space $E$, the following conditions are equivalent:*

1. *$ck(H) = 0$.*
2. *$k(H) = 0$.*
3. *$H$ is weakly relatively countably compact.*
4. *$H$ is weakly relatively compact.*

Krein's theorem also holds for Fréchet spaces but since the metric is not a norm it seems that the inequality obtained in Theorem 5.3 should not hold here. So the natural problem here is to find a continuous function $F : [0, 1] \to [0, 1]$ such that $F(0) = 0$ and $k(coH) \le F(k(H))$. The following theorem is a positive answer to this theorem.

**Theorem 5.7** [4] *For a bounded set $H$ in a Fréchet space $E$ the following inequality holds*

$$k(coH) < (2^{n+1} - 2)k(H) + \frac{1}{2^n}$$

*for all $n \in \mathbb{N}$. Consequently*

$$k(coH) \le \sqrt{k(H)}(3 - 2\sqrt{k(H)}).$$

*Proof* [Sketch of the proof] First we have to prove that $\gamma_n(H) = \gamma_n(coH)$. For this we can use the same ideas that were used in the proof of [9, Theorem 13] or the ideas of the proof of [7, Theorem 3.3]. Using the definition of $d$ and that the function $f(x) = x/(1+x)$ is strictly increasing we get that

$$k(H') \leq \frac{2^n - 1}{2^n} \frac{k_n(H')}{1 + k_n(H')} + \frac{1}{2^n},$$

for all $H' \subset E$. In particular

$$k(coH) \leq \frac{2^n - 1}{2^n} \frac{k_n(coH)}{1 + k_n(coH)} + \frac{1}{2^n}.$$

By Theorem 5.6 we get that

$$k_n(coH) \leq \gamma_n(coH) = \gamma_n(H) \leq 2k_n(coH),$$

and then since $f(x) = x/(1+x)$ is strictly increasing

$$k(coH) \leq \frac{2^n - 1}{2^n} \frac{k_n(coH)}{1 + k_n(coH)} + \frac{1}{2^n} \leq \frac{2k_n(H)}{1 + 2k_n(H)} + \frac{1}{2^n}.$$

To finish the proof we have to prove that

$$\frac{1}{2^n} \frac{2k_n(c)}{1 + 2k_n(H)} < 2k(H).$$

For this we use again the definition of $d$ and that $f(x)$ is strictly increasing. □

## 5.4 Distances to Spaces of Continuous Functions

In this section we collect several quantitative versions of classical theorems about compactness in spaces of continuous functions. First we provide a formula to measure distances to spaces of continuous functions from a normal space to $\mathbb{R}$ via oscillations, Theorem 5.9. This result appears in [6, Proposition 1.18] with proof in paracompact spaces but the authors also say that this result holds in normal spaces. We need the following theorem.

**Theorem 5.8** ([13], Theorem 12.16) *Let $X$ be a normal space and let $f_1 \leq f_2$ be to real functions on $X$ such that $f_1$ is upper semicontinuous and $f_2$ is lower semicontinuous. Then, there exists a continuous function $f \in C(X)$ such that $f_1(x) \leq f(x) \leq f_2(x)$ for all $x \in X$.*

**Definition 5.4** Let $X$ be a topological. The oscillation of a bounded function $f \in \mathbb{R}^X$ at the point $x \in X$ is defined by

$$osc(f,x) = \inf_U \sup_{y,z \in U} |f(x) - f(y)|,$$

where the infimum is taken over the neighborhoods $U$ of $x$ in $X$.

**Theorem 5.9** [6, Proposition 1.18] *Let $X$ be a normal space. If $f \in \mathbb{R}^X$, then*

$$d(f, C(X)) = \frac{1}{2} osc(f)$$

*where $osc(f) = \sup_{x \in X} osc(f,x)$ and*

$$osc(f,x) := \inf_U \{ \sup_{y,z \in U} |f(y) - f(z)| : U \subset X \text{ open}, x \in U\}.$$

*Proof* Let us prove that $d(f, C(X)) \geq \frac{1}{2} osc(f)$. If $d(f, C(X))$ is infinite, clearly the inequality holds. Suppose that $d = d(f, C(X))$ is finite. Fix $\varepsilon > 0$ and $x \in X$. There exist $g \in C(X)$ such that $d(f,g) \leq d + \varepsilon/3$. Since $g$ is continuous there is an open neighborhood $U$ of $x$ such that $diam(g(U)) < \varepsilon/3$. Then, if $y, z \in U$,

$$d(f(y), f(z)) \leq d(f(y), g(y)) + d(g(y), g(z)) + d(g(z), f(z)) < 2d + \varepsilon$$

so $osc(f,x) < 2d + \varepsilon$. Since we can do it for all $\varepsilon > 0$ we get that $osc(f,x) \leq 2d$.

Let us prove that $d(f, C(X)) \leq \frac{1}{2} osc(f)$. If $osc(f) = +\infty$ clearly the inequality holds so suppose that $\delta = \frac{1}{2} osc(f)$ is finite. For $x \in X$ denote by $\mathscr{U}_x$ the set of neighborhoods of $x$. Put

$$\mathscr{V}_x = \{U \in \mathscr{U}_x : diam(f(U)) < osc(f) + 1\}.$$

Clearly $\mathscr{V}_x$ is a basis of the neighborhoods of $x$ and for each $U \in \mathscr{V}_x$, $f|_U$ is upper and lower bounded. Now

$$\begin{aligned} 2\delta \geq osc(f,x) &= \inf_{U \in \mathscr{U}_x} diam(f(U)) = \inf_{U \in \mathscr{V}_x} diam(f(U)) \\ &= \inf_{U \in \mathscr{V}_x} \sup_{y,z \in U} (f(y) - f(z)) \\ &\geq \inf_{U,V \in \mathscr{V}_x} \sup_{y \in U, z \in V} (f(y) - f(z)) \\ &= \inf_{U \in \mathscr{V}_x} \sup_{y \in U} f(y) - \sup_{U \in \mathscr{V}_x} \inf_{z \in U} f(z). \end{aligned}$$

So if we define

$$f_1(x) = \inf_{U \in \mathscr{V}_x} \sup_{z \in U} f(z) - \delta$$

$$f_2(x) = \sup_{U \in \mathscr{V}} \inf_{z \in U} f(z) + \delta$$

then $f_1 \leq f_2$. It is easy to check that $f_1$ is upper semi-continuous and $f_2$ is lower semi-continuous. By Theorem 5.8, there is a continuous function $h \in C(X)$ such that

$$f_1(x) \leq h(x) \leq f_2(x)$$

for $x \in X$. Clearly

$$f_2(x) - \delta \leq f(x) \leq f_1(x) + \delta$$

so

$$h(x) - \delta \leq f_2(x) - \delta \leq f(x) \leq f_1(x) + \delta \leq h(x) + \delta$$

so $d(f, g) \leq \delta$ and this proves the inequality. □

Previous result characterizes the normality of $X$:

**Corollary 5.4** *Let X be a topological space. The following statements are equivalent:*

1. *X is normal.*
2. *For each $f \in \mathbb{R}^X$ there is $g \in C(X)$ such that $d(f, g) = \frac{1}{2}osc(f)$.*
3. *$d(f, C(X)) = \frac{1}{2}osc(f)$ for each function $f \in \mathbb{R}^X$.*

*Proof* By the previous result (and proof) we only have to prove that the third condition implies the first one. For this, if $A$ and $B$ are disjoint closed subsets of $X$, define $f = \chi_B - \chi_A$. Then $osc(f) \leq 1$ so by hypothesis, $d(f, C(X)) = \frac{1}{2}osc(f) \leq \frac{1}{2}$. Then we can pick a continuous function $g \in C(X)$ such that $d(f, g) < 1$. If we define $U = g^{-1}(-\infty, 0)$ and $V = g^{-1}(0, +\infty)$, then $U$ and $V$ are disjoint open subsets of $X$, $A \subset U$ and $B \subset V$ so $X$ is normal. □

**Definition 5.5** Let $K$ be a compact topological space and let $H$ be a uniformly bounded subset of $C(K)$. We define

$$ck(H) := \sup_{\varphi \in H^{\mathbb{N}}} d(clust_{\mathbb{R}^K}(\varphi), C(K))$$

and

$$\gamma_K(H) := \sup\{|\lim_n \lim_m f_m(x_n) - \lim_m \lim_n f_m(x_n)| : (f_m) \subset H, (x_n) \subset K\},$$

assuming the involved limits exist.

By definition $\inf \emptyset = +\infty$. Observe that if $H$ is a $\tau_p$-relatively countably compact subset of $C(K)$ then $ck(H) = 0$. Using $\gamma_K$, the usual notion of interchange of limits [12] can be defined as follows: $H$ interchanges limits with $K$ if, and only if $\gamma_K(H) = 0$.

Grothendieck proved in [12] that $H$ is relatively compact in $(C(K), \tau_p)$ if, and only if, $H$ interchanges limits with $K$. The following result establishes the relations between $\gamma_K(H)$, $ck(H)$ and $d(\overline{H}^{\mathbb{R}^K}, C(K))$.

**Theorem 5.10** [1, 7] *Let $K$ be a compact space and $H \subset C(K)$ a uniformly bounded set, then*

$$ck(H) \leq \hat{d}(\overline{H}^{\mathbb{R}^K}, C(K)) \leq \gamma_K(H) \leq 2ck(H).$$

The following theorem is a quantitative version of the Krein-Šmulyan theorem and combined with Theorem 5.5 implies Theorem 5.3.

**Theorem 5.11** [7] *Let $K$ be a compact topological space and let $H$ be a uniformly bounded subset of $\mathbb{R}^K$. Then*

$$\gamma_K(H) = \gamma_K(co(H))$$

*and as a consequence for $H \subset C(K)$ we obtain that*

$$\hat{d}(\overline{co(H)}^{\mathbb{R}^K}, C(K)) \leq 2\hat{d}(\overline{H}^{\mathbb{R}^K}, C(K))$$

*and if $H \subset \mathbb{R}^K$ is uniformly bounded then*

$$\hat{d}(\overline{co(H)}^{\mathbb{R}^K}, C(K)) \leq 5\hat{d}(\overline{H}^{\mathbb{R}^K}, C(K)).$$

Now we are going to prove Lemma 5.4. This lemma is a very useful tool to study quantitative versions of angelicity in spaces of continuous functions and was proved in [1] in a more general case, when $Z$ is a separable metric space. But since this result is a quantitative version of Theorem 5.2 we prefer to include here the compact metric space version and its proof.

**Lemma 5.3** *Suppose that $(Z, d)$ is a compact metric space and let $X$ be a set. Given functions $f_1, \ldots, f_n \in Z^X$, $D \subset X$ and $\varepsilon > 0$ there is a finite subset $L \subset D$ such that for every $x \in D$*

$$\min_{y \in L} \max_{1 \leq k \leq n} d(f_k(y), f_k(x)) < \varepsilon.$$

*Proof* The metric

$$d_\infty\big((t_k), (s_k)\big) := \sup_{1 \leq k \leq n} d(t_k, s_k),$$

$(t_k), (s_k) \in Z^n$, defines the product topology of the space $Z^n$. Define

$$H = \{(f_1(x), f_2(x), \ldots, f_n(x)) : x \in D\}.$$

Now we only have to apply the compactness of $(Z^n, d)$. □

**Lemma 5.4** [1] *Let $(Z, d)$ be a compact metric space, $X$ a set and $H$ a subset of the space $(Z^X, \tau_p)$ and $\varepsilon \geq 0$. We assume that:*

*(i) There is $\Sigma \subset \mathbb{N}^{\mathbb{N}}$ and a family $\{A_\alpha : \alpha \in \Sigma\}$ of non-void subsets of the set $X$ such that $X = \bigcup\{A_\alpha : \alpha \in \Sigma\}$.*

*(ii) For every $\alpha = (a_1, a_2, \ldots) \in \Sigma$ the set $H$ $\varepsilon$-interchanges limits in $Z$ with every sequence $(x_n)_n$ in $X$ that is eventually in each set*

$$C_{\alpha|m} = \bigcup\{A_\beta : \beta \in \Sigma, b = (b_m) \text{ and } b_j = a_j \text{ for } j = 1, 2, \ldots, m\},$$

*for $m \in \mathbb{N}$.*

*Then for any $f \in \overline{H}^{Z^X}$ there exists a sequence $(f_n)_{n\in\mathbb{N}}$ in $H$ such that*

$$\sup_{x\in X} d(g(x), f(x)) \leq \varepsilon$$

*for any cluster point $g$ of $(f_n)_{n\in\mathbb{N}}$ in $Z^X$.*

*Proof* Define $f_0 := f$. Since $\mathrm{F}(\Sigma)$ is countable and infinite there is a bijection $\varphi : \mathbb{N} \to \mathrm{F}(\Sigma)$. We define $D_n := C_{\varphi(n)}$ for each $n \in \mathbb{N}$. We claim that there are a sequence of functions $f_0, f_1, \ldots, f_n, \ldots$ and a sequence of finite sets $L_1, L_2, \ldots, L_n, \ldots$ with the properties:

- for each $n \in \mathbb{N}$ and every $x \in D_n$ we have

$$\min_{y\in L_n} \max_{0\leq k<n} d(f_k(y), f_k(x)) < \frac{1}{n}; \tag{5.2}$$

- for each $n \in \mathbb{N}$ the function $f_n$ belongs to $H$ and

$$d(f_n(y), f_0(y)) < \frac{1}{n} \text{ for every } y \in \bigcup\{L_m : 1 \leq m \leq n\}. \tag{5.3}$$

We prove the existence of the above sequences of functions and sets by recurrence. *FIRST STEP.* Applying Lemma 5.3 to $D := D_1$ and $f_0$ we obtain a finite subset $L_1$ of $D_1$ such that

$$\min_{y\in L_1} d(f_0(y), f_0(x)) < 1 \text{ for every } x \in D_1.$$

Since $f \in \overline{H}^{Z^X}$, there is $f_1 \in H$ such that

$$\max_{y\in L_1} d(f_1(y), f_0(y)) < 1.$$

*INDUCTION STEP.* Assuming we have produced $f_1, f_2, \ldots, f_n$ and $L_1, L_2, \ldots, L_n$ satisfying (5.2) and (5.3) we use Lemma 5.3 for $D := D_{n+1}$ and $f_0, f_1, \ldots, f_n$ to obtain $L_{n+1} \subset D_{n+1}$ satisfying

$$\inf_{y\in L_{n+1}} \max_{0\le k<n+1} d(f_k(y), f_k(x)) < \frac{1}{n+1} \text{ for every } x \in D_{n+1}.$$

Once again, since $f \in \overline{H}^{Z^X}$ we can take a function $f_{n+1} \in H$ satisfying

$$d(f_{n+1}(y), f_0(y)) < \frac{1}{n+1} \text{for every } y \in \bigcup\{L_m : 1 \le m \le n+1\}.$$

The constructed sequences $f_0, f_1, \ldots, f_n, \ldots$ and $L_1, L_2, \ldots, L_n, \ldots$ satisfy (5.2) and (5.3).

We shall prove now that $(f_n)_{n\in\mathbb{N}}$ has the property required in the thesis in the lemma: fix a cluster point $g$ of $(f_n)_n$ in $Z^X$ and fix a point $x \in X$ and let us prove that $d(g(x), f(x)) \le \varepsilon$. We note first that inequality (5.3) implies that

$$\lim_n f_n(y) = f(y) \text{ for every } y \in L = \bigcup_{n\in\mathbb{N}} L_n. \tag{5.4}$$

Now, we pick $\alpha = (a_1, a_2, \ldots) \in \Sigma$ such that $x \in A_\alpha$ and define

$$P := \varphi^{-1}(\{\alpha|n : n \in \mathbb{N}\}) \subset \mathbb{N}.$$

$P$ is an infinite subset because $\varphi$ is a bijection. Since the point $x \in \bigcap_{p\in P} D_p$, (5.2) for each $p \in P$ we can pick $y_p \in L_p$ with the property

$$d(f_k(y_p), f_k(x)) < \frac{1}{p} \text{ for } 0 \le k < p. \tag{5.5}$$

Being $P$ infinite we can and do fix $p_1 < p_2 < \cdots < p_j < \cdots \nearrow +\infty$ a strictly increasing sequence in $P$. We claim that the sequence $(y_{p_j})_j$ is eventually in $C_{\alpha|n}$ for every $n \in \mathbb{N}$. Indeed, for a given $n \in \mathbb{N}$ take $p_{j(n)}$ an element of the sequence $(p_j)_j$, with $p_{j(n)} > \varphi^{-1}(\alpha|i)$, $i = 1, 2, \ldots, n$. Therefore, if $j > j(n)$ then $p_j \ne \varphi^{-1}(\alpha|i)$ for $i = 1, 2, \ldots, n$ and consequently $\varphi(p_j) = \alpha|n(p_j)$ for some $n(p_j) > n$. The latter implies

$$y_{p_j} \in D_{p_j} = C_{\alpha|n(p_j)} \subset C_{\alpha|n}, \text{ for } j > j(n),$$

proving that $(y_{p_j})_j$ is eventually in each $C_{\alpha|n}$.

Observe also that (5.5) implies that

$$\lim_j f_k(y_{p_j}) = f_k(x) \text{ for } k = 0, 1, 2, \ldots. \tag{5.6}$$

Since $g(x)$ is a cluster point of $(f_n(x))_n$ in the metric space $(Z, d)$ we can choose a subsequence $(f_{n_k})_k$ of $(f_n)_n$ such that $\lim_k f_{n_k}(x) = g(x)$. With all the above we have

$$\lim_k \lim_j f_{n_k}(y_{p_j}) \overset{(5.6)}{=} \lim_k f_{n_k}(x) = g(x),$$

$$\lim_j \lim_k f_{n_k}(y_{p_j}) \overset{(5.4)}{=} \lim_j f(y_{p_j}) \overset{(5.6)}{=} f(x).$$

Being the sequence $(y_{p_j})_j$ eventually in every $C_{\alpha|n}$ the assumption (ii) in the lemma ensures us that $H$ $\varepsilon$-interchanges limits with $(y_{p_j})_j$, consequently

$$d(g(x), f(x)) = d(\lim_k \lim_j f_{n_k}(y_{p_j}), \lim_j \lim_k f_{n_k}(y_{p_j})) \leq \varepsilon,$$

and the proof is over. □

**Theorem 5.12** [3] Let $X$ be a web-compact space with a representation $D = \bigcup\{A_\alpha : \alpha \in \Sigma\}$ with $X = \overline{D}$. Let $(Z, d)$ be a compact metric space and $H \subset Z^X$. Then for each $f \in \overline{H}$ (the closure in $Z^X$) there exists a sequence $(f_n)_n$ in $H$ such that

$$\sup_{x \in D} d(f(x), g(x)) \leq 2ck(H) + 2\hat{d}(H, C(X, Z)) \leq 4ck(H)$$

for any cluster point $g$ of $(f_n)_n$ in $Z^X$.

*Proof* (Sketch of the proof) Let $\varepsilon := ck(H) + \hat{d}(H, C(X, Z))$ and let $\tilde{H} = \{f|_D : f \in H\}$. Then we have to prove that condition (ii) in Lemma 5.4 holds for $D$ and $\tilde{H}$.

Now let $f \in \overline{H}$ (the closure in $Z^X$). Since $f|_D \in \overline{\tilde{H}}$ (the closure in $Z^D$), by Lemma 5.4 there exists a sequence $(g_n)_n$ in $\tilde{H}$ such that $\sup_{x \in D} d(f(x), h(x)) \leq 2\varepsilon$ for each cluster point $h$ of $(g_n)_n$ in $Z^X$. For each $g_n$ there exists $f_n \in H$ such that $f_n|_D = g_n$. If $g$ is a cluster point of $(f_n)_n$ then $g|_D$ is cluster point of $(g_n)_n$ so $\sup_{x \in D} d(f(x), g(x)) \leq 2\varepsilon$. This yields the first inequality. The other inequality is trivial. □

**Corollary 5.5** [1, 3] Let $X$ be a strongly web-compact space, $(Z, d)$ a compact metric space, and let $H \subset Z^X$. Then

$$ck(H) \leq \hat{d}(\overline{H}, C(X, Z)) \leq 3ck(H) + 2\hat{d}(H, C(X, Z))) \leq 5ck(H).$$

*Proof* We only have to prove the middle inequality: We may assume that $ck(H) < \infty$. Fix $t > 0$ with $ck(H) < t$ and $f \in \overline{H}$. Fix

$$\varepsilon := ck(H) + \hat{d}(H, C(X, Z)).$$

By Theorem 5.12 there exists a sequence $(f_n)_n$ in $H$ such that $\sup_{x \in X} d(f(x), g(x)) \leq 2\varepsilon$ for any cluster point $g$ of $(f_n)_n$ in $Z^X$. Since $ck(H) < t$, for this sequence $(f_n)_n$ there exists a cluster point $g$ of $(f_n)_n$ such that $d(g, C(X, Z)) < t$. This yields the inequality. □

**Proposition 5.2** [3] *Let $X$ be a web-compact space with a representation $D = \bigcup\{A_\alpha : \alpha \in \Sigma\}$ with $X = \overline{D}$. Let $(Z, d)$ be a compact metric space, let $H \subset Z^X$ and let $f \in \overline{H}$ (the closure in $Z^X$). Then for each $\delta > 0$ and $x \in X$ there exists $U \subset X$ a neighborhood of $x$ such that*

$$d(f(x), f(d)) < 4ck(H) + 2\hat{d}(H, C(X, Z)) + \delta$$

*for every $d \in U \cap D$.*

*Proof* (Sketch of the proof) Fik $x \in X$ and define $H' = \{j \in H : d(j(x), f(x)) < 4^{-1}\delta\}$. Since $H'$ is the intersection of $H$ and an open neigbourhood of $f$ in $Z^X$, then $f \in \overline{H'}$. Now we obtain the sequence given by Theorem 5.12 and pick a cluster point $g$ of this sequence such that such that $d(g, C(X, Z)) < ck(H) + 4^{-1}\delta$. Then we can pick $h \in C(X, Z)$ such that

$$d(g(z), h(z)) < ck(H) + \delta/4$$

for all $z \in X$. Using the continuity of $h$ we can find $U \subset X$ a neighborhood of $x$ such that $d(h(x), h(z)) < \delta/4$. We can check that

$$d(f(x), f(d)) < 4ck(H) + 2\hat{d}(H, C(X, Z)) + \delta$$

for every $d \in U \cap D$. □

**Theorem 5.13** [3] *Let $X$ be a web-compact space, $(Z, d)$ be a compact metric space and $H \subset Z^X$. Then for each $f \in \overline{H}$ (the closure in $Z^X$) there exists a sequence $(f_n)_n$ in $H$ such that*

$$\sup_{x \in X} d(f(x), g(x)) \leq 10ck(H) + 6\hat{d}(H, C(X, Z)) \leq 16ck(H)$$

*for any cluster point $g$ of $(f_n)_n$ in $Z^X$.*

*Proof* Let $(f_n)$ be the sequence obtained from Theorem 5.12 and fix $x \in X$, $g$ a cluster point of $(f_n)_n$ and $\delta > 0$. Since $f, g \in \overline{H}$ by Proposition 5.2, there exist $U, V \subset X$ neighborhoods of $x$ such that

$$d(f(x), f(d)) < 4ck(H) + 2\hat{d}(H, C(X, Z))) + \delta$$

for every $d \in U \cap D$ and

$$d(g(x), g(d)) < 4ck(H) + 2\hat{d}(H, C(X, Z))) + \delta$$

for every $d \in V \cap D$. Pick $d \in D \cap U \cap V$, then

$$d(f(x), g(x)) \le d(f(x), f(d)) + d(f(d), g(d)) + d(g(d), g(x))$$
$$< 10ck(H) + 6\hat{d}(H, C(X, Z)) + 2\delta. \qquad \square$$

The following corollary follows from Theorem 5.13 like Corollary 5.5 from Theorem 5.12.

**Corollary 5.6** [3] *Let $X$ be a web-compact space, $(Z, d)$ a separable metric space and let $H \subset Z^X$ be a $\tau_p$-relatively compact set. Then*

$$ck(H) \le \hat{d}(\overline{H}, C(X, Z)) \le 11ck(H) + 6\hat{d}(H, C(X, Z))) \le 17ck(H).$$

**Corollary 5.7** *Let $X$ be a web-compact space, $(Z, d)$ a separable metric space and let $H \subset C(X, Z)$ be a $\tau_p$-relatively compact set in $X^Z$. The following conditions are equivalent:*

1. $ck(H) = 0$.
2. *$H$ is a relatively countably compact subset of $C(X, Z)$.*
3. *$H$ is a relatively compact subset of $C(X, Z)$.*

## References


1. Angosto, C., Cascales, B.: The quantitative difference between countable compactness and compactness. J. Math. Anal. Appl. **343**(1), 479–491 (2008)
2. Angosto, C., Cascales, B.: Measures of weak noncompactness in Banach spaces. Topology Appl. **156**, 1412–1421 (2009)
3. Angosto, C., Kąkol, J., López-Pellicer, M.: A quantitative approach to weak compactness in Fréchet spaces and spaces $C(X)$. J. Math. Anal. Appl. **403**(1), 13–22 (2013)
4. Angosto, C., Kąkol, J., Kubzdela, A., López-Pellicer, M.: A quantitative version of Krein's theorems for Fréchet spaces. Arch. Math. (Basel) **101**(1), 65–77 (2013)
5. Astala, K., Tylly, H.: Seminorms related to weak compactness to Tauberian operators. Math. Proc. Cambridge Philos. Soc. **107**, 367–375 (1990)
6. Benyamini, Y., Lindenstrauss, J.: Geometric nonlinear functional analysis. Vol. 1. American Mathematical Society Colloquium Publications, 48, American Mathematical Society, Providence, RI Springer-Verlag, New York (2000)
7. Cascales, B., Marciszewski, W., Raja, M.: Distance to spaces of continuous functions. Topology Appl. **153**, 2303–2319 (2006)
8. Eberlein, F.W.: Weak compactness in Banach spaces. Proc. Nat. Acad. Sci. U.S.A. **33**, 51–53 (1947)
9. Fabian, M., Hájek, P., Montesinos, V., Zizler, V.: A quantitative version of Krein's theorem. Rev. Mat. Iberoam. **21**, 237–248 (2005)
10. Floret, K.: Weakly compact sets. Lecture Notes in Mathematics, vol. 801, Springer, Berlin (1980)
11. Granero, A.S.: An extension of the Krein-Šmulian theorem. Rev. Mat. Iberoam. **22**, 93–110 (2006)
12. Grothendieck, A.: Critères de compacité dans les espaces fonctionnels généraux. Amer. J. Math. **74**, 168–186 (1952)
13. Jameson, G.J.O.: Topology and normed spaces. Chapman and Hall, London (1974)
14. Kelley, J.L., Namioka, I.: Linear topological spaces. Graduate Texts in Mathematics vol. 36, Springer-Verlag, New York (1976)

15. Orihuela, J.: Pointwise compactness in spaces of continuous functions. J. London Math. Soc. **36**, 143–152 (1987)
16. Pol, R.: On a question of H. H. Corson and some related problems. Fund. Math. **109**(2), 143–154 (1980)
17. Pryce, J.D.: A device of R. J. Whitley's applied to pointwise compactness in spaces of continuous functions. Proc. London Math. Soc. **23**(3), 532–546 (1971)
18. Šmulian, V.: Über lineare topologische Räume. Rec. Math. (Mat. Sbornik) N. S. 7(49), 425–448 (1940)

# Chapter 6
# Two Classes of Metrizable Spaces $\ell_c$-Invariant

**Manuel López-Pellicer**

**Abstract** Many properties of a Tychonoff space $X$ have been characterized by properties of $C_p(X)$ or $C_c(X)$, the spaces of continuous real-valued functions on $X$ provided with the topology of pointwise convergence or with the compact-open topology, respectively. The question of Arhangel'skii about preservation of metrizability by $\ell_p$-equivalence in the class of first countable spaces has been partially answered by Valov in the class of those first countable spaces that are Čech complete. The preservation of complete metrizability by $\ell_p$-equivalence in the class of metrizable spaces has been obtained by Baars, de Groot and Pelant. The $\ell_c$-invariance of separable metrizability and separable complete metrizability (i.e., Polish spaces) in the class of spaces of pointwise countable type has been considered very recently by Kąkol, López-Pellicer and Okunev. These two $\ell_c$-invariant properties were the aim of the talk given by the author in the First Meeting in Topology and Functional Analysis, dedicated to Professor Jerzy Kąkol on the occasion of his sixty birthday, September 27–28, in Elche (Spain). Here it is this talk with the proofs of properties needed to obtain these two $\ell_c$-invariant properties. As additional motivation for non specialist readers three classical $C(X)$ theorems related with $\ell$-equivalence's questions are also included.

**Keywords** Čech-complete space · Compact resolution · $C_p(X)$ and $L_p(X)$-spaces · $\ell_p$-, $\ell_c$- and $t$-equivalence · Metrizability · Polish space · Submetrizability · Spaces of pointwise countable type · $\mu$-space · $\aleph_0$-space

**2000 MSC Classification:** 46E10, 54C35, 46A20, 46A50

M. López-Pellicer (✉)
Department of Applied Mathematics and IUMPA, Universitat Politècnica de València, 46022 València, Spain
e-mail: mlopezpe@mat.upv.es

J. C. Ferrando and M. López-Pellicer (eds.), *Descriptive Topology and Functional Analysis*, Springer Proceedings in Mathematics & Statistics 80, DOI: 10.1007/978-3-319-05224-3_6, 
© Springer International Publishing Switzerland 2014

## 6.1 Introduction

All topological spaces (spaces, in brief) are assumed to be completely regular and Hausdorff. We use terminology and notation as in [9, 12, 13]. $C(X)$ is the set of all continuous real-valued functions on $X$, $C_p(X)$ is $C(X)$ endowed with the topology of pointwise convergence (also named simple convergence) and $C_c(X)$ is $C(X)$ with the compact open topology, which is the topology of uniform convergence on the compact subsets of $X$.

The symbol $\omega$ denotes the smallest infinite ordinal with the discrete topology and with the usual order, so $\omega$ is both in topology and order isomorphic to the set $\mathbb{N}^+$ of all non-negative integers.

Nagata's theorem [17] (see also [5, Theorem 0.6.1] or Sect. 6.2.2) states that two spaces $X$ and $Y$ are homeomorphic if and only if the topological rings $C_p(X)$ and $C_p(Y)$ are isomorphic. The spaces $X$ and $Y$ are said to be *t-equivalent* if the spaces $C_p(X)$ and $C_p(Y)$ are homeomorphic. A property $\mathscr{P}$ is preserved by $t$-equivalence if whenever $X$ and $Y$ are $t$-equivalent and $X$ has the property $\mathscr{P}$ then $Y$ has also the property $\mathscr{P}$. From Velichko's theorem (see [5, Theorem I.2.1] or Sect. 6.2.3) it follows that *finite cardinality* is preserved by $t$-equivalence.

Two spaces $X$ and $Y$ are said to be *$\ell_p$-equivalent* if the spaces $C_p(X)$ and $C_p(Y)$ are linearly homeomorphic. A topological property $\mathscr{P}$ is invariant by $\ell_p$-equivalence if whenever $X$ has the property $\mathscr{P}$ and $X$ is $\ell_p$-equivalent to $Y$ then $Y$ has the property $\mathscr{P}$ too. Clearly, a property $\mathscr{P}$ is preserved by $\ell_p$-equivalence if and only if $\mathscr{P}$ *admits a description in terms of the linear topological structure of* $C_p(X)$. There are many known results about preservation and non-preservation of various topological properties by $\ell_p$-equivalence; see, e.g., [4–6, 18, 25, 26]. For example, the properties of being *hemicompact*, $\aleph_0$, *Lindelöf*, *Lindelöf-*$\Sigma$, $K$*-analytic* and *analytic* are preserved by $\ell_p$-equivalence. On the other hand, *metrizability*, *local compactness*, *countable weight*, *normality* and *paracompactness* are not $\ell_p$-invariant.

The spaces $X$ and $Y$ are *$\ell_c$-equivalent* if the spaces $C_c(X)$ and $C_c(Y)$ are linearly homeomorphic. A topological property $\mathscr{S}$ is $\ell_c$-invariant if for each space $Y$ $\ell_c$-equivalent to a space $X$ with the property $\mathscr{S}$ then $Y$ has also the property $\mathscr{S}$. This is the case if and only if $\mathscr{S}$ *is characterized by a property of* $C_c(X)$. For instance, Nachbin -Shirota theorem ([16, 21], see also Theorem 6.1) states that *$\mu$-spaces* are preserved by $\ell_c$-equivalence, because $X$ is $\mu$-space if and only if $C_c(X)$ is barrelled. Let us notice that in the class of $\mu$-spaces the $\ell_p$-equivalence implies $\ell_c$-equivalence (see Sect. 6.2.6).

The space $X$ *$\ell_c$-covers* a space $Y$ if there is a continuous open linear map from $C_c(X)$ onto $C_c(Y)$. If $X$ and $Y$ are $\ell_c$-equivalent, then each of the two covers the other.

Arhangel'skii asked in [3, Problem 20] *if metrizability is preserved by $\ell_p$-equivalence in the class of first countable spaces.* Baars, de Groot and Pelant proved that *complete metrizability is preserved by $\ell_p$-equivalence in the class of metrizable spaces* [6, Theorem 3.3], given an alternative proof in the class of separable metrizable spaces by using Christensen's Theorem 6.4 [6, Theorems 5.1 and 5.3]. Later on, based on results in [24], Valov proved that a Čech-complete and first countable space $Y$ is metrizable when it is $\ell_p$-equivalent to a metrizable space $X$ [25, Corollary 4.6].

In the general class of spaces of pointwise countable type the preservation by $\ell_c$-equivalence of separable metrizability as well as the $\ell_c$-invariance of Polish spaces have been considered very recently by Kąkol, López-Pellicer and Okunev in [11]. These two $\ell_c$-invariant properties were the aim of the talk given by the author in the First Meeting in Topology and Functional Analysis, dedicated to Professor Jerzy Kąkol on the occasion of his sixty birthday, September 27–28, held in Elche (Spain). Here we complete the talk by collecting together all results needed to obtain, in Sect. 6.6.3, the two aforementioned $\ell_c$-invariant properties. For the sake of completeness all proofs are developed. We have added three classical $C(X)$ theorems related to $\ell$-equivalence as a motivation for readers non specialists in $\ell$-equivalence. By these reasons, Sect. 6.2 contains the classical Nagata, Velichko and Nachbin-Shirota $C(X)$ theorems, Sect. 6.3 give some results of the Michael paper [15] and Sect. 6.5 collect some results on compact resolutions, including Christensen's theorem in Sect. 6.5.2.

The class of spaces of pointwise countable type contains the first countable spaces and the locally compact spaces among others. The $\ell_c$-invariance of separable metrizability and of separable topologically complete metrizability for the spaces of pointwise countable type is consequence of its $\ell_c$-invariance in the subclass of first countable spaces. Therefore, as in [11], we present first the $\ell_c$-invariance of these two properties in the class of first countable spaces in Sects. 6.4 and 6.5.

## 6.2 Nagata, Velichko and Nachbin-Shirota Theorems in $C(X)$ Spaces

If $D$ is a subset of a topological space $X$ and $S$ is a subset of $\mathbb{R}$ then $W(D, S) := \{f \in C(X) : f(D) \subset S\}$.

### 6.2.1 *The Spaces $C_p(X)$ and $L_p(X)$*

Recall that $C_p(X)$ is the space of continuous real-valued functions on $X$ endowed with the topology of pointwise convergence. If $\mathscr{F}$ is the family of finite subsets of $X$ then the family $\{W(F, (-\varepsilon, \varepsilon)) : F \in \mathscr{F}, \varepsilon > 0\}$ is a base of neighborhoods of 0 in $C_p(X)$.

$L_p(X)$ is the topological dual $C_p(X)'$ of $C_p(X)$ with the topology induced by $C_p(C_p(X))$. This topology, denoted by $\sigma\left(C_p(X)', C_p(X)\right)$, is named the weak*-topology on $C_p(X)'$. A net $(\varphi_i : i \in I)$ converges to $\varphi$ in $L_p(X)$ if and only if $(\varphi_i(f) : i \in I)$ converges to $\varphi(f)$ for each $f \in C_p(X)$. The well known fact that a net $(x_i : i \in I)$ in $X$ converges to $x$ if and only if the limit of $(f(x_i) : i \in I)$ is $f(x)$, for each $f \in C(X)$, provides the standard embedding $\hat{\Delta}_X : X \to C_p(C_p(X))$ defined by

$\hat{\Delta}_X(x) = \hat{x}$, being $\hat{x}(g) = g(x)$ for each $g \in C(X)$. Clearly $\hat{\Delta}_X(X) \subset L_p(X)$ and if $\hat{X}$ is the set $\hat{\Delta}_X(X)$ with the topology induced by $C_p(C_p(X))$ then $\hat{\Delta}_X : X \to \hat{X}$ is a homeomorphism.

The space $L_p(X)$ is the linear hull of $\hat{X}$, (see [5, Proposition 0.5.7]), because if $\phi \in L_p(X)$ there exists a finite subset $F := \{x_1, x_2, \ldots, x_n\}$ in $X$, with $x_i \neq x_j$ if $i \neq j$, and a real number $\varepsilon > 0$ such that

$$\phi\left(W(F, (-\varepsilon, \varepsilon)\right) \subset [-1, 1].$$

Since for each $i \in \{1, 2, \ldots, n\}$ there exists $g_i \in C(X)$ with $g_i(x_i) = 1$ and $g_i(x_j) = 0$ if $j \neq i$, then for each $\lambda > 0$ and $f \in C(X)$ the function $g := f - \sum_{1 \leq i \leq n} f(x_i) g_i$ verifies that $\lambda g \in W(F, (-\varepsilon, \varepsilon))$, and therefore $\phi(g) = 0$. From

$$\begin{aligned} 0 = \phi(g) = \phi(f) - \sum_{1 \leq i \leq n} f(x_i)\phi(g_i) = \\ = \phi(f) - \sum_{1 \leq i \leq n} \phi(g_i)\hat{x}_i(f) = \\ = \left(\phi - \sum_{1 \leq i \leq n} \phi(g_i)\hat{x}_i\right)(f) \end{aligned}$$

it follows that $\phi = \sum_{1 \leq i \leq n} \phi(g_i)\hat{x}_i$.

In [5, Proposition 0.5.9] it is also stated that $\hat{X}$ is a closed subset of $L_p(X)$. This property follows from the observation that if $\psi \in \overline{\hat{X}}^{L_p(X)} \setminus \hat{X}$ then $\psi = \Sigma_{1 \leq i \leq n} \lambda_i \hat{x}_i$, with $x_i \in X$, $\lambda_i \in \mathbb{R}$, $1 \leqslant i \leqslant n$, and there exists $f \in C_p(L_p(X))$ such that $f(\psi) = 1$ and $f(\hat{x}_i) = 0$, for each $1 \leqslant i \leqslant n$. The subset $\hat{A} := \left\{\hat{x} \in \hat{X} : f(\hat{x}) \geqslant 1/2\right\}$ verifies that $\psi \in \overline{\hat{A}}^{L_p(X)}$ and the function $g \in C_p(X)$ such that $g(x) := f(\hat{x})$, for each $x \in X$, verifies that $\{\hat{x}(g) : \hat{x} \in \hat{A}\} = f(\hat{A}) \subset [1/2, \infty]$. Therefore $\psi(g) \geqslant 1/2$. But this inequality contradicts that

$$\psi(g) = \left(\Sigma_{1 \leq i \leq n} \lambda_i \hat{x}_i\right)(g) = \Sigma_{1 \leq i \leq n} \lambda_i g(x_i) = \Sigma_{1 \leq i \leq n} \lambda_i f(\hat{x}_i) = 0.$$

### 6.2.2 Nagata Theorem [17]

Let $\check{X}$ be the topological subspace of $L_p(X)$ whose points are the nonzero continuous multiplicative linear forms on $C_p(X)$. Clearly $\hat{X} \subset \check{X} \subset L_p(X) \setminus \{0\}$. Nagata proved that $\hat{X} = \check{X}$ by the following two claims [17]:

If $\varphi = \sum \left\{\lambda_i \hat{x}_i : 1 \leq i \leq n\right\}$, with $x_j \neq x_k$ and $\lambda_j \lambda_k \neq 0$, then $\varphi \notin \check{X}$, because for the functions $f_j$ and $f_k$ in $C(X)$ such that $f_j(x_j) = \lambda_j^{-1}$, $f_k(x_k) = \lambda_k^{-1}$ and $f_i(x_m) = 0$

for the remaining values of $i$ in $\{j, k\}$ and $m$ in $\{1, 2, \ldots, n\}$, we have that $\varphi(f_j) = \varphi(f_k) = 1$ and $\varphi(f_j \times f_k) = 0$.

If $\varphi = \lambda\hat{x}$, with $0 \neq \lambda \neq 1$ then $\varphi \notin \check{X}$, since the function $f$ identically equal to 1 verifies that $\varphi(f^2) = \varphi(f) = \lambda \neq \lambda^2 = [\varphi(f)]^2$.

Then the equality $\hat{X} = \check{X}$, the homeomorphism $\hat{\Delta}_X : X \to \hat{X}$ and the trivial fact that if the topological rings $C_p(X)$ and $C_p(Y)$ are topologically isomorphic then $\check{X}$ and $\check{Y}$ are homeomorphic imply Nagata's theorem, which states that if the topological rings $C_p(X)$ and $C_p(Y)$ are topologically isomorphic then $X$ and $Y$ are homeomorphic ([17] see also [5, Theorem 0.6.1]).

### *6.2.3 Velichko Theorem [22]*

Recall that a topological space $X$ is a $P$-space if the union of an increasing sequence of closed sets is closed. If $X$ is a $P$-space and $\{f_i : i \in \mathbb{N}\}$ is a sequence of real continuous functions with pairwise disjoint supports then $\Sigma\{f_i : i \in \mathbb{N}\}$ is a continuous function. Velichko theorem states that $X$ is finite if and only if $C_p(X)$ is $\sigma$-compact [22].

Suppose that $C_p(X) = \cup\{K_i : i \in \mathbb{N}\}$, each $K_i$ being a countably compact subset of $C_p(X)$. Velichko's theorem follows from the next two claims stating firstly that $X$ is a $P$-space and then that $X$ must be finite.

In fact, if $(F_n)_n$ were an increasing sequence of closed subsets of $X$ and there existed a point $x \in \overline{\cup_{n\in\omega} F_n} \setminus \cup_{n\in\omega} F_n$, then

$$\left\{f \in K_i : f(x) = 0,\ f\left(\cup_n F_n\right) \subset \mathbb{R}\setminus ]-2^{-i}, 2^{-i}[\right\} = \emptyset$$

for each $i \in \mathbb{N}$. By the countably compactness of $K_i$, there would exist $n_i$ such that

$$\left\{f \in K_i : f(x) = 0,\ f\left(F_{n_i}\right) \subset \mathbb{R}\setminus ]-2^{-i}, 2^{-i}[\right\} = \emptyset.$$

Then if $\varphi_i \in C_p(X)$, $\varphi_i(X) \subset [0, 2^{-i}]$, $\varphi_i(x) = 0$ and $\varphi_i(F_{n_i}) = \{2^{-i}\}$ we would have that $\varphi_i \notin K_i$ and hence $\Sigma_n\varphi_n \notin \cup\{K_i : i \in \mathbb{N}\} = C_p(X)$. As this relation contradicts the continuity of $\Sigma_n\varphi_n$ we have that $X$ is a $P$-space.

If $\{x_n : n \in \omega\}$ were an infinite subset of the $P$-space $X$ then, by induction, each $x_n$ would have a neighborhood $V_n$ such that $\overline{V_n} \cap \overline{V_m} = \emptyset$ when $n \neq m$. For each $n \in \omega$ let $\alpha_n = 1 + \sup\{|f(x_n)| : f \in K_n\}$ and let $g_n \in C_p(X)$ be such that $g_n(X) \subset [0, \alpha_n]$, $g_n(x_n) = \alpha_n$ and $g_n(X\setminus V_n) = \{0\}$. The contradiction that the continuous function $g = \Sigma_n g_n \notin \cup\{K_i : i \in \mathbb{N}\} = C_p(X)$ proves that $X$ is finite.

Conversely, if the cardinal of $X$ is $n$ then $C_p(X)$ is the $\sigma$-compact space $\mathbb{R}^n$. Therefore the following conditions are equivalent:

1. $C_p(X)$ is $\sigma$-compact.
2. $C_p(X)$ is $\sigma$-countably compact.
3. $X$ is finite.

The equivalence between 1 and 3 was obtained by Velichko [5, Theorem I.2.1] and the equivalence between 2 and 3 was due to Tkachuk and Shakhmatov in [22].

### 6.2.4 The Support of a Linear Continuous Functional on $C_c(X)$

Let $\mathscr{K}$ be the family of compact subsets of $X$. In the space $C_c(X)$ of continuous real-valued functions on $X$ endowed with the compact open topology, the family $\{W(K, (-\varepsilon, \varepsilon)) : K \in \mathscr{K}, \varepsilon > 0\}$ is a base of neighborhoods of 0 in $C_c(X)$. Let us observe the next Claim, as easy as helpful:

*Claim* If the sets $K \in \mathscr{K}$ and $D \subset X$ verify inclusion

$$W(K, (-\varepsilon, \varepsilon)) \subset W(D, [-m, m]),$$

with $m > 0$, then $D \subset K$.

If $x \in D \backslash K$ then there exists $f \in C(X)$ such that $f(K) = \{0\}$ and $f(x) = 2m$. Hence $f \in W(K, (-\varepsilon, \varepsilon)) \backslash W(D, [-m, m])$. This contradiction implies the Claim.

The name *compact open topology* comes from the fact that if $\mathscr{B}$ is a base of the topology of $\mathbb{R}$ then the family of sets $\{W(K, V) : K \in \mathscr{K}, V \in \mathscr{B}\}$ is a subbase of $C_c(X)$.

Recall that a subset in a locally convex space $E$ is a *barrel* if it is absorbing, closed and absolutely convex. If $V$ is a subset of a locally convex space $E$ its polar is the set $V^o := \{f \in E' : f(V) \subset [-1, 1]\}$ and if $V$ is a barrel then $V = V^{oo}$ [13, 20.8 (5)]. $E$ is *barrelled* if every barrel is a neighborhood of 0, or, equivalently, if each bounded subset of $(E', \sigma(E', E))$ is equicontinuous, where the weak* topology $\sigma(E', E)$ is the topology induced on the topological dual $E'$ of $E$ by $C_p(E)$ [13, 21.2 and 27]. Hence a locally convex space $E$ is barrelled if and only if its topology coincides with the strong topology $\beta(E, E')$, which is the topology of the uniform convergence on the bounded subsets of $(E', \sigma(E', E))$ [13, 21.2 (2)].

By Baire's category theorem it follows that Banach and Fréchet spaces are barrelled spaces. Then, since the set of the real continuous bounded functions in $X$ with the supremum norm is a Banach space, if $V$ is a barrel in $C_c(X)$ there exists a $\delta_V > 0$ such that $W(X, [-\delta_V, \delta_V]\} \subset V$. If additionally there exists in $X$ a subset $K$ such that $W(K, \{0\}\} \subset V$ then

$$W(K, [-\delta_V/2, \delta_V/2]) \subset V \tag{6.1}$$

because from $f(K) \subset [-\delta_V/2, \delta_V/2]$ it follows that the functions $g := \sup(f, \delta_V/2) + \inf(f, -\delta_V/2)$ and $f - g$ verify that $2g \in W(K, \{0\}\}(\subset V)$ and $2(f - g) \in W(X, [-\delta_V, \delta_V]\}(\subset V)$. Therefore $f = g + (f - g) \in 2^{-1}V + 2^{-1}V = V$. From (6.1) follows the next sufficient condition for a barrel to be a neighborhood of zero.

*Remark 6.1* If a barrel $V$ in $C_c(X)$ contains a set $W(K, \{0\})$, with $K$ compact, then $V$ is a neighborhood of zero in $C_c(X)$.

For each $\varphi \in C_c(X)'$ the barrel $V := \{f \in C_c(X) : |\varphi(f)| \leqslant 1\}$ is a neighborhood of 0 in $C_c(X)$ and then there exists a compact subset $K$ in $X$ and a positive number $\varepsilon$ such that $W(K, (-\varepsilon, \varepsilon)) \subset V$ and, in particular, $W(K, \{0\}) \subset \varphi^{-1}\{0\}$. Therefore the family $\mathscr{K}_\varphi := \{K_i : i \in I\}$ of compact subsets of $X$ such that $W(K_i, \{0\}) \subset \varphi^{-1}\{0\}$ for each $i \in I$ is nonvoid and it verifies the following two claims:

*Claim* $K := \cap\{K_i : i \in I\} \in \mathscr{K}_\varphi$.

If this Claim were not true then there would exist $f \in C_c(X)$ with $f(K) = \{0\}$ and $\varphi(f) = 2$. Let $\delta_V$ be the number considered in (6.1) for the barrel $V := \{f \in C_c(X) : |\varphi(f)| \leqslant 1\}$. Then $\{x \in X : |f(x)| < \delta_V/2\}$ is a neighborhood of $K$ and, by compactness, there exists $K_i \subset \{x \in X : |f(x)| < \delta_V/2\}$. From $f \in W(K_i, [-\delta_V/2, \delta_V/2]\}$ we deduce by (6.1) that $f \in V$ obtaining the contradiction $|\varphi(f)| \leqslant 1$. Therefore $K \in \mathscr{K}_\varphi$ and $K$ is the least compact subset of $X$ such that $W(K, \{0\}) \subset \varphi^{-1}\{0\}$. It is said that $K$ is the support $s(\varphi)$ of $\varphi$.

*Claim* If $K \in \mathscr{K}_\varphi$ and $F$ is a closed subset of $X$ with the property that $W(F, \{0\}) \subset \varphi^{-1}(0)$ then $K \cap F \in \mathscr{K}_\varphi$.

If this were not true then there would exist $f \in W(K \cap F, \{0\})$ such that $\varphi(f) = 2$. Again let $\delta_V$ be the number considered in (6.1) for the barrel $V := \{f \in C_c(X) : |\varphi(f)| \leqslant 1\}$. For the set $K_{\delta_V/2} := \{x \in K : |f(x)| \geqslant \delta_V/2\}$ there exists $n \in C_c(X)$ such that $n(X) \subset [0, 1]$, $n(K_{\delta_V/2}) = 0$ and $n(F) = \{1\}$. Then $f - f \times n \in W(F, \{0\})$ implies that $\varphi(f - f \times n) = 0$ and then $\varphi(f \times n) = \varphi(f) = 2$. From $(f \times n) \in W(K, [-\delta_V/2, \delta_V/2]\}$ and (6.1) follows that $f \times n \subset V$ and we get the contradiction $|\varphi(f \times n)| \leqslant 1$.

The relation $K \cap F \in \mathscr{K}_\varphi$ and the minimality of the $s(\varphi)$ implies that $s(\varphi) \subset K \cap F$.

### *6.2.5 Nachbin-Shirota Barrelledness Theorem [16, 21]*

Examples of properties of $X$ that may be characterized by properties of $C_c(X)$ may be seen in [12, Theorems 2.10, 2.11, 2.13 and 2.14 and Propositions 2.15 and 2.16]. For instance the space $X$ is realcompact if and only if $C_c(X)$ is an inductive limit of normed (or Banach) spaces [12, Proposition 2.16]. In the next theorem, obtained independently by Nachbin and Shirota in [16] and [21], it was proved that $C_c(X)$ is barrelled if and only $X$ is a $\mu$-space (see also [12, Proposition 2.15] and [20, Theorem 10.1.20]).

Recall that a set $A$ in a space $X$ is *functionally bounded* if every continuous real-valued function on $X$ is bounded on $A$. A space $X$ is a $\mu$-space if every closed functionally bounded subspace of $X$ is compact. As a realcompact space $X$ embeds into a closed subset of $\mathbb{R}^{C(X)}$ by the map defined by $\varphi(x) := \{f(x) : f \in C(X)\}$ then each functionally bounded subset of a realcompact space $X$ is relatively compact and therefore $X$ is a $\mu$-space [20, Observation 10.1.19]. In particular, Lindelöf spaces are $\mu$-spaces [12, Proposition 3.12]. Examples of $\mu$-spaces which are not realcompact can be found in [16] and [21].

**Theorem 6.1** *The topological space $X$ is a $\mu$-space if and only if $C_c(X)$ is barrelled.*

*Proof* Let us suppose that $X$ is a $\mu$-space. If $V$ is a barrel in $C_c(X)$ and

$$K := \overline{\cup\{s(\varphi) : \varphi \in V^{o}\}}$$

then from $\{f \in C(X) : f(K) = \{0\}\} \subset V^{oo} = V$ and Remark 6.1 it follows that in order to prove the barrelledness of $C_c(X)$ it suffices to show that $K$ is compact.

If $K$ were a non compact subset of the $\mu$-space $X$ then there would exist $f \in C_c(X)$ unbounded on $K$. By an easy induction, we may obtain an increasing sequence $(n_i)_i$ of natural numbers, a sequence $(\varphi_i)_i$ in $V^o$ and a sequence $(g_i)$ in $C_c(X)$, with $\varphi_i(g_i) = 1$, such that for each $i \in \mathbb{N}$ the sets $A_i = \{x \in K : |f(x)| > n_i\}$ verify that $s(\varphi_i) \cap A_i \neq \emptyset$, $s(\varphi_j) \cap A_i = \emptyset$, when $j < i$, and $g_i(X \backslash A_i) = \{0\}$ . In fact, for $n_1 = 1$ there exists $\varphi_1 \in V^o$ such that $s(\varphi_1) \cap A_1 \neq \emptyset$. Recall that if $g \in C_c(X)$ and $g(X \backslash A_1) = \{0\}$ would imply $\varphi_1(g) = 0$ then, by Claim 6.2.4, $s(\varphi_1) \subset s(\varphi_1) \cap (X \backslash A_1)$, in contradiction with $s(\varphi_1) \cap A_1 \neq \emptyset$. Therefore there exists $g_1 \in C_c(X)$ such that $g_1(X \backslash A_1) = \{0\}$ and $\varphi_1(g_1) = 1$. Now fix $n_2$ $(> n_1)$ such that $s(\varphi_1) \cap A_2 = \emptyset$ and then choose $\varphi_2 \in V^o$ with $s(\varphi_2) \cap A_2 \neq \emptyset$; as in preceding case there exists $g_2 \in C_c(X)$ such that $g_2(X \backslash A_2) = \{0\}$ and $\varphi_2(g_2) = 1$. The induction follows in an obvious way.

The function $g := \sum_{i=1}^{\infty} c_i g_i \in C_c(X)$ for any choice of coefficients $c_i$, $i \in \mathbb{N}$, because $\left(\sum_{i=j+1}^{\infty} c_i g_i\right)(X \backslash A_{j+1}) = \{0\}$. Then the relation $\varphi_j(g) = \sum_{i=1}^{i=j} c_i \varphi_j(g_i) = \sum_{i=1}^{i=j-1} c_i \varphi_j(g_i) + c_j$ enables to select $(c_i)_i$ such that $\varphi_j(g) = j + 1$. As each $\varphi_j \in V^o$ we get that $g \notin iV$, for each $i \in \mathbb{N}$, contradicting that $V$ is absorbing. Therefore $K$ is compact and $C_c(X)$ is barrelled by Remark 6.1.

To prove the converse let $D$ be a closed noncompact subset in $X$. If the barrel $W(D, [-1, +1])$ were a neighborhood of zero, then there would exist a compact subset $K$ and a positive number $\varepsilon$ such that $W(K, (-\varepsilon, \varepsilon)) \subset W(D, [-1, +1])$. Then Claim 6.2.4 implies that $D \subset K$, which contradicts the noncompactness of $D$. Therefore if $C_c(X)$ is barrelled the set $W(D, [-1, +1])$ fails to be absorbent and hence there exists $f \in C_c(X)$ unbounded in $D$. This proves that $X$ is a $\mu$-space. □

### 6.2.6 $\mu$-Spaces and $\ell$-Equivalence

In the class of $\mu$-spaces we have that if $h : C_p(X) \to C_p(Y)$ is a linear homeomorphism then $h : C_c(X) \to C_c(Y)$ is also a linear homeomorphism, hence $\ell_p$-equivalence implies $\ell_c$-equivalence. This result follows from the following facts:

If $h : C_p(X) \to C_p(Y)$ is a linear homeomorphism then its dual map $h' : L_p(Y) \to L_p(X)$ is also a linear homeomorphism by [13, 20.4.6]. This implies that a subset

$B$ of $L_p(Y)$ is bounded if and only if $h'(B)$ is bounded and then by [13, 21.2] we have that $h : (C(X), \beta(C_p(X), L_p(X))) \to (C(Y), \beta(C_p(Y), L_p(Y)))$ is also a linear homeomorphism.

If $Z$ is a $\mu$-space and $\tau_c(Z)$ is the topology of $C_c(Z)$ then $\tau_c(Z) = \beta(C_p(Z), L_p(Z))$ because:

1. The homeomorphism from $Z$ onto the subspace $\hat{Z}$ of $L_p(Z)$ imply that $\tau_c(Z) \leqslant \beta(C_p(Z), L_p(Z))$.
2. The identity map is a continuous immersion from $C_c(Z)$ onto $C_p(Z)$ and then $\beta(C_p(Z), L_p(Z)) \leqslant \beta(C_c(Z), C_c(Z)')$.
3. Finally, Theorem 6.1 implies that $\beta(C_c(Z), C_c(Z)') = \tau_c(Z)$.

In [4, Theorems 1 and 3], by a combination of Milyutin's and Pestov's results, it is proved that $[0, 1]$ and $[0, 1] \times [0, 1]$ are $\ell_c$-equivalent but not $\ell_p$-equivalent.

## 6.3 Michael's Results on $\aleph_0$-Spaces

All the propositions of this section were obtained by E. Michael in [15].

### *6.3.1 Elementary Properties. Heritability by k-Equivalence*

A family $\mathscr{N}$ of subsets of a space $X$ is called a $k$-network in $X$ if, whenever $K \subset U$, with $K$ compact and $U$ open in $X$, then $K \subset \cup\{F : F \in \mathscr{F}\} \subset U$ for some finite family $\mathscr{F} \subset \mathscr{N}$. A topological space $X$ is called $\aleph_0$-space if it is regular and it has a countable $k$-network. First-countable or locally compact $\aleph_0$-spaces are separable and metrizable. Any subspace of an $\aleph_0$-space is an $\aleph_0$-space and the countable product of a family of $\aleph_0$-spaces is an $\aleph_0$-space. Every $\aleph_0$-space $X$ is Lindelöf (hence normal [12, Lemma 6.1]), hereditary separable and every open subset is an $F_\sigma$. These direct properties are proved in [15].

Two Hausdorff topological spaces $X$ and $Y$ are $k$-equivalent if there exists a bijection $f : X \to Y$ such that $f$ and $f^{-1}$ preserve the compact subsets. More precisely, it is said that $X$ and $Y$ are $k$-equivalent by the bijection $f$. Two Hausdorff topologies $\tau_1$ and $\tau_2$ on a set $X$ are *k-equivalent* if they yield the same compact subsets. Let $(X, \tau)$ be a topological space; from Tychonoff's product theorem it follows that the topology $\tau$ is $k$-equivalent to the supremum $\tau_k$ of the family of all topologies $k$-equivalent to $\tau$. The topology $\tau_k$ is the finest topology $k$-equivalent to $\tau$ and it is said that $\tau_k$ is the $k$-topology associated to $\tau$ and that $X_k := (X, \tau_k)$ is the $k$-space associated to $(X, \tau)$.

A topological space $(X, \tau)$ is a $k$-space if $\tau = \tau_k$ and $(X, \tau)$ is a $k$-space if and only if a subset $A$ is open when for each compact subset $C$ the intersection $C \cap A$ is an open subset in $C$ with the induced topology.

A function $f : (X, \tau) \to Y$ is $k$-continuous if its restriction to every compact subset of $X$ is continuous. This happens if and only if $f : (X, \tau_k) \to Y$ is continuous.

It is obvious that if the Hausdorff topological spaces $(X, \tau)$ and $Y$ are $k$-equivalent by the bijection $f$, $(X, \tau)$ is a $k$-space and $f^{-1}$ is continuous then $f$ is a homeomophism, because $f : (X, \tau_k) \to Y$ is continuous and $\tau = \tau_k$.

**Proposition 6.1** ([15, Proposition 8.2]) *If $\mathscr{J}_1$ and $\mathscr{J}_2$ are two k-equivalent topologies on X then $(X, \mathscr{J}_2)$ has a countable k-network if and only if $(X, \mathscr{J}_1)$ does.*

*Proof* Let $\mathscr{P}$ be a countable $k$-network for $\mathscr{J}_1$ closed under finite intersections and suppose that $\mathscr{P}$ is not a countable $k$-network for $\mathscr{J}_2$. Then there exists a compact subset $C$ contained in a $\mathscr{J}_2$-open subset $U_2$ such that, if $(P_n)_n$ is an enumeration of the elements of $\mathscr{P}$ containing $C$, then for each $n$ there exists $x_n \in P'_n \backslash U_2$, with $P'_n = \cap \{P_m : 1 \leqslant m \leqslant n\}$. If $C \subset U$ and $U$ is a $\mathscr{J}_1$ open there exists $P'_{n(U)}$ such that $C \subset P'_{n(U)} \subset U$. Therefore $\{x_m : n(U) \leqslant m\} \subset U$ and from this inclusion it follows that $C \cup \{x_n : n \in \omega\}$ is a compact set. Let $U_1$ be the $\mathscr{J}_1$ open set such that

$$U_1 \cap [C \cup \{x_n : n \in \omega\}] = U_2 \cap [C \cup \{x_n : n \in \omega\}] .$$

From this equality it follows a contradiction because $\{x_m : n(U_1) \leqslant m\} \subset U_1$ and $U_2 \cap [C \cup \{x_n : n \in \omega\}] = C$. □

### 6.3.2 $C_c(X)$ $\aleph_0$-Spaces

In [15, Theorem 9.3] Michael proves that if $X$ is a Hausdorff space with countable $k$-network (named pseudobase in [15]) and if $Y$ is and $\aleph_0$-space then $C_c(X, Y)$ is and $\aleph_0$-space. The below Theorem 6.2 is the particular case of this result needed to proof Theorem 6.3.

**Lemma 6.1** ([15, Lemma 9.1]) *Let $M$ be a compact subset of $X$. Then the mapping $\psi : M \times C_c(X) \to C_c(X)$ such that $\psi(x,f) = f(x)$ is continuous.*

*Proof* Fix $(x,f) \in M \times C_c(X)$. Let $V$ be a neighborhood of $f(x)$ and $D$ be a closed neighborhood of $x$ in $M$ such that $f(D) \subset V$. Then $D$ and $W(D, V)$ verify that $\psi((D \times W(D, V))) \subset V$. Therefore $\psi$ is continuous in $(x,f)$. □

**Lemma 6.2** ([15, Lemma 9.2]) *Let $K$ be a compact subset of $C_c(X)$ and let $U$ be an open subset of $\mathbb{R}$. If $X$ is a k-space then $V := \{x \in X : K(x) \subset U\}$ is open.*

*Proof* It is enough to prove that for each compact subset $M$ of $X$ the set

$$V \cap M = \{x \in M : K(x) \subset U\}$$

is open in $M$. Let $x$ be a point in $V \cap M$. Then, by Lemma 6.1, for each $f \in K$ there exists a $M$-neighborhood $D_f$ of $x$ and a neighborhood $W_f$ of $f$ such that $W_f(D_f) \subset U$. By compactness there exists a finite set $\{f_1, f_2, \ldots, f_n\} \subset K$ such that $K \subset \cup \{W_{f_i} : 1 \leqslant i \leqslant n\}$. This implies that $K\left(\cap D_{f_i}\right) \subset U$ and then $\cap D_{f_i} \subset V \cap M$. This proves that $V \cap M$ is open in $M$. □

To proof the next Lemma 6.3 we need to observe that if a normal space $X$ is the union of a finite family $\{A_i : 1 \leqslant i \leqslant n\}$ of open subsets, then $X$ is covered by a family $\{B_i : 1 \leqslant i \leqslant n\}$ of open subsets such that $\overline{B}_i \subset A_i$, for each $1 \leqslant i \leqslant n$. In fact, as $A_1$ is an open neighborhood of the closed set $X \setminus \cup\{A_i : 2 \leqslant i \leqslant n\}$ then, by normality, there exists an open set $B_1$ such that $X \setminus \cup\{A_i : 2 \leqslant i \leqslant n\} \subset B_1 \subset \overline{B}_1 \subset A_1$. Hence $X = B_1 \cup \{A_i : 2 \leqslant i \leqslant n\}$. After $n$-steps of an obvious induction process we get the open covering $X = \cup\{B_i : 1 \leqslant i \leqslant n\}$, with $\overline{B}_i \subset A_i$, for $1 \leqslant i \leqslant n$.

**Lemma 6.3** ([15, Lemma 5.1]) *Let $\mathscr{S}$ be a subbase of the Hausdorff topological space X and let $\mathscr{P}$ be a collection of subsets of X such that for each compact set C and each open set $U \in \mathscr{S}$ there exists $P \in \mathscr{P}$ such that $C \subset P \subset U$. Then X has a countable k-network.*

*Proof* Let $\mathscr{R}$ be the family of finite unions of finite intersections of elements of $\mathscr{P}$. We will prove that the countable family $\mathscr{R}$ is a $k$-network for $X$. Let $C$ be a compact subset of $X$ and let $U$ be an open neighborhood of $C$. The proof is trivial if $U$ is a finite intersection $\cap\{S_i : 1 \leqslant i \leqslant n\}$ of elements of $\mathscr{S}$, because then for each $i$ there exist $P_i$ such that $C \subset P_i \subset S_i$. Hence $C \subset \cap\{P_i : 1 \leqslant i \leqslant n\} \subset \cap\{S_i : 1 \leqslant i \leqslant n\} = U$ and $R := \cap\{P_i : 1 \leqslant i \leqslant n\} \in \mathscr{R}$. To finish the proof recall that the base $\mathscr{B}$ generated by $\mathscr{S}$ is the family formed by the finite intersections of elements of $\mathscr{S}$. By compactness we may suppose that $C \subset \cup\{A_j : 1 \leqslant j \leqslant p\}$, with $A_j \in \mathscr{B}$ and $A_j \subset U$ for each $1 \leqslant j \leqslant p$. By the observation preceding this lemma we deduce that $C = \cup\{\overline{B}_j : 1 \leqslant j \leqslant p\}$, where each $\overline{B}_j$ is a compact set contained in $A_j$. We have proved that there exists $R_j \in \mathscr{R}$ such that $\overline{B}_j \subset R_j \subset A_j$ for each $1 \leqslant j \leqslant p$. From these relations it follows that the set $R = \cup\{R_j : 1 \leqslant j \leqslant p\} \in \mathscr{R}$ and verifies the inclusions $C \subset R \subset U$. □

**Theorem 6.2** ([15, Theorem 9.3]) *If X is an $\aleph_0$-space then $C_c(X)$ is an $\aleph_0$-space.*

*Proof* By the Proposition 6.1 we know that $kX$ is an $\aleph_0$-space. As $C_c(X)$ is a subspace of $C_c(kX)$ it is enough to prove the theorem in the case that $X$ is a $k$-space. Let $\{Q_m : m \in \omega\}$ be a countable $k$-network in $X$ and let $\mathscr{B} = \{B_n : n \in \omega\}$ be a base of the topology of $\mathbb{R}$. By Lemma 6.3 it is enough to prove that there exists a countable family $\mathscr{P}$ of subsets of $C_c(X)$ such that for each compact subset $K$ of $C_c(X)$, each compact subset $M$ of $X$ and each $B_n \in \mathscr{B}$ such that $K \subset W(M, B_n)$ there exists $P \in \mathscr{P}$ that verifies $K \subset P \subset W(M, B_n)$. Clearly $M \subset \{x \in X : K(x) \subset B_n\}$ and then, by Lemma 6.2 there exists $Q_m$ such that $M \subset Q_m \subset \{x \in X : K(x) \subset B_n\}$. Hence $K \subset W(Q_m, B_n) \subset W(M, B_n)$ and then $\mathscr{P} := \{W(Q_m, B_n) : (m, n) \in \omega^2\}$ is the countable family we are looking for. □

## 6.4 Preservation of $\aleph_0$-Spaces by $\ell_c$-Equivalence

The main result in this section is that $\aleph_0$-spaces are preserved by $\ell_c$-equivalence. It was obtained in [12, Theorem 21].

As $C_p(C_p(X))$ is a topological subspace of $C_p(C_c(X))$ we have that the map $\hat{\Delta}_X : X \to C_p(C_c(X))$, defined by $\hat{\Delta}_X(x) := \hat{x}$ with $\hat{x}(f) = f(x)$ for each $f \in C(X)$, is an embedding. Then $X$ is homeomorphic to the space $\hat{X}$, which is the set $\hat{\Delta}_X(X)$ endowed with the topology induced by $C_p(C_c(X))$.

In general, the map $\tilde{\Delta}_X : X \to C_c(C_c(X))$ defined by $\tilde{\Delta}_X(x) = \tilde{x}$, where also $\tilde{x}(f) = f(x)$ for each $f \in C(X)$, is not continuous because the topology induced in $\tilde{\Delta}_X(X)$ by $C_c(C_c(X))$ is finer that the topology induced by $C_p(C_c(X))$. Therefore the map $\tilde{\Delta}_X$ is continuous if and only if it is an embedding. The space $\tilde{X}$ is the set $\tilde{\Delta}_X(X)$ equipped with the topology induced by $C_c(C_c(X))$.

**Lemma 6.4** ([12, Lemma 20]) *The map $\tilde{\Delta}_X$ is k-continuous. Therefore the topological spaces $\hat{X}$ and $\tilde{X}$ are k-equivalents by the map i defined by $i(\hat{x}) = \tilde{x}$ for each $x \in X$ and if X is a k-space then the map $\tilde{\Delta}_X$ is an embedding.*

*Proof* Let $M$ be a compact subset of $X$ and $F = C_c(C_c(X)) \setminus W(K, V)$, where $K$ is a compact subset of $C_c(X)$ and $V$ is an open subset of $\mathbb{R}$. Let $x$ be the limit of a convergent net $(x_l : l \in L)$, with each $x_l \in M \cap \tilde{\Delta}_X^{-1}(F)$. For each $l \in L$ fix $f_l \in K$ with $f_l(x_l) \notin V$. By compactness we may suppose, taking a subnet if necessary, that the net $(f_l : l \in L)$ converges uniformly on $M$ to some $f \in K$. Then $(f_l(x_l) : l \in L)$ converges to $f(x)$ and the relation $f_l(x_l) \notin V$, for each $l \in l$, implies $f(x) \notin V$. Therefore $x \in M \cap \tilde{\Delta}_X^{-1}(F)$. This proves that the map $\tilde{\Delta}_X$ is $k$-continuous, and then the remaining assertions are obvious. □

The first part of the next theorem extends Theorem 21 in [11]. It is well known that the $\aleph_0$-space property in not preserved by open maps.

**Theorem 6.3** *If X is an $\aleph_0$-space and there is a continuous linear mapping h from $C_c(X)$ onto $C_c(Y)$, then Y is an $\aleph_0$-space. In particular the property of being $\aleph_0$-space is preserved by $\ell_c$-equivalence.*

*Proof* The injective dual mapping $h^* : C(C_c(Y)) \to C(C_c(X))$ verifies that $h^* : C_c(C_c(Y)) \to C_c(C_c(X))$ is continuous and that $h^* : C_p(C_c(Y)) \to C_p(C_c(X))$ is an embedding. We endow $h^*(\tilde{Y})$ and $h^*(\hat{Y})$ with the topologies induced by $C_c(C_c(X))$ and $C_p(C_c(X))$, respectively. Let $j : h^*(\hat{Y}) \to h^*(\tilde{Y})$ be the map defined by $j(h^*(\hat{y})) = h^*(\tilde{y})$, for each $y \in Y$. Since $j$ is the restriction to $h^*(\hat{Y})$ of the natural embedding of $C_p(C_c(X))$ onto $C_c(C_c(X))$, then the inverse map $j^{-1}$ is continuous. By Lemma 6.4, $\hat{Y}$ and $\tilde{Y}$ are $k$-equivalent by the map $i : \hat{Y} \to \tilde{Y}$ such that $i(\hat{y}) = \tilde{y}$ for each $y \in Y$. By definition $h^* i = jh^*|_{\hat{Y}}$.

If $M$ is a compact subset of $\hat{Y}$ then, by $k$-equivalence and continuity, we have that $h^* i(M)$ is a compact subset of $h^*(\tilde{Y})$. If $K$ is a compact subset of $h^*(\tilde{Y})$ then by continuity $\left(jh^*|_{\hat{Y}}\right)^{-1}(K)$ is a compact subset of $\hat{Y}$. Therefore $h^*(\tilde{Y})$ is $k$-equivalent to $\hat{Y}$ and, by homeomorphism is also $k$-equivalent to $Y$.

From Michael's Theorem 6.2 it follows that $C_c(C_c(X))$ is an $\aleph_0$-space. So its subspace $h^*(\tilde{Y})$ is an $\aleph_0$-space. Proposition 6.1 implies that $Y$ is an $\aleph_0$-space. □

Let $\mathfrak{D}$ be the class of first countable or locally compact spaces. From this theorem and Michael results asserting that $\aleph_0$-spaces in $\mathfrak{D}$ are metrizable [15], it follows the next corollary (the first countable case is in [11, Carollary 22]).

**Corollary 6.1** *If $X$ is an $\aleph_0$-space, $Y \in \mathfrak{D}$ and there exists a continuous linear map $h$ from $C_c(X)$ onto $C_c(Y)$ , then $Y$ is metrizable and separable. In particular, the property of being metrizable and separable is preserved by $\ell_c$-equivalence in the class $\mathfrak{D}$.*

As a regular space is second countable if and only if it is metrizable and separable we get that this corollary applies when $X$ is second countable, providing and extension of Arhangel'skii's [3, Theorem 16].

In the next section we are going to prove that the property of being a Polish space is also invariant by $\ell_c$-equivalence in this class $\mathfrak{D}$, extending [11, Corollary 23].

## 6.5 Preservation of Polish Spaces by $\ell_c$-Equivalence

Recall that a topological space $X$ is a *Polish space* if it is separable and completely metrizable.

### *6.5.1 Compact Resolutions and $\mathfrak{G}$-Bases*

A family $C = \{C_{n_1 \cdots n_{p-1} n_p} : n_i \in \omega, 1 \leqslant i \leqslant p\}$ of subsets of $X$ is a *web* if $X = \bigcup_{n_1} C_{n_1}$ and $C_{n_1 \cdots n_{p-1}} = \bigcup_{n_p} C_{n_1 \cdots n_{p-1} n_p}$, for each $p \geqslant 2$.

A *resolution* $X$ is a family $\mathscr{K} = \{K_\alpha : \alpha \in \omega^\omega\}$ of subsets of $X$ such that $X = \bigcup_\alpha K_\alpha$ and $K_\alpha \subset K_\beta$ if $\alpha \leqslant \beta$. For $\eta = (n_i : i \in \omega) \in \omega^\omega$ and $p \in \omega$ we write that $\eta|p := (n_i : 1 \leqslant i \leqslant p)$ and then the sets $C_{n_1 \cdots n_{p-1} n_p} = \bigcup \{A_\alpha : \alpha \in \omega^\omega, \alpha|p = \eta|p\}$ determine the web $\mathscr{C}$ associate to the resolution $\mathscr{K}$. A resolution $\{K_\alpha : \alpha \in \omega^\omega\}$ in a topological space $X$ is compact if each $K_\alpha$ is compact. If, additionally, for each compact subset $K$ of $X$ there exists $K_\alpha \in \mathscr{K}$ such that $K \subset K_\alpha$ then it is said that $\mathscr{K}$ is a *compact resolution swallowing compact sets.*

The name resolution appears in [8]. In [23] a space endowed with a compact resolution is called a *space dominated by irrationals.*

A base $\{U_\alpha : \alpha \in \omega^\omega\}$ of neighborhoods of zero in the topological vector space $E$ is a *$\mathfrak{G}$-base* if $U_\beta \subset U_\alpha$ whenever $\alpha \leq \beta$ [12, Lemma 15.2 (iii)]. If $f$ is a linear continuous open map from the topological vector space $E$ onto the topological vector space $F$ and $\{U_\alpha : \alpha \in \omega^\omega\}$ is a $\mathfrak{G}$- base in $E$ then $\{f(U_\alpha) : \alpha \in \omega^\omega\}$ is a $\mathfrak{G}$-base in $F$.

The following proposition given in [10, Theorem 2] furnishes a relation between resolutions in $X$ and $\mathfrak{G}$-bases in $C_c(X)$.

**Proposition 6.2** ([10, Theorem 2]) *A completely regular topological space $X$ has a compact resolution $\mathscr{K} = \{K_\alpha : \alpha \in \omega^\omega\}$ swallowing compact sets if and only if $C_c(X)$ has a $\mathfrak{G}$-base.*

*Proof* If $\mathscr{K} = \{K_\alpha : \alpha \in \omega^\omega\}$ is a compact resolution on $X$ swallowing compact sets then for each $K \in \mathscr{K}$ and $\varepsilon > 0$ there exists $K_\alpha$, with $\alpha = (a_n)_n \in \omega^\omega$, such that $K \subset K_\alpha$ and $a_1^{-1} < \varepsilon$. Let $U_\alpha := W(K_\alpha, (-a_1^{-1}, a_1^{-1}))$. As $U_\alpha \subset W(K, \varepsilon)$ and $U_\beta \subset U_\alpha$ when $\alpha \leq \beta$ we have that $\mathscr{U} = \{U_\alpha : \alpha \in \omega^\omega\}$ is a $\mathfrak{G}$-base in $C_c(X)$.

Conversely, let $\mathscr{U} = \{U_\alpha : \alpha \in \omega^\omega\}$ be a $\mathfrak{G}$-base of $C_c(X)$. For each $\alpha = (a_n)_n \in \omega^\omega$ there exists $K \in \mathscr{K}$ and $\varepsilon > 0$ such that $W(K, (-\varepsilon, \varepsilon)) \subset U_\alpha$. As the set $K_\alpha := \{x \in X : |f(x)| \leq a_1 \text{ for all } f \in U_\alpha\}$ verifies that $U_\alpha \subset W(K_\alpha, [-a_1, a_1])$ then $W(K, (-\varepsilon, \varepsilon)) \subset W(K_\alpha, [-a_1, a_1])$ and Claim in 6.2.4 imply that $K_\alpha \subset K$. Therefore the closed set $K_\alpha$ is compact. If $D$ is a compact subset of $X$ then there exists $\alpha = (a_n)_n \in \omega^\omega$ such that $U_\alpha \subset W(D, [-1, 1]\}$. Therefore for each $x \in D$ and each $f \in U_\alpha$ we have that $|f(x)| \leq 1$ ($\leq a_1$), hence, by definition, $x \in K_\alpha$. The obtained inclusion $D \subset K_\alpha$ and and the fact $K_\alpha \subset K_\beta$ when $\alpha \leqslant \beta$ in $\omega^\omega$ imply that $\mathscr{K} = \{K_\alpha : \alpha \in \omega^\omega\}$ is a compact resolution swallowing compact subsets of $X$. □

### 6.5.2 Christensen Theorem

Let $A$ be an uncountable subset of $X$ endowed with a resolution $\mathscr{K} = \{K_\alpha : \alpha \in \omega^\omega\}$ and let $\mathscr{C}$ be the web associate to $\mathscr{K}$. By induction we get a sequence $\eta \in \omega^\omega$ such that $A \cap C_{\eta|p}$ is uncountable for each $p \in \omega$. Therefore there exists an infinite subset $\{x_p : p \in \omega\}$ in $X$ and a sequence $\{\beta(p) : p \in \omega\}$ in $\omega^\omega$ such that $x_p \in K_{\beta(p)}$ and $\beta(p)|p = \eta|p$ for each $p \in \omega$. If $\alpha := \sup\{\beta(p) : p \in \omega\}$ we have that $\{x_p : p \in \omega\} \subset K_\alpha$. This implies that if $(X, d)$ is a metric space with a compact resolution $\mathscr{K}$ and $A_n$ is a maximal subset of $X$ such that $d(x, y) \geqslant n^{-1}$, for each $(x, y) \in A_n^2$ with $x \neq y$, then $A_n$ is countable, because the intersection of $A_n$ with every compact subset must be finite. The maximality of each $A_n$ implies that $\cup\{A_n : n \in \omega\}$ is a dense subset of the space $(X, d)$. Therefore a metric space with a compact resolution is separable.

A separable and complete metric space $(X, d)$ has a compact resolution $\mathscr{K}$ swallowing compact sets, because if $\{x_n : n \in \omega\}$ is a dense subset of $(X, d)$, $B(x, \varepsilon)$ is the closed ball of center $x$ and radius $\varepsilon$ and for $\alpha = (a_n)_n \in \omega^\omega$, $K_{a_n} := \cup\left\{B(x_m; n^{-1}) : m \leqslant a_n\right\}$ and $K_\alpha := \cap\left\{K_{a_n} : n \in \omega\right\}$ we get that $K_\alpha$ is compact, because is closed and precompact, and if $K$ is a compact subset of $X$ then, by compact definition, for each $n$ there exists $b_n \in \omega$ such that $K \subset K_{b_n}$. From this inclusion it follows that $K \subset K_\beta$, with $\beta := (b_n : n \in \omega)$. Then $\{K_\alpha : \alpha \in \omega^\omega\}$ is a compact resolution in $X$ swallowing compact subsets.

This proves that each Polish space $X$ has a compact resolution swallowing compact sets. To proof Theorem 6.5 we need the converse of this result, which is a deep theorem due to Christensen [7]. A detailed proof of Christensen's theorem may be found in [12, Theorem 6.1] and the essential parts of this proof are below in Theorem 6.4.

If $F$ is a subset of a metric space $(X, d)$ then the equality $d(x, F) = \inf_{y \in F} d(x, y)$ defines a real continuous function on $X$, because from $d(x, F) \leqslant d(x, y) \leqslant d(x, z)$

$+ d(z, y)$ for each $y \in F$ it follows that $d(x, F) \leqslant d(x, z) + d(z, F)$, and then, by symmetry, $|d(z, F) - d(x, F)| \leqslant d(z, x)$. The continuity of $d(x, F)$ is the key in the proof of the next lemma.

**Lemma 6.5** ([9, 4.3.22 Lemma]) *Every $G_\delta$-set $A$ in a metric space $(X, d)$ is topologically homeomorphic to a closed subspace of the cartesian product $X \times \mathbb{R}^\omega$. Therefore every $G_\delta$-set $A$ in a complete metrizable space $(X, d)$ is topologically complete.*

*Proof* $X \backslash A$ is the union of a sequence $(F_n)_{n\in\omega}$ of closed subsets of $X$. The continuity of each $d(x, F_n)$ implies that the injective map $f$ from $A$ into $X \times \mathbb{R}^\omega$ defined by $f(x) = \left(x, \left([d(x, F_n)]^{-1}, n \in \omega\right)\right)$ is a homeomorphism from $A$ onto $f(A)$. If there exists a sequence $(f(x_p))_{p\in\omega}$ in $f(A)$ that converges to $(x, (r_n, n \in \omega))$ then $x = \lim_p x_p$ and $\lim_p [d(x_p, F_n)]^{-1} = r_n$ for each $n \in \omega$. If $x \notin A$ then there exists $m \in \omega$ such that $x \in F_m$ and we obtain the contradiction $\lim_p \left[d(x_p, F_m)\right]^{-1} = \infty$. Therefore $f(A)$ is closed. If $(X, d)$ is complete then $A$ is homeomorphic to the closed subspace $f(A)$ of the complete metric space $(X \times \mathbb{R}^\omega, \rho)$, being $\rho$ the usual metric associated to a countable product of metric spaces. Hence $A$ is complete with the metric $\rho'$ induced by $f$ (i.e., $\rho'(x, y) := \rho(f(x), f(y))$ for each $x, y \in A$). The lemma follows from the fact that the topologies induced by $d$ and $\rho'$ on $A$ are the same. □

**Theorem 6.4** ([12, Theorem 6.1, Christensen Theorem]) *A metrizable topological space $(X, d)$ has a compact resolution swallowing compact sets if and only if $(X, d)$ is a Polish space.*

*Proof* Let's suppose that $(X, d)$ has a compact resolution $\mathscr{K} = \{K_\alpha : \alpha \in \omega^\omega\}$ swallowing compact subsets and let $\{x_{n_1} : n_1 \in \omega\}$ be a dense subset in $X$.

For each $x_{n_1}$ there exists an open ball $B_{n_1} = B(x_{n_1}, p_1(n_1)^{-1})$, with $p_1(n_1) \in \omega$, such that each compact subset $K$ of $B_{n_1}$ is contained in a $K_\alpha$ with $\alpha|1 = p_1(n_1)$. In fact, if this were not true then for each $m \in \omega$ there would exist a compact subset $K_m$ in $B(x_{n_1}, m^{-1})$ not contained in every $K_\alpha$ with $\alpha|1 = m$. As there exists $\beta \in \omega^\omega$ such that the compact set $\{x_{n_1}\} \cup \{\cup K_m : m \in \omega\}$ is contained in $K_\beta$, we get contradiction for $m = \beta|1$. Let $p_1 : \omega \to \omega$ be the map whose values are $p_1(n_1)$, $n_1 \in \omega$.

Let $\{x_{n_1 n_2} : n_2 \in \omega\}$ be a dense subset in $B_{n_1}$. Again for each $x_{n_1 n_2}$ there exists an open ball $B_{n_1 n_2} = B(x_{n_1 n_2}, p_2(n_1, n_2)^{-1})$ contained in $B_{n_1}$ with $p_2(n_1, n_2) \in \omega$, such that each compact subset $K$ of $B_{n_1 n_2}$ is contained in a $K_\alpha$ with $\alpha|2 = (p_1(n_1), p_2(n_1, n_2))$. If this were not true then for each $m$ such that $B(x_{n_1 n_2}, m^{-1}) \subset B_{n_1}$ there would exist a compact subset $K_m$ in $B(x_{n_1 n_2}, m^{-1})$ not contained in each $K_\alpha$ with $\alpha|2 = (p_1(n_1), m)$. The set $H = \{x_{n_1 n_2}\} \cup \left\{\cup K_m : m \in \omega, B(x_{n_1 n_2}, m^{-1}) \subset B_{n_1}\right\}$ is a compact subset of $B_{n_1}$ and we have proved that there exists $\beta \in \omega^\omega$ with $\beta|1 = p_1(n_1)$ such that $H \subset K_\beta$. We may suppose that $\beta|2 = (p_1(n_1), m)$ verifies that $B(x_{n_1 n_2}, m^{-1}) \subset B_{n_1}$ and then the inclusion $K_m \subset K_\beta$ is a contradiction. We denote by $p_2$ the map defined in $\omega^2$ and generated by the natural numbers $p_2(n_1, n_2)$ obtained in this second step of this induction process.

By a clear induction process we obtain an open web

$$\mathscr{B} = \left\{B_{n_1 \cdots n_{q-1} n_q} : n_i \in \omega, 1 \leqslant i \leqslant q\right\}$$

and a sequence of mappings $(p_i)_i$ such that for each sequence $\eta = (n_i)_i \in \omega^\omega$ and each sequence $(x_q)_q$, with $x_q \in B_{n_1 \cdots n_{q-1} n_q}$ for each $q \in \omega$, there exists a sequence $(\alpha(q))_q$ in $\omega^\omega$ such that $x_q \in K_{\alpha(q)}$ and

$$\alpha(q)|q = (p_1(n_1), p_2(n_1, n_2), \ldots, p_q(n_1 \cdots n_{q-1} n_q)).$$

From $\sup_q \alpha(q) = \beta \in \omega^\omega$ it follows that the sequence $(x_q)_q$ is contained in the compact set $K_\beta$ and therefore it has an adherent point in $K_\beta$.

Let $\left(\tilde{X}, \tilde{d}\right)$ be the $\tilde{d}$-completion of $(X, d)$. For each $B_{n_1 \cdots n_{q-1} n_q}$ let $A_{n_1 \cdots n_{q-1} n_q}$ be an open subset of $\widetilde{X}$ such that

$$A_{n_1 \cdots n_{q-1} n_q} \cap X = B_{n_1 \cdots n_{q-1} n_q}$$

and let

$$M := \cup \left\{ \cap \{A_{n_1 \cdots n_{q-1} n_q}, q \in \omega\}, \eta = (n_i)_i \in \omega^\omega \right\}.$$

By web's definition it follows that $X \subset M$. For each $y \in M$ we may choose $\eta = (n_i)_i \in \omega^\omega$ such that $y \in A_{n_1 \cdots n_{q-1} n_q}$ for each $q \in \mathbb{N}$. Then, for each $q \in \mathbb{N}$ there exists

$$x_q \in B_{n_1 \cdots n_{q-1} n_q}$$

such that

$$d(y, x_q) < q^{-1}.$$

We have proved that the sequence $(x_q)_g$ has an adherent point in $X$. Therefore $y = \lim_q x_q \in X$ and then $X = M$. Clearly $\widetilde{X} \backslash X = \widetilde{X} \backslash M$ is the union of the closed set

$$\widetilde{X} \backslash \left\{ \cup [A_{n_1} : n_1 \in \omega] \right\},$$

and the sets

$$A_{n_1 \cdots n_{q-1} n_q} \backslash \left\{ \cup [A_{n_1 \cdots n_{q-1} n_q n_{q+1}} : n_{q+1} \in \omega] \right\}$$

for each $q \in \omega$, $(n_1 \cdots n_{q-1} n_q) \in \omega^q$. By the separability of $X$ we note that each set $A_{n_1 \cdots n_{q-1} n_q}$ is a countable union of closed sets. Hence $\widetilde{X} \backslash X$ is a $F_\sigma$- subset of $\widetilde{X}$. Therefore $X$ is a $G_\delta$ subset of the complete space $\left(\tilde{X}, \tilde{d}\right)$ and, by Lemma 6.5, $(X, d)$ is a Polish space. □

Let us notice that the web $\mathscr{B} = \left\{ B_{n_1 \cdots n_{q-1} n_q} : n_i \in \omega, 1 \leqslant i \leqslant q \right\}$ in Theorem 6.4 was constructed in such a way that for each $\eta = (n_i : i \in \omega) \in \omega^\omega$ and each $x_q \in B_{n_1 \cdots n_{q-1} n_q}$, with $q \in \omega$, we have that the sequence $(x_q)_q$ has a cluster point in

$X$. This condition reminds Orihuela's definition of *web-compact spaces* [19] that has many interesting applications (see also [12, Sects. 4.3 and 5.1]).

### 6.5.3 Preservation of Polish Spaces by $\ell_c$-Equivalence in the Class $\mathfrak{D}$

The next theorem on the preservation of Polish spaces under $\ell_c$-equivalence in the class $\mathfrak{D}$ of first-countable or locally compact spaces extends Pelant's result [4, Theorem 3.27]. It completes [11, Corollary 23] for locally compact spaces.

**Theorem 6.5** *Let $X$ be a Polish space and $Y \in \mathfrak{D}$. If $X$ $\ell_c$-cover $Y$ then $Y$ is a Polish space. Therefore the property of being Polish space is invariant by $\ell_c$-equivalence in the class $\mathfrak{D}$.*

*Proof* The space $Y$ is $\aleph_0$-space by Theorem 6.3 and then $Y$ is metrizable by [15]. As $X$ has a compact resolution swallowing compact sets then $C_c(X)$ has a $\mathfrak{G}$-base by Proposition 6.2. Also $C_c(Y)$ has a $\mathfrak{G}$-base, because there exits an open continuous linear map from $C_c(X)$ onto $C_c(Y)$. The Proposition 6.2 implies that $Y$ has a compact resolution swallowing compact sets and then Christensen's Theorem 6.4 implies that $Y$ is a Polish space. □

## 6.6 Extension to Spaces of Pointwise Countable Type

Corollary 6.1 and Theorem 6.5 stated that second countability and the property of being Polish space are invariants by $\ell_c$-equivalence in the class $\mathfrak{D}$ of first countable or locally compact spaces. The $\ell_c$-invariance of Polish spaces was extended in [11, Corollary 24] to the class of pointwise countable type spaces. This result (see Corollary 6.3) and an extension of preservation of second countability by $\ell_c$-equivalence in the class of pointwise countable type spaces (see Corollary 6.2) are provided in this section.

### 6.6.1 Spaces of Pointwise Countable Type

A space $Y$ is *of pointwise countable type* [5, Chap. 0, Sect. 2] if for every $y \in Y$ there exists a compact set $K$ that contains $y$ such that $K$ has a countable base of neighborhoods in $Y$. From this definition it follows that the class of spaces of pointwise countable type contains the first countable and the locally compact spaces.

Let us notice that if $Y$ is a Čech-complete space then $Y$ is a space of pointwise countable type. In fact, as $Y = \cap O_n$, being each $O_n$ an open subset of the Stone-Čech compactification $\beta Y$ of $Y$, then for each $y \in Y$ and each $n$ there exists in $\beta Y$ a closed neighborhood $A_n$ of $y$ such that $A_n \subset O_n$. The set $K := \cap A_n$ is a compact subset of $X$. We may suppose that $A_{n+1} \subset A_n$ for each $n \in \omega$. Clearly $y \in K$ and by compactness if $A$ is an open neighborhood of $K$ in $\beta Y$ there exists $n \in \omega$ such that $A_n \subset A$. Then $\{A_n \cap X : n \in \omega\}$ is a countable base of neighborhoods of $K$ and hence $Y$ is a space of pointwise countable type.

The next Propositions 6.3 and 6.4 are the Exercise 3.3.I of [9] and the Theorem 5.6.2 of [14]. We recall that the proofs are given for the sake of completeness.

**Proposition 6.3** *Every space X of pointwise countable type is a k-space.*

*Proof* Let us suppose that $X$ is a topological space of pointwise countable type that it is not $k$-space. Then there exists a point $x \in \overline{A} \backslash A$, being $A$ a subset such that the intersection of $A$ with every compact subset of $X$ is closed. By hypothesis the point $x$ is contained in a compact subset $K$ that has a decreasing base of neighborhoods $(V_n)_n$. Since the point $x$ does not belong to the closed set $A \cap K$ then there exists a closed neighborhood $V$ of $x$ such that $V \cap A \cap K = \emptyset$.

For each $n$ there exists $x_n \in V \cap V_n \cap A$. If $U$ is a neighborhood of $K$ there exists $m$ such that $K \subset V_m \subset U$. Therefore from $K \cup \{x_n : n \geqslant m\} \subset U$ it follows that $K \cup \{x_n : n \in \omega\}$ is a compact subset of $X$. Hence the set $V \cap A \cap (K \cup \{x_n : n \in \omega\})$ is closed and equal to $\{x_n : n \in \omega\}$. This implies that the set $X \backslash \{x_n : n \in \omega\}$ is open and contains $K$. Therefore there exists $q \in \omega$ such that $K \subset V_q \subset X \backslash \{x_n : n \in \omega\}$. This inclusion yields the contradiction $x_q \in X \backslash \{x_n : n \in \omega\}$, so $X$ is a $k$-space. □

## 6.6.2 Two Results on Submetrizability

A topological space $X$ is submetrizable if there exists a metric space $M$ and a continuous bijective map $\Phi : X \rightarrow M$. It is clear that if a topological space $Y$ has a dense $\sigma$-compact subset $A$, i.e. $A = \cup\{K_n : n \in \omega\}$, $\overline{A} = Y$ and each $K_n$ compact, then $C_c(Y)$ is submetrizable, because the topology in $C(Y)$ of the uniform convergence on each $K_n$, $n \in \omega$, is metrizable and weaker than the topology of $C_c(Y)$. It is not so obvious to proof the next proposition ([14, Theorem 5.6.2]).

**Proposition 6.4** *If $X$ is submetrizable then $C_c(X)$ has a dense $\sigma$-compact subset.*

*Proof* By hypothesis there exists a metric space $M$ and a continuous bijective mapping $\Phi : X \rightarrow M$. Then the mapping $\Phi^* : C_c(M) \rightarrow C_c(X)$ defined by $\Phi^*(g) = g \cdot \Phi$ is continuous. From the fact that the restriction of $\Phi$ to a compact subset $K$ of $X$ is a homeomorphism it follows that $\Phi^*(C_c(M))$ is a dense subset of $C_c(X)$; in fact, if $f \in C_c(X)$ and $K$ is a compact subset of $X$ then the map $h$ defined on $\Phi(K)$ by $h(\Phi(x)) = f(x)$ is continuous and hence it has a continuous extension $g \in C_c(M)$. For each $x \in K$ we have that $(\Phi^*(g))(x) = h(\Phi(x)) = f(x)$ and then from $\sup\{|f(x) - (\Phi^*(g))(x)| : x \in K\} = 0$ it follows that $\Phi^*(C_c(M))$ is a dense subset of $C_c(X)$. Therefore it is enough to prove this proposition when $X$ is metrizable.

The paracompactness of the metric space $X$ implies that for each $n \in \omega$ there exists a locally finite partition of the unit $\mathscr{F}_n$ subordinated to the open subsets of diameter less than $1/n$. This means that $f(X) \subset [0, 1]$, for each $f \in \mathscr{F}_n$, being $\{f^{-1}((0, 1]) : f \in \mathscr{F}_n\}$ a locally finite family of sets with diameter less than $1/n$ and such that $\sum\{f(x) : f \in \mathscr{F}_n\} = 1$ for each $x \in X$. The locally finite condition implies that for each point $x$ there exists a neighborhood $V(x)$ and a finite subset $F_n(x)$ of $\mathscr{F}_n$ such that $f(V(x)) = \{0\}$ for each $f \in \mathscr{F}_n \backslash F_n(x)$. Therefore each net $\mathscr{N}$ in $\mathscr{F}_n$ has a subnet that converges uniformly on $V(x)$, hence if $K$ is a compact subset of $X$ the net $\mathscr{N}$ has also a subnet that converges uniformly in $K$. By Tychonov theorem each net $\mathscr{N}$ in $\mathscr{F}_n$ has a subnet that converges uniformly in each compact subset $K$ of $X$ to a real function $g$. Since $X$ is a $k$-space we get that $g \in C(X)$. Hence $\mathscr{F}_n$ is a relatively compact subset of $C_c(X)$.

Let $e \in C_c(X)$ be such that $e(X) = \{1\}$. From the continuity of the sum, products (of two functions as well as the product of a function by scalar) and the compactness of finite subsets it follows that the sets

$$\mathscr{F}'_n := \{rf : -1 \leqslant r \leqslant 1,\ f \in \mathscr{F}_n \cup \{e\}\}$$

$$\mathscr{G}_n := \Big\{f_1 f_2 \cdots f_k : 1 \leqslant k \leqslant n,\ \{f_i : 1 \leqslant i \leqslant k\} \subset \cup\{\mathscr{F}'_j : 1 \leqslant j \leqslant n\}\Big\}$$

$$\mathscr{H}_n := \{f_1 + f_2 + \cdots + f_k : 1 \leqslant k \leqslant n,\ \{f_i : 1 \leqslant i \leqslant k\} \subset \mathscr{G}_n\}$$

are relatively compact subsets of $C_c(X)$. $\mathscr{H} = \cup\{\mathscr{H}_n : n \in \omega\}$ is a subalgebra of $C_c(X)$ since from $f \in \mathscr{H}_m$, $g \in \mathscr{H}_n$ and $t \in \mathbb{R}\backslash\{0\}$ with $|t| \leqslant p$ it follows that $f + g \in \mathscr{H}_{m+n}$, $fg \in \mathscr{H}_{mn}$ and $tf = (tp^{-1})pf \in \mathscr{H}_{mp}$.

If $x \neq y$ are two points of $X$ then there exists $n$ such that $d(x, y) > n^{-1}$ and since $\mathscr{F}_n$ is a partition of the unit there exists $f \in \mathscr{F}_n$ such that $f(x) \neq 0$. By construction, the diameter of $f^{-1}((0, 1])$ is less than $1/n$ and therefore $f(y) = 0$. Then $\mathscr{H}$ separates points of $X$ and the Stone-Weierstrass theorem assures that the algebra $\mathscr{H}$ is dense in $C_c(X)$. Therefore $\cup\{\overline{\mathscr{H}_n}^{C_c(X)} : n \in \omega\}$ is a dense $\sigma$-compact subset of $C_c(X)$. □

The *transitivity of character for compact sets* implies that if $F_1$ and $F_2$ are compact subsets of a topological space $Y$ such that $F_1 \subset F_2$, $F_1$ has countable character in $F_2$ and $F_2$ has countable character in $Y$ then $F_1$ has countable character in $Y$, see [1, Proposition 3.3]. In particular, a submetrizable space $Y$ of pointwise countable type is first countable. To proof this property apply the transitivity of character for a compact set $F_1 := \{y_0\}$ and a compact set $F_2$ of countable character in $Y$ that contains $y_0$. Nevertheless an easy direct proof of this property is given in Proposition 6.5.

**Proposition 6.5** *Let $Y$ be a submetrizable space of pointwise countable type. Then $Y$ is first countable.*

*Proof* Each $y \in Y$ is contained in a compact subset $K$ which has a decreasing base $(V_n)_n$ of closed neighborhoods. As $K$ is metrizable there exists in $Y$ a decreasing sequence $(W_n)_n$ of closed neighborhoods of $y$ such that $(W_n \cap K)_n$ is a base of neighborhoods of $y$ in $K$. We may suppose that $W_n \subset V_n$ for each $n \in \omega$. Therefore if $V$ is an open neighborhood of $y$ there exist $m$ such that $W_m \cap K \subset V$.

Let us suppose that for each $n \in \omega$ there exists $x_n \in W_n \backslash V$. Then from $x_p \in V_n$ for each $p \geqslant n$ it follows that $K \cup \{x_n : n \in \omega\}$ is compact. If $x$ is an adherent point of $(x_n)_n$ we have that $x \in W_n$ for each $n \in \omega$, and then $x \in \bigcap \{V_n : n \in \omega\} = K$. Therefore $x \in W_m \cap K \subset V$. But the relations $x_n \in W_n \backslash V \subset E \backslash V$ for each $n \in \omega$ yield the contradiction $x \in E \backslash V$. Hence there exists $W_p \subset V$, and the sequence $(W_n)_n$ is a base of neighborhoods of $y$. □

### 6.6.3 Preservation of Second Countable and Polish Spaces by $\ell_c$-Equivalence for Spaces of Pointwise Countable Type

**Proposition 6.6** ([11, Proposition 18]) *Let $X$ be a submetrizable space and let $Y$ be a space of pointwise countable type. If there exists a continuous map from $C_c(X)$ onto $C_c(Y)$ then $Y$ is submetrizable and first countable.*

*Proof* By Proposition 6.4 the space $C_c(X)$ has a dense $\sigma$-compact subset. Hence by continuity $C_c(Y)$ also has a dense $\sigma$-compact subset, which implies that $C_c(C_c(Y))$ is submetrizable. From Proposition 6.3 it follows that $Y$ is a $k$-space and then, by Lemma 6.4, $Y$ is embeds in $C_c(C_c(Y))$. Therefore $Y$ is a submetrizable space. From Proposition 6.5 we get that $Y$ is first countable. □

The next corollary extends [11, Corollary 22] for spaces of pointwise countable type.

**Corollary 6.2** *If $X$ is metrizable and separable, $Y$ is a space of pointwise countable type and $h$ is a continuous linear mapping from $C_c(X)$ onto $C_c(Y)$, then $Y$ is metrizable and separable. Therefore second countability is preserved by $\ell_c$-equivalence for spaces of pointwise countable type.*

*Proof* From Proposition 6.6 it follows that $Y$ is first countable. Then from Corollary 6.1 it is deduced the statement of this corollary. □

The property that Polish spaces are preserved by $\ell_c$-equivalence in the spaces of countable type may be obtained by a combination of results of Baars, de Groot, Pelant and Valov on preservation of complete metrizability by $\ell_p$-equivalence in the class of metrizable spaces [6, 25]. The next corollary is the simple proof given in [11, Corollay 24].

**Corollary 6.3** ([11, Corollay 24]) *Let $X$ be a Polish space that $\ell_c$-covers the space $Y$. If $Y$ is of pointwise countable type then $Y$ is a Polish space. Thus the property of being a Polish space is preserved by $\ell_c$-equivalence for spaces of pointwise countable type.*

*Proof* From Proposition 6.6 it follows that $Y$ is first countable. So Theorem 6.5 applies. □

**Acknowledgments** Supported by Generalitat Valenciana, Conselleria d'Educació Cultura i Esport, Spain, Grant PROMETEO/2013/058.

## References

1. Arhangel'skii, A.V.: Bicompact sets and the topology of spaces. Trudy Moskov. Mat. Obšč. **13**, 3–55 (1965) (in Russian)
2. Arhangel'skii, A.V.: On linear homomorphisms of function spaces. Dokl. Akad. Nauk SSSR. **264**, 1289–1292 (1982)
3. Arhangel'skii, A.V.: A survey of $C_p$-theory. Questions Answers Gen. Topology. **5**, 1–109 (1987)
4. Arhangel'skii, A.V.: General Topology III. Encyclopaedia of Mathematical Sciences, 51. Springer, Berlin (1991)
5. Arhangel'skii, A.V.: Topological Function Spaces. Mathematics and its Applications 78. Kluwer Academic Publishers, Dordrecht (1992)
6. Baars, J., de Groot, J., Pelant, J.: Function spaces of completely metrizable spaces. Trans. Amer. Math. Soc. **340**, 871–883 (1993)
7. Christensen, J.P.R.: Topology and Borel Structure. North-Holland Mathematics Studies 10, North-Holland, Amsterdam (1974)
8. Drewnowski, L.: Resolutions of topological linear spaces and continuity of linear maps. J. Math. Anal. Appl. **335**, 1177–1195 (2007)
9. Engelking, R.: General Topology. Monografie Matematiczne 60. PWN, Warszawa (1975)
10. Ferrando, J.C., Kąkol, J.: On precompact sets in spaces $C_c(X)$. Georgian Math. J. **20**, 247–254 (2013)
11. Kąkol, J., López-Pellicer, M., Okunev, O.: Compact covers and functions spaces. J. Math. Anal. Appl. **411**, 372–380 (2014)
12. Kąkol, J., Kubiś, W., López-Pellicer, M.: Descriptive Topology in Selected Topics of Functional Analysis. Developments in Mathematics 24. Springer, New York (2011)
13. Köthe, G.: Topological Vector Spaces. Springer, Berlin (1969)
14. McCoy, R.A., Ntantu, I.: Topological Properties of Spaces of Continuous Functions. Lecture Notes in Mathematics 1315. Springer (1988)
15. Michael, E.: $\aleph_0$-spaces. J. Math. Mech. **15**, 983–1002 (1966)
16. Nachbin, L.: Topological vector spaces of continuous functions. Proc. Nat. Acad. Sci. **40**, 471–474 (1945)
17. Nagata, J.: On lattices of functions on topological spaces and of functions on uniform spaces. Osaka Math. J. **1**, 166–181 (1949)
18. Okunev, O.: A relation between spaces implied by their $t$-equivalence. Topology Appl. **158**, 2158–2164 (2011)
19. Orihuela, J.: Pointwise compactness in spaces of continuous functions. J. Lond. Math. Soc. **36**, 143–152 (1987)
20. Pérez Carreras, P.; Bonet, J.: Barrelled Locally Convex Spaces. North-Holland Mathematics Studies 131. North-Holland, Amsterdam (1987)
21. Shirota, T.: On locally convex vector spaces of continuous functions. Proc. Japan Acad. **30**, 294–298 (1954)
22. Tkachuk, V.V., Shakhmatov, D.B.: When is space $C_p(X)$ $\sigma$-countably compact? Moscow Univ. Math. Bull. **42**, 23–26 (1987)
23. Tkachuk, V.V.: A space $C_p(X)$ is dominated by irrationals if and only if it is $K$-analytic. Acta Math. Hungar. **107**, 253–265 (2005)

24. Uspenski, V.: On the topology of a free locally convex space. Soviet Math. Dokl. **27**, 781–785 (1983)
25. Valov, V.: Function spaces. Topology Appl. **81**, 1–22 (1997)
26. Velichko, N.V.: The Lindelöf property is *l*-invariant. Topology Appl. **89**, 277–283 (1998)

# Chapter 7
# Characteristics of the Mackey Topology for Abelian Topological Groups

**José Manuel Díaz Nieto and Elena Martín Peinador**

*The present chapter, written for a joyful event, has suddenly changed its sign: the first author died a few days after its submission. Hiking the mountains of Gredos, near his homeland Ávila, and with a big experience and passion in this sport, José Manuel flew to scale new heights beyond the Mathematics. The chapter contains a part of his Doctoral Thesis, which was to be defended around the coming November.*

**Abstract** This chapter is inspired on the Mackey-Arens Theorem, and consists on a thorough study of its validity in the class of locally quasi-convex abelian topological groups. If $G$ is an abelian group and $H$ is a group of characters which separates points of $G$, the pair $(G, H)$ is said to be a dual pair. Any group topology on $G$ which has $H$ as its group of continuous characters is said to be compatible with the pair $(G, H)$ or with the group duality $(G, H)$. If the starting group $G$ is already a topological group, a natural duality is obtained taking $H$ as the group of its continuous characters. The family of all locally quasi-convex topologies defined on an abelian group $G$, with a fixed common character group $H$, was studied for the first time in [9]. It is a problem not solved yet if the supremum of a family of locally quasi-convex compatible topologies on an abelian topological group $G$ is again a compatible topology. If it is, then it is called the Mackey topology for $G$. A locally quasi-convex group $G$ is said to be a Mackey group whenever it carries the Mackey topology. Locally quasi-convex topologies can be characterized in terms of the families of equicontinuous subsets that they produce in the corresponding dual group. We have adopted this point of view and we have defined a grading of Mackey-type properties for abelian topological groups. We also study the stability of these properties through quotients.

J. M. D. Nieto · E. M. Peinador (✉)
Departamento de Geometría y Topología, Facultad de Matemáticas,
Universidad Complutense de Madrid, 28040 Madrid, Spain
e-mail: peinador@mat.ucm.es

J. C. Ferrando and M. López-Pellicer (eds.), *Descriptive Topology and Functional Analysis*, 117
Springer Proceedings in Mathematics & Statistics 80, DOI: 10.1007/978-3-319-05224-3_7,
© Springer International Publishing Switzerland 2014

**Keywords** Mackey topology on a topological group · Equicontinuous subset · Locally quasi-convex group · Precompact group · Bohr compactification

**MSC 2000 Classification:** Primary 22A10, 22D35, Secondary 20K45

## 7.1 Introduction

The Mackey Theory is a well known topic in Functional Analysis. If $(E, \tau)$ denotes a real locally convex topological vector space and $E'$ its dual space, the Mackey topology is defined as the finest locally convex topology on $E$ producing the same continuous linear functionals as $(E, \tau)$. The existence of the Mackey topology and an external description of it, are provided by the Mackey-Arens Theorem. Any vector topology $\nu$ on $E$ admitting $E'$ as dual space is called a compatible topology for $(E, \tau)$ or just for the pair $(E, E^{'})$. Thus the Mackey topology on $E$ is the finest locally convex compatible topology for $(E, E^{'})$, and it can be explicitly described as the topology of uniform convergence on the family of all absolutely convex weakly compact subsets of $E'$.

J. Kakol was the first to realize that local convexity is an essential requirement in the Theorem of Mackey-Arens. In [14] it is proved that in some classes $\mathscr{R}$ of topological vector spaces, for a fixed $(X, \tau) \in \mathscr{R}$ it is not always possible to obtain a vector space topology $\mu$ on $X$ such that $(X, \mu) \in \mathscr{R}$ and it is the finest compatible with $\tau$. Thus, if the class of locally convex spaces is substituted by some other class of topological vector spaces $\mathscr{R}$, the analogous to the Mackey-Arens Theorem does not necessarily hold in $\mathscr{R}$.

A different context to extend Mackey-Arens Theorem is provided by the locally quasi-convex groups. Inspired on the Hahn-Banach Theorem, Vilenkin defined the notion of quasi-convex subset of an abelian topological group [19]. He also introduced the locally quasi-convex groups as those abelian topological groups which admit a basis of quasi-convex zero neighborhoods. On the other hand Banaszczyk proved that a topological vector space is locally convex if and only if it is locally quasi-convex as an abelian topological group [5]. This result permits to conceive the locally quasi-convex groups as a class which extends that of locally convex spaces. The quasi-convexity is by no means as easy to handle as convexity: it requires hard calculations, even if the group involved is the underlying group of a topological vector space. Nevertheless there is some parallellism between duality defined in the category $\mathscr{LCS}$ of locally convex spaces and in the category $\mathscr{LQG}$ of locally quasi-convex groups.

In [9] the Mackey theory for locally convex spaces was generalized—as far as possible—to locally quasi-convex groups. It was also proved that certain classes of groups like the locally compact abelian groups or the complete metrizable locally quasi-convex groups are Mackey groups. Later on, in [3] and in [10] it was proved that completeness cannot be dropped in the previous sentence. The main open problem in this topic is the existence of a Mackey topology in a topological group $G$, without further constrains on $G$.

In $\mathscr{L}\mathscr{Q}\mathscr{G}$ the dualizing object is the circle group $\mathbb{T}$, that is, the group of complex numbers of modulus one, endowed with its natural topology induced from $\mathbb{C}$. The homomorphisms from an abelian group $G$ into $\mathbb{T}$, are usually called characters on $G$. The set of characters on $G$ with respect to the pointwise operation is a group, frequently designated as the algebraic dual of $G$ and represented by the symbol $G^*$. If $(G, \tau)$ is an abelian topological group, the continuous characters on $G$ constitute a subgroup of $G^*$: it will be denoted by $(G, \tau)^\wedge$ (or simply $G^\wedge$), and called the dual of $G$. A group topology $\nu$ on a topological abelian group $(G, \tau)$ is said to be *compatible* for $(G, \tau)$—or for the pair $(G, G^\wedge)$—if $(G, \nu)^\wedge = G^\wedge$, in other words if it gives rise to the same dual group as the original topology.

Varopoulos was the first to consider duality in the framework of abelian groups. In [18], a group duality is defined as a pair $(G, H)$ where $G$ is an abelian group and $H$ is a group of characters on $G$ that separates the points of $G$. The groups $G$ and $H$ play a symmetric role, since the elements of $G$ can be considered as characters on $H$, just identifying each point $x$ of $G$ to the evaluation $\tilde{x} : H \to \mathbb{T}$ which carries $\phi \in H$ into $\phi(x)$.

Although the above definition of duality is purely algebraic, it gives rise to two standard topologies: $\sigma(G, H)$ and $\sigma(H, G)$. The first one is the topology on $G$ of pointwise convergence on the elements of $H$ and $\sigma(H, G)$ is the topology on $H$ of pointwise convergence on the elements of $G$ considered as evaluation mappings. It is proved in [18] that $(G, \sigma(G, H))$ has dual group precisely $H$ (Corollary, p. 481) and now it makes sense to consider those topologies on $G$ which admit $H$ as dual group, and call them compatible for the duality $(G, H)$. The symmetric situation for $(H, \sigma(H, G))$ is described in an identical way.

The definition of a duality $(G, H)$ can be done also if $H$ does not separate the points of $G$. If this is the case, one can replace the group $G$ by the quotient group $G/H^\perp$, where $H^\perp := \bigcap_{\phi \in H} \phi^{-1}(\{0\})$ is the von Neumann kernel. The duality $(G/H^\perp, H)$ is now separating, and loosely speaking, the pair $(G/H^\perp, H)$ carries all the information about the duality $(G, H)$. In our future considerations we do not require the duality $(G, H)$ to be separating and we will write $< G, H >$ instead of $(G, H)$ to stress the fact that we speak of duality in its most general context.

For a topological group $(G, \tau)$ there is a standard natural duality obtained by taking $H$ as $G^\wedge$, the dual group of $G$. As said above, *a topology on $G$ compatible with $\tau$* means a topology compatible with $(G, G^\wedge)$.

Trying to emulate the Mackey theory, Varopoulos considers for a fixed group duality $(G, H)$, the family $\mathscr{P}C$ of all locally precompact compatible topologies on $G$. The supremum of $\mathscr{P}C$ in the lattice of all topologies, say $\nu(G, H)$, is again a compatible topology, which we will call the Varopoulos topology. However $\nu(G, H)$ cannot be considered a candidate to be the Mackey topology in this new setting of abelian topological groups.

We outline next an example given in [9], which might convince the reader of our point of view. Take in the place of $G$ a topological vector space $E$ and let $E^\wedge$ denote the dual of $E$ as a topological group. The topology $\nu(E, E^\wedge)$ coincides with $\sigma(E, E')$, the weak topology on $E$ as a vector space. Clearly $\sigma(E, E')$ is seldom the Mackey topology on $E$, as a locally convex topological vector space. For instance, if $E$ is an infinite dimensional Banach space, its Mackey topology is the original one while $\sigma(E, E')$ is not even metrizable.

The authors of [9] had the idea to substitute the above family $\mathscr{P}C$ by the family LQCC of all locally quasi-convex topologies in $G$ which are compatible with $(G, H)$. The framework provided by locally quasi-convex groups, briefly LQC groups, is more suitable for a generalization of the Mackey theory for abelian (topological) groups. However, it is not known for the general case (that is, without any restrictions on the duality) if the supremum of a family of topologies in LQCC is inside that class. It is not hard to prove that the supremum of any family of LQC topologies on a topological group $G$ is again a locally quasi-convex topology, but it is not known so far if the supremum of a family of compatible topologies is compatible.

Following the track of the Mackey-Arens Theorem, two candidates arise to be defined as the Mackey topology of an abelian topological group $(G, \tau)$. The first option would be to take the finest among all the topologies which are locally quasi-convex and **compatible** with $\tau$, if such an object exists.

The existence of the Mackey topology in this sense (compatibility of the supremum of locally quasi-convex compatible topologies) is proved in [9] for some particular classes of topological groups and the general case is mentioned as an open problem.

The second candidate to be the Mackey topology on $G$ is the topology $\tau_{\mathscr{Q}}$ of uniform convergence on the family $\mathscr{Q}$ of all the quasi-convex and compact subsets of $G^\wedge_p$, where the latter is the dual group of $G$ endowed with the $\sigma(G^\wedge, G)$-topology. In [9] it is asked if these two approaches are equivalent. A negative answer is provided in [6], by means of examples of topological groups $G$, for which the finest LQC compatible topology does not coincide with the topology of uniform convergence on the weakly compact and quasi-convex subsets of the dual $G^\wedge$. More precisely, Example (4.2) in [6] describes a topological group which has non-continuous characters, i.e. $G^\wedge \neq G^*$, and nevertheless the topology of uniform convergence on the family $\mathscr{Q}$ is discrete, so $(G, \tau_{\mathscr{Q}})^\wedge = G^*$. Thus, the Mackey-Arens Theorem does not admit a counterpart for abelian locally quasi-convex groups, at least in the most natural way to generalize it.

With the just mentioned Example, it was natural to seek a family $\mathscr{H}$ of subsets of the dual $G^\wedge$ of a topological group $G$, such that the inverse polars of the members of $\mathscr{H}$ were essentially a basis of neighbourhoods for the greatest compatible topology in $G$. The authors of [6] pointed out a family $\mathscr{H}$ which could have such a property. This approach has suggested us to define a grading in the property "to be a Mackey group", and this will be done in Sects. 7.6 and 7.7 of the present chapter. The Mackey topology for a topological group $(G, \tau)$ (as used subsequently, for instance in [3], [10] and some other papers) is the greatest locally quasi-convex compatible topology, provided it exists. Accordingly, a topological group $(G, \tau)$ is said to be a Mackey group if its original topology $\tau$ is the Mackey topology. Whenever $\tau_{\mathscr{Q}}$ is compatible, it is precisely the finest with this condition, and therefore the Mackey topology.

We describe now the contents of the present chapter. Sects. 7.2 and 7.3 are introductory, explaining notation and preliminaries. In Sect. 7.4 we introduce some definitions and technical lemmata, relating different topologies in a quotient group $G/H$ with the corresponding of the original group $G$. These results will be used in Sects. 7.6 and 7.7.

In Sect. 7.5 we prove that the quotients of g-barrelled (pre-Mackey) are essentially g-barrelled (pre-Mackey) and therefore Mackey.

The main results are in Sects. 7.6, 7.7 and 7.8. In Sect. 7.6 we provide a necessary and sufficient condition for the existence of the Mackey topology for an abelian topological group $G$, and also for the quotient group $G/H$, where $H$ is any subgroup of $G$. In order to describe these results, we introduce next some notation.

For an abelian topological group $G$ and for a fixed subset $B \subset G^\wedge$, let $\tau_{\{B\}}$ be the topology on $G$ of uniform convergence on $B$. It is a group topology on $G$, which has as a basis of zero-neighborhoods the family

$$\mathscr{B} := \{\cap_{\varphi \in B}\varphi^{-1}(\mathbb{T}_n) : n \in \mathbb{N}\}, \text{ where } \mathbb{T}_n := [-1/4n, 1/4n] + \mathbb{Z} \subset \mathbb{T}.$$

The subset $B$ is said to "determine equicontinuity" for $(G, G^\wedge)$ if $(G, \tau_{\{B\}})^\wedge \subseteq G^\wedge$. If $\mathscr{M}$ is the family of all subsets determining equicontinuity, then $\tau_{\mathscr{M}} := sup_{B \in \mathscr{M}} \tau_{\{B\}}$ is the topology defined by uniform convergence on the sets of the family $\mathscr{M}$. We prove that the above mentioned family $\mathscr{H}$ obtained in [6] is equivalent to the family $\mathscr{M}$. However, the topology $\tau_{\mathscr{M}}$ is not the supremum of all locally quasi convex compatible topologies in $(G, \tau)$, as the authors of [6] thought. We characterize the case when $\tau_{\mathscr{M}}$ is a compatible topology (Theorem 7.9), being then $\tau_{\mathscr{M}}$ the Mackey topology corresponding to $(G, G^\wedge)$. We have called $(\mathscr{M})$-Mackey these kind of groups which are Mackey groups in a stronger sense. Thus, compatibility of $\tau_{\mathscr{M}}$ is a sufficient condition for the existence of the Mackey topology, but we do not know if it is also necessary. We end the section with the proof that the quotients of $(\mathscr{M})$-Mackey groups are essentially $(\mathscr{M})$-Mackey groups.

A more accurated family must be defined in order to obtain a necessary and sufficient condition for the existence of the Mackey topology. Let $\mathscr{S} \subset \mathscr{M}$ be the subfamily formed by those sets $B \in \mathscr{M}$ such that $\tau_{\{B\}} \vee \sigma(G, G^\wedge)$ is a topology compatible for $G$. We denote by $\tau_{\mathscr{S}} = sup_{B \in \mathscr{S}} \tau_{\{B\}}$ the topology defined by uniform convergence on the sets of $\mathscr{S}$. Clearly $\tau_{\mathscr{S}} \subset \tau_{\mathscr{M}}$, and $\tau_{\mathscr{S}}$ is the supremum of all compatible topologies (Theorem 7.15). Thus, compatibility of $\tau_{\mathscr{S}}$ is the backbone for the existence of the Mackey topology. In Theorem 7.14, we give a necessary and sufficient condition for it, therefore for the existence of the Mackey topology.

Finally we prove that if $(G, \tau)$ is a Mackey group (i.e. $\tau = \tau_{\mathscr{S}}$) and $H$ any subgroup of $G$, then $G/H$ is also a Mackey group, provided it is a locally quasi-convex group.

In Sect. 7.8 we obtain the following two consequences of our previous results. First, a countable group of bounded exponent does not admit a nondiscrete reflexive group topology. This result has been recently proved in [4] by other means. In the second place a more impressive result: the group of rationals $\mathbb{Q}$ with its usual topology is not a Mackey group.

## 7.2 Notation and Preliminaries

The chapter deals with abelian groups, therefore we omit the term "abelian" in the sequel and we use additive notation. The symbol 0 denotes the neutral element of a group $G$. For a group $G$, a natural number $n \in \mathbb{N}$ and a subset $A \subset G$, we define two new subsets:

$nA := \{nx : x \in A\}$, $\frac{1}{n}A := \{x \in G : nx \in A\}$.

Let $A$ and $B$ be subsets of the group $G$, we define

$A + B := \{a + b : a \in A, b \in B\} \subset G$.

We denote by $\mathbb{R}$ the additive group of real numbers endowed with the euclidean topology, by $\mathbb{T}$ the quotient topological group $\mathbb{R}/\mathbb{Z}$, and let

$\mathbb{T}_+ := [-1/4, 1/4] + \mathbb{Z} \subset \mathbb{T}$,

$\mathbb{T}_n := [-1/4n, 1/4n] + \mathbb{Z} \subset \mathbb{T}$, where $n \in \mathbb{N}$.

It is easy to prove that $\mathbb{T}_n = \bigcap_{m=1}^{n} \frac{1}{m}\mathbb{T}_+$.

For a group $G$, the symbol $H \leq G$ means that $H$ is a subgroup of $G$. If $(G, \tau)$ is a topological group, $H \leq (G, \tau)$ means that $H$ is a subgroup of $G$ endowed with the relative topology $\tau|_H$.

A **character** on a group $G$ is a homomorphism from $G$ to $\mathbb{T}$.

The set of all characters of $G$ endowed with the pointwise operation is a group which will be denoted by $G^* := Hom(G, \mathbb{T})$. If $G$ is a topological group, the **dual group** of $G$, denoted by $G^\wedge$, is the subgroup of $G^*$ formed by the continuous characters. Thus $G^\wedge := CHom(G, \mathbb{T}) \leq G^*$.

A subgroup $H$ of a topological group $G$ is called **dually embedded** in $G$ if every continuous character of $H$ can be extended to a continuous character of $G$. The subgroup $H$ is said to be **dually closed** if for every $x \in G \setminus H$ there is an element $\psi$ in $G^\wedge$ such that $\psi(x) \neq 0$ and $\psi(H) = \{0\}$.

**Definition 7.1** Let $G$ be a topological group and $\lambda$ another group topology on $G$. The topology $\lambda$ is said to be **compatible** with the original topology of $G$ if $G$ and $(G, \lambda)$ have the same dual group, i.e. if $(G, \lambda)^\wedge = G^\wedge$. Similarly, $\lambda$ is said to be **subcompatible** if $(G, \lambda)^\wedge \leq G^\wedge$.

The topological group $G$ is said to be **maximally almost periodic (MAP)** if the dual group $G^\wedge \leq G^*$ is a separating subgroup, which means that $G^\wedge$ separates the points of $G$. In this case we will say that the duality $<G, G^\wedge>$ is a separating duality.

The **Bohr topology** of a topological group $G$, denoted by $\sigma(G, G^\wedge)$, is the weak topology induced by $G^\wedge$ on $G$. It is a group topology, and clearly the topological group $(G, \sigma(G, G^\wedge))$ is Hausdorff if and only if $G$ is MAP. In the dual group $G^\wedge$, we denote by $\sigma(G^\wedge, G)$ the topology of pointwise convergence, and we write $G_p^\wedge := (G^\wedge, \sigma(G^\wedge, G))$. Note that $G_p^\wedge$ is always Hausdorff.

For topologies $\lambda$ and $\tau$ defined on a set $X$, the symbol $\tau \vee \lambda$ will denote the **supremum topology**, i.e. the topology generated by $\tau \cup \lambda$. If $X = G$ is an abelian group, and $\lambda$ and $\tau$ are group topologies, the supremum $\tau \vee \lambda$ is also a group topology with basis of 0-neighbourhoods given by $\{V \cap W : V \in \mathcal{B}_\tau, W \in \mathcal{B}_\lambda\}$, where $\mathcal{B}_\tau$, $\mathcal{B}_\lambda$ are basis of 0-neighborhoods in $\tau$ and $\lambda$ respectively.

Let $G$ be a topological group. For a subset $A \subset G$, the **polar set** of $A$ in $G^*$ is defined as $A^{\blacktriangleright} := \{\chi \in G^* : \chi(C) \subset \mathbb{T}_+\}$, and the polar set of $A$ (in $G^\wedge$) as $A^{\triangleright} := A^{\blacktriangleright} \cap G^\wedge$. If $H$ is a subgroup of $G$, the **annihilator** of $H$ (in $G^\wedge$) is the subgroup of $G^\wedge$ defined by $H^\perp := \{\chi \in G^\wedge : \chi(H) = \{0\}\}$. Clearly $H^\perp = H^{\triangleright}$ and $H^\perp \leq G^\wedge$. Sometimes we will write $H^{\perp G^*}$, for the annihilator group of $H$ in $G^*$, hence $H^{\perp G^*} = H^{\blacktriangleright}$ and clearly $H^\perp = H^{\perp G^*} \cap G^\wedge$. It is well-known and easy to prove that if $U$ is a 0-neighborhood, then $U^{\triangleright} = U^{\blacktriangleright}$.

For a topological group $G$ and a subgroup $H \leq G$, $G/H$ denotes the **quotient group** endowed with the quotient topology. It will be explicitly stated if some other topology is considered on the algebraic quotient group $G/H$. The projection $G \longrightarrow G/H$ is a continuous and open homomorphism. Since we may canonically identify $(G/H)^*$ with the subgroup $H^{\perp G^*} \leq G^*$, it is clear that the dual group of the quotient topological group $(G/H)^\wedge$ may be identified with $H^\perp$, since:

$$(G/H)^\wedge = (G/H)^* \cap G^\wedge = H^{\perp G^*} \cap G^\wedge = H^\perp \leq G^\wedge.$$

Thus the Bohr topology of the quotient group $G/H$ is denoted by $\sigma(G/H, H^\perp)$. It can be also proved that $< G/H, H^\perp >$ is a separating duality whenever $H$ is dually closed in $G$. In other words: if $H$ is Bohr closed, then $G/H$ is MAP being this fact independent of whether $G$ is MAP or not.

For a subset $B \subset G^\wedge$ we define the **inverse polar** of B (in G) as

$$B^{\triangleleft} := \{x \in G : \chi(x) \in \mathbb{T}_+ \text{ for each } \chi \in B\} = \bigcap_{\chi \in B} \chi^{-1}(\mathbb{T}_+).$$

We also define
$(B, \mathbb{T}_n) := \{x \in G : \chi(x) \in \mathbb{T}_n, \forall \chi \in B\} = \bigcap_{\chi \in B} \chi^{-1}(\mathbb{T}_n)$, $n \in \mathbb{N}$,
$(B, \varepsilon) := \{x \in G : \chi(x) \in [-\varepsilon, \varepsilon] + \mathbb{Z}, \forall \chi \in B\}$, $\varepsilon > 0$.
The following formulae are easy to check:
$(\bigcup_{i \in I} B_i)^{\triangleleft} = \bigcap_{i \in I} (B_i)^{\triangleleft}$, $(mB)^{\triangleleft} = \frac{1}{m} B^{\triangleleft}$, for $m \in \mathbb{N}$
and $(B, \mathbb{T}_n) = \bigcap_{m=1}^{n} (mB)^{\triangleleft} = (\bigcup_{m=1}^{n} mB)^{\triangleleft}$, where $n \in \mathbb{N}$.

A nonempty subset $A$ of a topological group $G$ is said to be **quasi-convex** if for every $x \in G \setminus A$ there exists a continuous character $\chi \in G^\wedge$ so that $\chi \in A^{\triangleright}$ and $\chi(x) \notin \mathbb{T}_+$. It is straightforward to prove that the intersection of any family of quasi-convex subsets is quasi-convex. Hence the quasi-convex hull $q(A)$ of a subset $A \subset G$ can be defined as the smallest quasi-convex subset containing $A$ and explicitly, $q(A)$ is the intersection of all quasi-convex subsets containing $A$. It is easy to see that $q(A) = A^{\triangleright\triangleleft}$, therefore $A$ is quasi-convex if and only if $A = A^{\triangleright\triangleleft}$. Any quasi-convex subset $A$ is closed in the Bohr topology $\sigma(G, G^\wedge)$, since $A = A^{\triangleright\triangleleft} = \bigcap_{\chi \in A^{\triangleright}} \chi^{-1}(\mathbb{T}_+)$. For $B \subset G^\wedge$, observe that $B^{\triangleleft} = (B^{\triangleleft})^{\triangleright\triangleleft}$ and then the inverse polar $B^{\triangleleft}$ is always quasi-convex.

In a similar way one can define quasi-convex subsets in $G_p^\wedge$. A subset $B \subset G^\wedge$ is quasi-convex if and only if $B = B^{\triangleleft\triangleright}$ and in this case $B$ is closed in $G_p^\wedge$. We also have that $A^{\triangleright}$ is quasi-convex in $G_p^\wedge$, for any $A \subset G$.

**Definition 7.2** A topological group $G$ is said to be **locally quasi-convex (LQC)** if there exists a neighborhood basis of 0 consisting of quasi-convex subsets.

If $G$ is a LQC group then $\mathscr{B} := \{U^{\triangleright\triangleleft} : U$ is a 0-neighborhood of $G\}$ is a neighborhood basis of 0 consisting of quasi-convex subsets. As observed in [7], the family $\mathscr{B}$ is always a basis of 0-neighborhoods for a group topology on $G$, which is strictly coarser than the original one if $G$ is not LQC.

Let $(G, \tau)$ be a topological group and $\mathscr{B}$ a neighborhood basis of 0 for $(G, \tau)$. From [16] it can be deduced that the family of **$\tau$-equicontinuous subsets** of $G^\wedge$ is:

$$\mathscr{E}(\tau) := \{B \subset G^\wedge : B \text{ is } \tau\text{-equicontinuous }\} = \bigcup_{U\in\mathscr{B}} \mathscr{P}(U^\triangleright) = \bigcup_{U\in\mathscr{B}} \mathscr{P}(U^\blacktriangleright), \tag{7.1}$$

where $\mathscr{P}(A)$ is the power-set of the set $A$. It is known that a character $\chi \in G^*$ is continuous, i.e. $\chi \in G^\wedge$, if and only if there exists a 0-neighborhood $U$ so that $\chi \in U^\blacktriangleright$. Consequently, $U^\triangleright = U^\blacktriangleright$ for every 0-neighborhood $U$. Since $U^\blacktriangleright$ is closed in the compact group $G^*_p = (G^*, \sigma(G^*, G))$, we obtain that $U^\triangleright = U^\blacktriangleright$ is compact and quasi-convex in $G^\wedge_p \leq G^*_p$.

**Definition 7.3** For a topological group $(G, \tau)$, the **locally quasi-convex modification** (LQC modification) of $\tau$ is a new group topology $\tau_{lqc}$ defined as the finest among all LQC topologies on $G$ coarser than $\tau$. We denote $(G, \tau_{lqc})$ by $(G, \tau)_{lqc}$.

*Remark 7.1* The family $\mathscr{B} := \{U^{\triangleright\triangleleft} : U \text{ is a } 0 - \text{neighborhood for } (G, \tau)\}$ is a neighborhood basis of 0 for $(G, \tau)_{lqc}$. Clearly $(G, \tau)$ is LQC iff $(G, \tau)_{lqc} = (G, \tau)$. The LQC modification appears for the first time in [7].

**Proposition 7.1** *The topological groups $(G, \tau)$ and $(G, \tau)_{lqc}$ admit the same dual group, that is $(G, \tau)^\wedge = (G, \tau_{lqc})^\wedge$. Furthermore, the $\tau$-equicontinuous subsets of $G^\wedge$ coincide with the $\tau_{lqc}$-equicontinuous. Thus we can write: $\mathscr{E}(\tau) = \mathscr{E}(\tau_{lqc})$.*

*Proof* From $\tau_{lqc} \subset \tau$, we obtain $(G, \tau_{lqc})^\wedge \leq (G, \tau)^\wedge$.

Conversely, if $\chi \in (G, \tau)^\wedge$, there exists a 0-neighborhood $U$ for $\tau$ so that $\chi \in U^\triangleright$. Since $(U^\triangleright)^{\triangleleft\triangleright} = U^\triangleright$, $\chi \in (U^{\triangleright\triangleleft})^\triangleright$. As $U^{\triangleright\triangleleft}$ is a 0-neighborhood for $\tau_{lqc}$, then $\chi \in (G, \tau_{lqc})^\wedge$.

The second assertion is derived from the equality $(U^\triangleright)^{\triangleleft\triangleright} = (U^{\triangleright\triangleleft})^\triangleright = U^\triangleright$. □

Thus every group topology gives rise, in a standard way, to a LQC topology which admits the same dual group. This fact allows to replace in some occasions the family of all compatible topologies for a fixed topological group $(G, \tau)$, by the family of all compatible LQC topologies for $(G, \tau)$.

## 7.3 Uniform Convergence Topologies on Abelian Groups

Let $G$ be an algebraic group, $G^* = Hom(G, \mathbb{T})$ the group of characters of $G$, and let $\mathcal{G} \subset \mathcal{P}(G^*)$ be a family of subsets of $G^*$. We define $\tau_{\mathcal{G}}$ as the **topology of uniform convergence on the family** $\mathcal{G}$. More explicitly, if $\tau_{\{B\}}$ is the topology of uniform convergence on the subset $B \subset G^*$, then $\tau_{\mathcal{G}} = sup_{B\in\mathcal{G}}\tau_{\{B\}}$. The topology $\tau_{\mathcal{G}}$ is a group topology on $G$, in fact we will see that $(G, \tau_{\mathcal{G}})$ is LQC.

**Definition 7.4** The above family $\mathcal{G}$ is **directed** if for every $A \in \mathcal{G}, B \in \mathcal{G}$ there exists $C \in \mathcal{G}$ so that $A \cup B \subset C$. It is **well directed** if it is directed and for every $n \in \mathbb{N}, A \in \mathcal{G}$ there exists $B \in \mathcal{G}$ so that $nA \subset B$.

The following two symbols express respectively the **directed envelope** and the **well directed envelope** of a fixed family $\mathcal{G}$:
$\widetilde{\mathcal{G}} := \{\cup_{i\in F} A_i : A_i \in \mathcal{G}, F \text{ finite}\}$,
$\overline{\mathcal{G}} := \{\cup_{i\in F}\, n_i A_i : A_i \in \mathcal{G}, F \text{ finite}, n_i \in \mathbb{N}\}$.
Clearly $\mathcal{G} \subset \widetilde{\mathcal{G}} \subset \overline{\mathcal{G}}$.

*Remark 7.2* For a topological group $(G, \tau)$, the family $\mathcal{E}(\tau)$ of $\tau$-equicontinuous subsets of $G^\wedge$ (see (7.1)) is well directed, furthermore $\mathcal{E}(\tau) = \overline{\mathcal{E}(\tau)}$. The proof is straightforward.

**Theorem 7.1** *With the above notation, the following statements hold:*

(1) *A neighborhood basis of* 0 *for* $(G, \tau_{\mathcal{G}})$ *is* $\mathcal{B}_1 = \{(B, \mathbb{T}_n) : B \in \widetilde{\mathcal{G}}, n \in \mathbb{N}\}$, *and also* $\mathcal{B}_2 = \{B^{\triangleleft} : B \in \overline{\mathcal{G}}\}$. *It is clear that* $\tau_{\mathcal{G}} = \tau_{\widetilde{\mathcal{G}}} = \tau_{\overline{\mathcal{G}}}$.
(2) $(G, \tau_{\mathcal{G}})$ *is a LQC group, not necessarily Hausdorff.*
(3) *The dual group is* $(G, \tau_{\mathcal{G}})^\wedge = \bigcup_{B\in\widetilde{\mathcal{G}}} \bigcup_{n\in\mathbb{N}} (B, \mathbb{T}_n)^{\blacktriangleright} = \bigcup_{B\in\overline{\mathcal{G}}} B^{\triangleleft\blacktriangleright}$.
(4) *Let* $\overline{\mathcal{G}}^{\triangleleft\blacktriangleright} := \{B^{\triangleleft\blacktriangleright} : B \in \overline{\mathcal{G}}\}$. *Then the family of* $\tau_{\mathcal{G}}$*-equicontinuous is* $\mathcal{E}(\tau_{\mathcal{G}}) = \bigcup_{B\in\overline{\mathcal{G}}} \mathcal{P}(B^{\triangleleft\blacktriangleright}) = \bigcup_{B\in\overline{\mathcal{G}}^{\triangleleft\blacktriangleright}} \mathcal{P}(B)$, *and* $\overline{\mathcal{G}}^{\triangleleft\blacktriangleright}$ *is a well directed family.*

*Proof* (1) and (2) are in [9]. For the proof of (3), recall that $U^{\triangleright} = U^{\blacktriangleright}$ holds for any 0-neighborhood $U$, and therefore the dual group can be expressed as:

$$(G, \tau_{\mathcal{G}})^\wedge = \bigcup_{U\in\mathcal{B}_1} U^{\blacktriangleright} = \bigcup_{V\in\mathcal{B}_2} V^{\blacktriangleright}$$

In order to prove 4), we input $\mathcal{B}_2$ to the expression for $\mathcal{E}(\tau)$ obtained in (7.1). □

**Corollary 7.1** *Let* $(G, \tau)$ *be a topological group. Then the LQC modification of* $\tau$ *is the topology of uniform convergence on the family of the* $\tau$*-equicontinuous subsets,* $\tau_{lqc} = \tau_{\mathcal{E}(\tau)}$, *and hence* $(G, \tau)$ *is LQC if and only if* $\tau = \tau_{\mathcal{E}(\tau)}$. *On the other hand* $(G, \tau)_{lqc}$ *is Hausdorff if and only if it is MAP.*

The Bohr topology of a topological group $G$ coincides with the topology of the uniform convergence on the well directed family $\mathcal{F} := \{F \subset G^\wedge : F \text{ finite }\}$ of the finite subsets of $G^\wedge$, so $\sigma(G, G^\wedge) = \tau_{\mathcal{F}}$ and $(G, \tau_{\mathcal{F}})$ is LQC.

**Definition 7.5** Let $(G, \tau)$ be a topological group. We denote by $\tau_g(G, G^\wedge)$ the least upper bound of the family of all LQC topologies compatible with $\tau$ and by $\eta_g(G, G^\wedge)$ the least upper bound of the family of all LQC topologies subcompatible with $\tau$. We can write:

$$\tau_g(G, G^\wedge) := \sup\{\lambda : (G, \lambda) \text{ is a LQC group and } (G, \lambda)^\wedge = G^\wedge\}. \tag{7.2}$$

$$\eta_g(G, G^\wedge) := \sup\{\lambda : (G, \lambda) \text{ is a LQC group and } (G, \lambda)^\wedge \leq G^\wedge\}. \tag{7.3}$$

In a similar way, for a subgroup $A \leq G^*$ we define:

$$\tau_g(G, A) := \sup\{\lambda : (G, \lambda) \text{ is a LQC group and } (G, \lambda)^\wedge = A\}. \tag{7.4}$$

$$\eta_g(G, A) := \sup\{\lambda : (G, \lambda) \text{ is a LQC group and } (G, \lambda)^\wedge \leq A\}. \tag{7.5}$$

By their definitions it is clear that $\tau_g \leq \eta_g$. We do not know if they are equal. To see that $\tau_g$ and $\eta_g$ are LQC topologies we need the following theorem.

**Theorem 7.2** *Let G be a group and let $\{\tau_i, i \in I\}$ be a family of LQC topologies on G. Then, $\tau := sup\{\tau_i : i \in I\}$ is a LQC topology on G. Moreover $\tau$ can be described as the topology of uniform convergence on the family $\mathscr{G} = \bigcup_{i \in I} \mathscr{E}(\tau_i)$.*

*Proof* The first assertion is in [9, Theorem 3.3]. For the second one, observe that a 0-neighborhood for the supremum topology $\tau$ can be written as

$$\bigcap_{i \in F} (B_i, \mathbb{T}_{n_i}) = \bigcap_{i \in F} \bigcap_{m_i=1}^{n_i} (m_i B_i)^{\triangleleft} = (\bigcup_{i \in F} \bigcup_{m_i=1}^{n_i} m_i B_i)^{\triangleleft}$$

where $B_i \in \mathscr{E}(\tau_i)$, $n_i \in \mathbb{N}$ for every $i \in F \subset I$, with $F$ finite.

Since $\bigcup_{i \in F} \bigcup_{m_i=1}^{n_i} m_i B_i \in \overline{\mathscr{G}}$ and any element of $\overline{\mathscr{G}}$ can be expressed in this way, the second assertion follows from Theorem 7.1. □

*Remark 7.3* In the last theorem if $\tau_i = \tau_{\mathscr{G}_i}$, where $\mathscr{G}_i$ is a family of subsets of $G^*$ for every $i \in I$, then the supremum topology is
$\tau = \sup\{\tau_i : i \in I\} = \tau_{\mathscr{G}}$, being now $\mathscr{G} = \bigcup_{i \in I} \mathscr{G}_i$.

The proof of this remark is totally analogous to the above theorem.

**Corollary 7.2** *Let $(G, \tau)$ be a topological group. Let us define the families*
$\mathscr{S} := \bigcup_{\lambda, (G,\lambda)^\wedge = G^\wedge} \mathscr{E}(\lambda)$ *and* $\mathscr{M} := \bigcup_{\lambda, (G,\lambda)^\wedge \leq G^\wedge} \mathscr{E}(\lambda)$.
*Then:* $(G, \tau_g(G, G^\wedge)) = (G, \tau_{\mathscr{S}})$ *and* $(G, \eta_g(G, G^\wedge)) = (G, \tau_{\mathscr{M}})$.

It is straightforward to check that $A \subset B$ and $B \in \mathscr{M}$ imply $A \in \mathscr{M}$. The same is true for $\mathscr{S}$. It is also clear that $\mathscr{S} \subset \mathscr{M}$.

Observe that $(G, \tau_g(G, G^\wedge))$ and $(G, \eta_g(G, G^\wedge))$ are LQC groups (Theorem 7.2), but the topologies $\tau_g$ and $\eta_g$ might not be compatible. This justifies the following definitions:

**Definition 7.6** (1) We will say that the **Mackey topology** exists for the duality $< G, G^\wedge >$ if $\tau_g(G, G^\wedge)$ is compatible with the duality $< G, G^\wedge >$. That is, if $(G, \tau_g(G, G^\wedge))^\wedge = G^\wedge$.

(2) A LQC group $(G, \tau)$ is a **Mackey group** if $(G, \tau) = (G, \tau_g(G, G^\wedge))$. In other words, a LQC group $(G, \tau)$ is said to be a Mackey group if for any locally quasi-convex topology $\nu$ on $G$ such that $(G, \nu)^\wedge = (G, \tau)^\wedge$, it holds $\nu \leq \tau$.

It is an open problem to know if the equality $(G, \tau_g(G, G^\wedge))^\wedge = G^\wedge$ holds for every topological group $G$. For some classes of topological groups an affirmative answer was obtained in [9].

It is clear that if the topological group $(G, \tau)$ is Mackey then $\mathscr{S} = \mathscr{E}(\tau)$ and $\tau = \tau_g(G, G^\wedge) = \tau_{\mathscr{S}}$. For a topological group $(G, \tau)$, it may happen that $\mathscr{M} = \mathscr{E}(\tau)$ and then $\tau = \eta_g(G, G^\wedge) = \tau_{\mathscr{M}}$. In this case $(G, \tau)$ is also a Mackey group, which justifies the following definition.

**Definition 7.7** Let us say that a topological group $(G, \tau)$ is $(\mathscr{M})$-Mackey if it is LQC and the family of its equicontinuous subsets $\mathscr{E}(\tau)$ coincides with the family $\mathscr{M}$, therefore $\tau = \tau_{\mathscr{M}} = \eta_g(G, G^\wedge)$.

Thus a $(\mathscr{M})$-Mackey group is Mackey in a stronger sense. In the following theorem we relate the topologies $\tau_g$ and $\eta_g$.

**Theorem 7.3** *Let G be a topological group. Then:*

$$\tau_{\mathscr{M}} = sup\{\lambda : (G, \lambda) \text{ is LQC with } (G, \lambda)^\wedge \leq G^\wedge\} = sup\{\tau_g(G, A) : A \leq G^\wedge\}.$$

*It may happen that $\tau_g(G, A)$ is not Hausdorff.*

*Proof* With the definition of the family $\mathscr{M}$ and Theorem 7.2 the first equality is clear. Now, for each $A \leq G^\wedge$, write $\mathscr{S}(A) := \bigcup_{\lambda, (G,\lambda)^\wedge = A} \mathscr{E}(\lambda) \subset \mathscr{M}$. By the definition of $\mathscr{M}$ we can set

$\mathscr{M} = \bigcup_{A \leq G^\wedge} \bigcup_{\lambda, (G,\lambda)^\wedge = A} \mathscr{E}(\lambda) = \bigcup_{A \leq G^\wedge} \mathscr{S}(A)$

Recall that $\tau_{\mathscr{S}(A)} = \tau_g(G, A)$ and therefore the second equality again follows from Remark 7.3. □

Let $(G, \tau)$ be a topological group. Besides the families $\mathscr{M}$ and $\mathscr{S}$ of Corollary 7.2, the following families will be also used in the sequel:

$\mathscr{F} := \{F \subset G^\wedge : F \text{ finite}\}$.

$\mathscr{Q} := \{B \subset G_p^\wedge : \exists K \text{ compact and quasi-convex in } G_p^\wedge \text{ with } B \subset K\} = \bigcup_K \mathscr{P}(K)$ with $K$ compact and quasi-convex subset of $G_p^\wedge$.

$\mathscr{K} := \{B \subset G_p^\wedge : \exists K \text{ compact in } G_p^\wedge \text{ with } B \subset K\} = \bigcup_K \mathscr{P}(K)$ with $K$ a compact subset of $G_p^\wedge$.

The families $\mathscr{F}$, $\mathscr{E}(\tau)$ and $\mathscr{K}$ are well directed, and from their definitions it follows that $\mathscr{F} \subset \mathscr{E}(\tau) \subset \mathscr{S} \subset \mathscr{M}$ and $\mathscr{Q} \subset \mathscr{K}$. Further, by Theorem 7.1, if $B \in \mathscr{M}$ then $B \subset B^{\triangleleft\blacktriangleright} \subset G^\wedge$ and clearly $B^{\triangleleft\blacktriangleright}$ is compact and quasi-convex in $G_p^\wedge$. Thus $\mathscr{M} \subset \mathscr{Q}$.

We define an order on the families of subsets of $G^*$, by establishing $\mathcal{G} < \mathcal{H}$ if $\tau_{\mathcal{G}} \subset \tau_{\mathcal{H}}$. If $\mathcal{G} \subset \mathcal{H}$ then $\mathcal{G} < \mathcal{H}$, and we will say $\mathcal{G} \approx \mathcal{H}$ when $\tau_{\mathcal{G}} = \tau_{\mathcal{H}}$. If $(G, \tau)$ is a LQC group then:

$$\mathcal{F} \subset \mathcal{E}(\tau) \subset \mathcal{S} \subset \mathcal{M} \subset \mathcal{Q} \subset \mathcal{K}.$$

Consequently,

$$\sigma(G, G^\wedge) = \tau_{\mathcal{F}} \subset \tau = \tau_{\mathcal{E}(\tau)} \subset \tau_{\mathcal{S}} = \tau_g(G, G^\wedge) \subset \tau_{\mathcal{M}} = \eta_g(G, G^\wedge) \subset \tau_{\mathcal{Q}} \subset \tau_{\mathcal{K}},$$

and $G^\wedge \le (G, \tau_{\mathcal{S}})^\wedge \le (G, \tau_{\mathcal{M}})^\wedge \le (G, \tau_{\mathcal{Q}})^\wedge \le (G, \tau_{\mathcal{K}})^\wedge$.

We now recall some definitions which are related with the families $\mathcal{Q}$ and $\mathcal{K}$ above defined.

**Definition 7.8** [9] A LQC group $(G, \tau)$ is said to be:

1. **g-barrelled**, if every compact subset of $G_p^\wedge$ is equicontinuous. Equivalently, if $\mathcal{E}(\tau) = \mathcal{K}$, and then $(G, \tau) = (G, \tau_{\mathcal{K}})$.
2. **pre-Mackey**, if every compact and quasi-convex subset of $G_p^\wedge$ is equicontinuous, or equivalently, if $\mathcal{E}(\tau) = \mathcal{Q}$, and then $(G, \tau) = (G, \tau_{\mathcal{Q}})$.

**Theorem 7.4** *[9] Let G be a LQC group. The following implications hold true: G is g-barrelled $\Longrightarrow$ G is pre-Mackey $\Longrightarrow$ G is Mackey.*

With Definition 7.7, and since $\tau_{\mathcal{M}} \subset \tau_{\mathcal{Q}}$, we may introduce a new implication: $G$ is pre-Mackey $\Longrightarrow$ $G$ is $(\mathcal{M})$-Mackey $\Longrightarrow$ $G$ is Mackey.

*Remark 7.4* The g-barrelled groups were introduced in [9] by the following statement: An abelian topological group $(G, \tau)$ is g-barrelled if every $\sigma(G^\wedge, G)$-compact subset is equicontinuous. With this definition it can be easily checked (through Proposition 7.1) that $(G, \tau)$ is g-barrelled iff $(G, \tau_{lqc})$ is g-barrelled. Since we only deal with LQC g-barrelled groups, we find convenient to define g-barrelled groups only in the class of LQC groups.

As proved in [9], the class of g-barrelled groups includes several well known types of groups like the Čech-complete, the metrizable hereditarily Baire, the separable Baire and the pseudocompact groups. Taking into account our modified definition, we should require further local quasi-convexity for the validity of this assertion, or else do the assertion for the LQC modification. For example, if $G$ is a metrizable and complete group, then $G_{lqc}$ is g-barrelled, however it may not be complete (Example 7.1).

In [6] there are examples of Mackey groups that are not g-barrelled neither pre-Mackey. These examples are $(\mathcal{M})$-Mackey groups, therefore $(G, \tau) = (G, \tau_{\mathcal{M}})$, that is $\mathcal{E}(\tau) = \mathcal{M} \neq \mathcal{Q}$.

In Definition 7.7 a new class of Mackey groups has been obtained by means of the condition $\mathcal{E}(\tau) = \mathcal{M}$ and clearly $\mathcal{M} \subset \mathcal{Q} \subset \mathcal{K}$. In the root paper [9] the groups $(G, \tau)$ for which $\mathcal{E}(\tau) = \mathcal{K}$ or $\mathcal{E}(\tau) = \mathcal{Q}$ were respectively called g-barrelled or pre-Mackey. In order to ensamble this new class in the existing schema it is convenient to introduce new names.

**Definition 7.9** Let us say that a topological group $(G, \tau)$ is $(\mathscr{K})$-Mackey or $(\mathscr{Q})$-Mackey if it is locally quasi-convex and the family of its equicontinuous subsets $\mathscr{E}(\tau)$ coincides respectively with $\mathscr{K}$ or $\mathscr{Q}$, therefore $\tau = \tau_{\mathscr{K}}$ or $\tau = \tau_{\mathscr{Q}}$.

From now on the g-barrelled (pre-Mackey) groups will be named $(\mathscr{K})$-Mackey ($(\mathscr{Q})$-Mackey) groups.

Observe that the conditions $\mathscr{E}(\tau) = \mathscr{K}$ and $\mathscr{E}(\tau) = \mathscr{Q}$—which define respectively when a group $(G, \tau)$ is $g$-barrelled or pre-Mackey—can be weakened to simply require $\mathscr{E}(\tau) = \mathscr{M}$ or $\mathscr{E}(\tau) = \mathscr{S}$. A topological group $G$ for which these equalities occur is also a Mackey group, and so we could establish a grading in "being a Mackey group". We shall use the term **Mackey-type properties** to indicate these sort of properties which in particular imply that $(G, \tau)$ is a Mackey group.

*Example 7.1* Consider the topological group $G = (l_p, \tau_p), p \in (0, 1)$, where $(l_p, \tau_p)$ is the topological vector space formed by the sequences $(x_n)_{n\in\mathbb{N}}$ in $\mathbb{R}$ such that $\sum_{n\in\mathbb{N}} |x_n|^p < \infty$. The topology $\tau_p$ is given by the metric
$\rho(x, y) = \sum_{n\in\mathbb{N}} |x_n - y_n|^p$, where $x = (x_n)_{n\in\mathbb{N}}$ and $y = (y_n)_{n\in\mathbb{N}}$.

It is known that $G$ is metrizable and complete. In [14] it is proved that $(l_p, \tau_p)$ is a topological vector space which is not locally convex. Therefore $G$ is not a LQC group [5].

The LQC modification $G_{lqc}$ is metrizable non complete. In fact the locally quasi convex modification of the topology $\tau_p$ coincides with the restriction to $l_p$ of $\tau_1$, the standard topology of $l_1$. Summarizing, $G_{lqc} = (l_p, \tau_1) \leq (l_1, \tau_1)$, and $G_{lqc}$ is dense in $l_1 \neq G_{lqc}$.

We conclude that $G_{lqc} = (l_p, \tau_1)$ is metrizable, non complete, $(\mathscr{K})$-Mackey (or g-barrelled), and then it is a Mackey group.

This example provides a metrizable complete group $G$, such that its LQC modification $G_{lqc}$ is a $(\mathscr{K})$-Mackey metrizable non complete group.

## 7.4 Quotient Groups of LQC Groups

In this section we first focus our attention on the Hausdorff property. If a topological group $G$ is non-Hausdorff, it is well known that $G/\overline{\{0\}}$ is a Hausdorff group very related to $G$. In a natural sense, it can be said that $G/\overline{\{0\}}$ is the Hausdorff-ication of $G$. We next study this operation in the context of LQC groups.

If a topological group $G$ is not MAP, then the closure of $\{0\}$ in the Bohr Topology of $G$ is $\overline{\{0\}} = (G^\wedge)^\perp$, the von Neumann kernel. It is straightforward to prove that a Hausdorff LQC group is MAP. It is frequent in the literature, to define LQC groups already in the class of Hausdorff abelian groups, to make sure that they are MAP groups, a convenient property. Clearly, a non-Hausdorff LQC group $G$ has nontrivial von Neumann kernel, $(G^\wedge)^\perp$. We next formulate explicitly some items essential for our future considerations.

**Definition 7.10** Let $(G,\lambda)$ be a LQC group and let $A := (G,\lambda)^\wedge$. We define a LQC and Haudorff group topology $h(\lambda)$ on $G/A^\perp$, by stating $(G/A^\perp, h(\lambda)) := (G,\lambda)/A^\perp$.

**Lemma 7.1** *The topological group $(G/A^\perp, h(\lambda))$ is LQC and Hausdorff. If the dual group of $G/A^\perp$ is identified with A, from the equality $(G/A^\perp, h(\lambda))^\wedge = A$, we obtain:*

$$\mathscr{E}(\lambda) = \mathscr{E}(h(\lambda)) \text{ and } h(\lambda) = \tau_{\mathscr{E}(\lambda)}.$$

*Proof* Since $< G/A^\perp, A >$ is a separating duality it is clear that $h(\lambda)$ is Hausdorff.

Let us prove that $(G,\lambda)/A^\perp$ is a LQC group. Fix $U$, a quasi-convex neighborhood of 0 for $(G,\lambda)$. Then, $A^\perp \subset U$ since the non-null elements of $A^\perp$ cannot be separated from 0 by continuous characters. It is easy to deduce that $(U+A^\perp)^\rhd = U^\rhd \cap A = U^\rhd$. In order to see that the 0-neighborhood $U + A^\perp$ of $(G/A^\perp, h(\lambda))$ is quasi-convex, take into account that:

$(U+A^\perp)^{\rhd\lhd G/A^\perp} = (U^\rhd \cap A)^{\lhd G/A^\perp} = (U^\rhd \cap A)^\lhd + A^\perp = U^{\rhd\lhd} + A^\perp = U + A^\perp \subset G/A^\perp$.

As $(U + A^\perp)^\rhd = U^\rhd$, with Eq. (7.1) we obtain that $\mathscr{E}(\lambda) = \mathscr{E}(h(\lambda))$.

Since $h(\lambda)$ is LQC and $\mathscr{E}(\lambda) = \mathscr{E}(h(\lambda))$, it follows from Corollary 7.1 that $h(\lambda) = \tau_{\mathscr{E}(\lambda)}$. □

Summarizing, by means of $h$ we have constructed a locally quasi-convex topology on the quotient of a LQC group $G$ by its von Neumann kernel, which loosely speaking has the same dual group as the original group $G$ and the same equicontinuous family. Now we try to do the inverse operation, that is, starting with a locally quasi-convex topology on a quotient group $G/H$, we construct a locally quasi-convex topology in $G$ so that both dual groups together with the corresponding equicontinuous families might be identified.

*Notation 7.1* If $(G/H, \nu)$ is a topological group and $h : G \longrightarrow G/H$ the canonical mapping, we denote by $h^{-1}(\nu)$ the inverse image (or initial) topology defined on $G$, whose neighborhood basis of 0 is $\mathscr{B}_{h^{-1}(\nu)} = \{h^{-1}(U) : U \text{ is a 0-neighborhood for } \nu\}$.

**Lemma 7.2** *Let $(G/H,\nu)$ be a LQC group (non necessarily Hausdorff), let $A := (G/H,\nu)^\wedge$ and $h : G \longrightarrow G/H$ the canonical mapping. Then $(G, h^{-1}(\nu))$ is a LQC group, where $(G, h^{-1}(\nu))^\wedge$ may be identified with A and $\mathscr{E}(h^{-1}(\nu)) = \mathscr{E}(\nu)$.*

*Proof* It is straightforward to prove that $(G, h^{-1}(\nu))$ is LQC. Now let $U$ be a 0-neighborhood for $\nu$, then

$(h^{-1}(U))^\blacktriangleright = (h^{-1}(U) + H)^\blacktriangleright = (h^{-1}(U))^\blacktriangleright \cap H^{\perp G^*}$.

After identifying $(G/H)^*$ with $H^{\perp G^*}$,

$(h^{-1}(U))^\blacktriangleright \cap H^{\perp G^*} = U^{\blacktriangleright (G/H)^*} = U^\rhd$.

Therefore $(G, h^{-1}(\nu))^\wedge = A$ and $\mathscr{E}(h^{-1}(\nu)) = \mathscr{E}(\nu)$.

The groups $(G/H,\nu)$ and $(G, h^{-1}(\nu))$ in Notation 7.1 and Lemma 7.2 are not necessarily Hausdorff. The operations $h$ and $h^{-1}$ performed respectively in Definition 7.10 and in Notation 7.1 are inverse to each other as we specify next.

**Lemma 7.3** *With the above definitions, the following assertions hold:*

(1) *If* $(G, \lambda)$ *is a LQC group and* $(G, \lambda)^\wedge = A$, *then* $(G, \lambda) = (G, h^{-1}(h(\lambda)))$.
(2) *If* $(G/H, \nu)$ *is a Hausdorff LQC group and* $(G/H, \nu)^\wedge = A$, *then:*

$$(G/H, \nu) = (G/H, h(h^{-1}(\nu)) = (G, h^{-1}(\nu))/H.$$

*Proof* The same argument proves (1) and (2). By Lemmas 7.1 and 7.2, it is only necessary to observe that all the topological groups involved in the equalities of (1) and (2) are LQC, with identical dual groups and identical equicontinuous families. In fact, (1) follows from

$$(G, \lambda)^\wedge = (G, h^{-1}(h(\lambda)))^\wedge \text{ and } \mathscr{E}(\lambda) = \mathscr{E}(h(\lambda)) = \mathscr{E}(h^{-1}(h(\lambda))),$$

and (2) follows from $(G/H, \nu)^\wedge = (G/H, h(h^{-1}(\nu)))^\wedge = ((G, h^{-1}(\nu))/H)^\wedge$ and from $\mathscr{E}(\nu) = \mathscr{E}(h^{-1}(\nu)) = \mathscr{E}(h(h^{-1}(\nu)))$. □

The next result will be useful in order to prove some of the main theorems. It also has interest on its own.

**Theorem 7.5** *Let* $H$ *be a subgroup of a topological group* $G$. *Let* $(G/H, \nu)$ *be a LQC group so that* $(G/H, \nu)^\wedge = A \le H^\perp \le G^\wedge$. *Then:*

(1) $(G/H, \nu \vee \sigma(G/H, H^\perp)) = (G, h^{-1}(\nu) \vee \sigma(G, G^\wedge))/H$.
(2) $(G/H, \nu \vee \sigma(G/H, H^\perp))^\wedge = H^\perp \Leftrightarrow (G, h^{-1}(\nu) \vee \sigma(G, G^\wedge))^\wedge = G^\wedge$.

*Proof* (1) Denote by $U$ a 0-neighborhood for $\nu$ and by $V$ a 0-neighborhood for $\sigma(G, G^\wedge)$. A 0-neighborhood for $(G, h^{-1}(\nu) \vee \sigma(G, G^\wedge))/H$ can be written as $h^{-1}(U) \cap V + H$. Clearly $h^{-1}(U) \cap V + H = U \cap h(V)$, and since $h(V)$ is a 0-neighborhood of $(G, \sigma(G, G^\wedge))/H = (G/H, \sigma(G/H, H^\perp))$ we can deduce that $(G, h^{-1}(\nu) \vee \sigma(G, G^\wedge))/H = (G/H, \nu \vee \sigma(G/H, H^\perp))$.

(2) The implication $\Leftarrow$) follows from (1). In order to prove the reverse implication, assume

$$(G, h^{-1}(\nu) \vee \sigma(G, G^\wedge))^\wedge = D. \qquad (\dagger)$$

It is easy to observe that
$(H, (h^{-1}(\nu) \vee \sigma(G, G^\wedge))|_H) = (H, \sigma(G, G^\wedge)|_H) = (H, \sigma(H, G^\wedge/H^\perp))$.
From $(\dagger)$ we have that $D \supset G^\wedge$ and $h^{-1}(\nu) \vee \sigma(G, G^\wedge) = h^{-1}(\nu) \vee \sigma(G, D)$.
As above:
$(h^{-1}(\nu) \vee \sigma(G, D))|_H = \sigma(G, D)|_H = \sigma(H, D/H^{\perp D})$, where $H^{\perp D} = H^{\perp G^*} \cap D$.
Then $\sigma(H, G^\wedge/H^\perp) = \sigma(H, D/H^{\perp D})$, Which implies that $D/H^{\perp D} = G^\wedge/H^\perp$.
From this and from $H^\perp = H^{\perp D} \cap G^\wedge$, by the Second Isomorphism Theorem (Noether) we deduce that $D = G^\wedge + H^{\perp D}$. By $(\dagger)$ applied to 1), we obtain that

$$(G/H, \nu \vee \sigma(G/H, H^\perp)) = (G, h^{-1}(\nu) \vee \sigma(G, D))/H = (G/H, \nu \vee \sigma(G/H, H^{\perp D})).$$

Since $(G/H, \nu \vee \sigma(G/H, H^{\perp}))^{\wedge} = H^{\perp}$ and $H^{\perp} \leq H^{\perp D}$, it yields that $H^{\perp} = H^{\perp D}$, and finally
$D = G^{\wedge} + H^{\perp D} = G^{\wedge} + H^{\perp} = G^{\wedge}$. □

## 7.5 Stability of Mackey-Type Properties with Respect to Quotients

As indicated in the title, in this section we deal with quotients of Mackey groups. We need to realize first that a LQC group might have quotients groups that are not LQC groups, [5, (5.3)]. This fact is not disturbing for our considerations on compatible topologies, since we can replace such a quotient by its locally quasi-convex modification, which has the same dual group.

Based upon Theorem 7.4 and the comments after it, together with the stability properties listed in the following Lemma, we had the conjecture that all the Mackey-type properties behave well with respect to the operation of taking quotients. We shall prove it in this and in subsequent sections.

**Lemma 7.4** *Let G be a topological group and H a closed subgroup of G. Then the following assertions hold:*

1. *If G is completely metrizable, then G/H is also completely metrizable.*
2. *If G is Čech-complete, then G/H is also Čech-complete.*
3. *If G is Baire separable, then G/H is also Baire separable.*

*Proof* The proofs of 1 and 2 are respectively [1, 4.3.26] and [20, (6.10, p. 47) ] . The proof of 3 follows from the facts that separability is preserved through continuous mappings, and the Baire property is preserved for quotients as can be easily derived from Exercise B in [15, Sect. 9]. □

For all the classes of groups considered in Lemma 7.4, the locally quasi-convex modification of the corresponding groups and quotient groups are $(\mathscr{K})$-Mackey (g-barrelled) groups. Thus, one suspects that the locally quasi-convex modification of the quotient of a $(\mathscr{K})$-Mackey ($g$-barrelled) group is again $(\mathscr{K})$-Mackey ($g$-barrelled). In order to have the tools to analyze these questions, we describe next the family $\mathscr{G}$ of equicontinuous subsets for the quotient group $G/H$ of a topological group $G$. Clearly, the LQC modification is given through: $(G/H)_{lqc} = (G/H, \tau_{\mathscr{G}})$.

**Lemma 7.5** *Let $(G, \tau)$ be a topological group and $H \leq G$ a fixed subgroup. If we denote the quotient group by $(G/H, \lambda) := (G, \tau)/H$, then the family of $\lambda$-equicontinuous subsets is $\mathscr{E}(\lambda) = \{B \cap H^{\perp} : B \in \mathscr{E}(\tau)\}$. Here we have identified $((G, \tau)/H)^{\wedge}$ with the subgroup $H^{\perp} \leq G^{\wedge}$.*

*Proof* If $\mathscr{B}$ is a neighborhood basis of 0 for $(G, \tau)$, then $\mathscr{B}' = \{U + H \subset G/H : U \in \mathscr{B}\}$ is a neighborhood basis of 0 for $(G/H, \lambda) = (G, \tau)/H$. In terms of the basis $\mathscr{B}'$, the family of $\lambda$-equicontinuous subsets is $\mathscr{E}(\lambda) = \bigcup_{V\in\mathscr{B}'} \mathscr{P}(V^{\rhd H^\perp})$. Let us describe $V^{\rhd H^\perp}$ for $V \in \mathscr{B}'$. To that end, fix $V = U + H \in \mathscr{B}'$ with $U \in \mathscr{B}$. We have:

$$\begin{aligned} V^{\rhd H^\perp} &= (U+H)^{\rhd H^\perp} = (\bigcup_{x\in U}\{x+H\})^{\rhd H^\perp} \\ &= \bigcap_{x\in U}\{x+H\}^{\rhd H^\perp} = \bigcap_{x\in U}(\{x\}^{\rhd} \cap H^\perp) \\ &= U^{\rhd} \cap H^\perp. \end{aligned}$$

Since $U^{\rhd} \in \mathscr{E}(\tau)$, and every $\tau$-equicontinuous subset is contained in the polar of a zero neighborhood, we obtain the proposed expression for $\mathscr{E}(\lambda)$. □

With this Lemma we are ready to prove the stability through quotients of the Mackey-type properties. In this section we study quotients of $(\mathscr{K})$-Mackey and $(\mathscr{Q})$-Mackey groups.

**Theorem 7.6** *Let G be a $(\mathscr{K})$-Mackey group and H a subgroup of G. Then the topological group $(G/H)_{lqc}$ is a $(\mathscr{K})$-Mackey group.*

*Proof* As $(G,\tau)$ is $(\mathscr{K})$-Mackey, $(G,\tau) = (G,\tau_{\mathscr{K}})$ where $\mathscr{E}(\tau) = \mathscr{K}$. Let us define now:

$(G/H,\lambda) := (G/H)_{lqc} = ((G,\tau_{\mathscr{K}})/H)_{lqc}$.

By Lemma 7.5 and Proposition 7.1, $\mathscr{E}(\lambda) = \{B \cap H^\perp : B \in \mathscr{E}(\tau) = \mathscr{K}\}$. Let $\mathscr{K}' := \{B \subset H^\perp : \exists K \text{ compact in } H^\perp \le G_p^\wedge \text{ with } B \subset K\} = \bigcup_K \mathscr{P}(K)$ where $K$ runs over the compact subsets of $H^\perp = (G/H)_p^\wedge$.

The family $\mathscr{K}'$ considered in $(G/H)_p^\wedge$ plays the same role as the family $\mathscr{K}$ in $G_p^\wedge$. It is clear that $\mathscr{K}' \subset \mathscr{K}$ and since for $B \in \mathscr{K}'$ we have $B \subset H^\perp$, it yields $\mathscr{K}' \subset \mathscr{E}(\lambda)$.

It also holds $\mathscr{E}(\lambda) \subset \mathscr{K}'$, and hence $\mathscr{K}' = \mathscr{E}(\lambda)$ and $(G/H)_{lqc} = (G/H, \tau_{\mathscr{K}'})$. The dual group is $(G/H,\tau_{\mathscr{K}'})^\wedge = (G/H)^\wedge = H^\perp$ concluding that $(G/H,\tau_{\mathscr{K}'})$ is $(\mathscr{K})$-Mackey. □

**Theorem 7.7** *Let G be a $(\mathscr{Q})$-Mackey group and H a subgroup of G. Then the topological group $(G/H)_{lqc}$ is a $(\mathscr{Q})$-Mackey group.*

*Proof* Since $(G,\tau)$ is a $(\mathscr{Q})$-Mackey group, $(G,\tau) = (G,\tau_{\mathscr{Q}})$ where $\mathscr{E}(\tau) = \mathscr{Q}$. Let now

$(G/H,\lambda) := (G/H)_{lqc} = ((G,\tau_{\mathscr{Q}})/H)_{lqc}$.

By Lemma 7.5, $\mathscr{E}(\lambda) = \{B \cap H^\perp : B \in \mathscr{E}(\tau) = \mathscr{Q}\}$ is the family of the $\lambda$-equicontinuous subsets. Define:

$\mathscr{Q}' := \{B \subset H^\perp : \exists K \text{ compact and quasi-convex in } H^\perp \le G_p^\wedge \text{ with } B \subset K\} = \bigcup_K \mathscr{P}(K)$ where $K$ is a compact and quasi-convex subset of $H^\perp = (G/H)_p^\wedge$.

The family $\mathscr{Q}'$ in $(G/H)_p^\wedge$ plays the same role as the family $\mathscr{Q}$ in $G_p^\wedge$.

From the fact that $H^\perp$ is a dually closed and dually embedded subgroup of $G_p^\wedge$, by [11, (2.2)] we obtain that any quasi-convex subset $K \subset H^\perp$ is also a quasi-convex subset of $G_p^\wedge$. Thus, $\mathscr{Q}' \subset \mathscr{Q}$. Since for $B \in \mathscr{Q}'$ we have $B \subset H^\perp$, $\mathscr{Q}' \subset \mathscr{E}(\lambda)$ derives from $\mathscr{Q} \subset \mathscr{E}(\tau)$. Clearly $\mathscr{E}(\lambda) \subset \mathscr{Q}'$ also holds, hence $\mathscr{Q}' = \mathscr{E}(\lambda)$ and $(G/H)_{lqc} = (G/H, \tau_{\mathscr{Q}'})$. The dual group is $(G/H, \tau_{\mathscr{Q}'})^\wedge = (G/H)^\wedge = H^\perp$, concluding that $(G/H, \tau_{\mathscr{Q}'})$ is a $(\mathscr{Q})$-Mackey group. □

We shall prove in Sects. 7.6 and 7.7 respectively, that being $\mathscr{M}$-Mackey, or in general Mackey, are also properties stable through quotients.

## 7.6 The Family $\mathscr{M}$ of Subsets Determining Equicontinuity

For a topological group $(G, \tau)$ the family $\mathscr{M}$ was explicitly defined in Corollary 7.2. We start this section giving new representations of it. By means of these representations, we characterize when $\tau_{\mathscr{M}}$ is compatible and consequently when $(G, \tau_{\mathscr{M}})$ is a Mackey group. In that case it will be moreover $(\mathscr{M})$-Mackey.

**Theorem 7.8** *Let $G$ be a topological group. In the duality $< G, G^\wedge >$, the family $\mathscr{M} := \bigcup_{\lambda,(G,\lambda)^\wedge \le G^\wedge} \mathscr{E}(\lambda)$ can also be represented as:*

(1) $\mathscr{M}_1 = \bigcup_{H,\lambda} \mathscr{E}(\lambda)$ *where $H$ and $\lambda$ run respectively over all dually closed subgroups of $G$ and over all the group topologies on $G/H$ such that $(G/H, \lambda)$ is MAP with $(G/H, \lambda)^\wedge = A \le H^\perp$.*

(1′) $\mathscr{M}_1' = \bigcup_{H,\rho} \mathscr{E}(\rho)$ *where $H$ and $\rho$ run respectively over all subgroups of $G$ and group topologies on the quotient $G/H$ such that $(G/H, \rho)^\wedge \le H^\perp$ (here $(G/H, \rho)$ is not necessarily MAP).*

(2) $\mathscr{M}_2 = \{B \subset G^\wedge : (G, \tau_{\{B\}})^\wedge \le G^\wedge\} = \{B \subset G^\wedge : (G/B^\perp, \tau_{\{B\}})^\wedge \le B^{\perp\perp}\}$. *Here $B^{\perp\perp} = \overline{< B >}^{G_p^\wedge}$ and $(G/B^\perp, \tau_{\{B\}})$ is Hausdorff and LQC, therefore MAP.*

(3) $\mathscr{M}_3 = \{B \subset G^\wedge : (B, \mathbb{T}_n)^\triangleright = (B, \mathbb{T}_n)^\blacktriangleright, \forall n \in \mathbb{N}\}$.

(3′) $\mathscr{M}'_3 = \{B \subset G^\wedge : (\bigcup_{m=1}^n mB)^{\triangleleft\triangleright} = (\bigcup_{m=1}^n mB)^{\triangleleft\blacktriangleright}, \forall n \in \mathbb{N}\}$.

(4) $\mathscr{M}_4 = \{B \subset G^\wedge : (B, \varepsilon)^\triangleright = (B, \varepsilon)^\blacktriangleright, \forall \varepsilon \in (0, 1/4]\}$.

*Proof* For (1) and (1′), it is clear that $\mathscr{M}_1 \subset \mathscr{M}_1'$. By Lemma 7.2 it is straightforward to see that $\mathscr{M}_1' \subset \mathscr{M}$. From Lemma 7.1 it follows that $\mathscr{M} \subset \mathscr{M}_1$, concluding that $\mathscr{M}_1 = \mathscr{M}_1' = \mathscr{M}$.

(2) Recall that $(G, \tau_{\{B\}})$ is the group $G$ with the topology of the uniform convergence on $B \subset G^\wedge$. Clearly $\mathscr{M}_2 \subset \mathscr{M}$. For the converse, take $B \in \mathscr{M}$. By the definition of $\mathscr{M}$ there exists a LQC topology on $G$, say $\lambda$, such that $B \in \mathscr{E}(\lambda)$ and $(G, \lambda)^\wedge \le G^\wedge$. Therefore $\tau_{\{B\}} \le \lambda$ and then $(G, \tau_{\{B\}})^\wedge \le (G, \lambda)^\wedge \le G^\wedge$, so $B \in \mathscr{M}_2$. With Lemmas 7.1 and 7.2 we obtain the second equality of (2).

(3) We check that $\mathscr{M}_3 = \mathscr{M}_2$. Take $B \in \mathscr{M}_2$. By Theorem 7.1 (3) we deduce that $(G, \tau_{\{B\}})^\wedge = \bigcup_{n\in\mathbb{N}} (B, \mathbb{T}_n)^\blacktriangleright$, and from here it follows that:

$$B \in \mathscr{M}_2 \Leftrightarrow \bigcup_{n\in\mathbb{N}} (B, \mathbb{T}_n)^{\blacktriangleright} \leq G^{\wedge} \Leftrightarrow (B, \mathbb{T}_n)^{\blacktriangleright} = (B, \mathbb{T}_n)^{\triangleright}, \forall n \in \mathbb{N} \Leftrightarrow B \in \mathscr{M}_3$$

(3′) follows from the fact that $(B, \mathbb{T}_n) = \bigcap_{m=1}^{n}(mB)^{\triangleleft} = (\bigcup_{m=1}^{n} mB)^{\triangleleft}$.

(4) Let us prove that $\mathscr{M}_4 = \mathscr{M}_3$. Since $(B, \mathbb{T}_n) = (B, 1/4n)$ where $n \in \mathbb{N}$, it is clear that $\mathscr{M}_4 \subset \mathscr{M}_3$. Let now $B \in \mathscr{M}_3$. For $\varepsilon \in (0, 1/4]$ there exists $n \in \mathbb{N}$ so that $\frac{1}{4n} \leq \varepsilon$, and hence $(B, \mathbb{T}_n) \subset (B, \varepsilon)$. This implies that $(B, \mathbb{T}_n)^{\blacktriangleright} \supset (B, \varepsilon)^{\blacktriangleright}$. From $(B, \mathbb{T}_n)^{\triangleright} = (B, \mathbb{T}_n)^{\blacktriangleright}$, we obtain $(B, \varepsilon)^{\triangleright} = (B, \varepsilon)^{\blacktriangleright}$. □

*Remark 7.5* The family $\mathscr{M}_4$ was already considered in [6]. The authors proved that if $\mathscr{M}_4$ is a well directed family, then there exists a Mackey topology in the duality $< G, G^{\wedge} >$. In Theorem 7.9 we prove that "well directed" can be weakened to "directed" in their statement. However we do not know if the existence of the Mackey topology implies that $\mathscr{M}$ must be a directed family, or equivalently if $G$ is Mackey implies that $G$ is $(\mathscr{M})$-Mackey.

**Theorem 7.9** *Let G be a topological group. The following assertions are equivalent:*

(1) *The group topology $\tau_{\mathscr{M}}$ is compatible (i.e. $(G, \tau_{\mathscr{M}})^{\wedge} = G^{\wedge}$ and $(G, \tau_{\mathscr{M}})$ is a $(\mathscr{M})$-Mackey group), and hence it is the Mackey topology in $< G, G^{\wedge} >$.*
(2) *The family $\mathscr{M}$ is directed (i.e. $\mathscr{M} = \widetilde{\mathscr{M}}$).*

*Proof* (1) $\Rightarrow$ (2). Since $(G, \tau_{\mathscr{M}})^{\wedge} = G^{\wedge}$, we have $\mathscr{E}(\tau_{\mathscr{M}}) \subset \mathscr{M}$. It is obvious that $\mathscr{M} \subset \mathscr{E}(\tau_{\mathscr{M}})$ and, by Remark 7.2, the family $\mathscr{M} = \mathscr{E}(\tau_{\mathscr{M}})$ is directed.

(2) $\Rightarrow$ (1). By Theorem 7.8. (2), if $B \in \mathscr{M}$ then:

$\mathscr{E}(\tau_{\{B\}}) = \bigcup_{n\in\mathbb{N}} \mathscr{P}((B, \mathbb{T}_n)^{\triangleright}) = \bigcup_{n\in\mathbb{N}} \mathscr{P}((\bigcup_{m=1}^{n} mB)^{\triangleleft\triangleright}) \subset \mathscr{M}$,

so $B^{\triangleleft\triangleright} \in \mathscr{M}$ and $nB \in \mathscr{M}, \forall n \in \mathbb{N}$.

Since $\mathscr{M} = \widetilde{\mathscr{M}}$ and $nB \in \mathscr{M}$ for every $B \in \mathscr{M}$, $n \in \mathbb{N}$, it follows that $\mathscr{M} = \widetilde{\mathscr{M}} = \overline{\mathscr{M}}$ is well directed and with Theorem 7.1 a neighborhood basis of 0 for $\tau_{\mathscr{M}}$ is $\mathscr{B}(\tau_{\mathscr{M}}) = \{B^{\triangleleft} : B \in \mathscr{M}\}$. For this basis we can set $(G, \tau_{\mathscr{M}})^{\wedge} = \bigcup_{B\in\mathscr{M}} B^{\triangleleft\blacktriangleright}$. By Theorem 7.8 (3′) $B^{\triangleleft\blacktriangleright} = B^{\triangleleft\triangleright}$, and $(G, \tau_{\mathscr{M}})^{\wedge} = \bigcup_{B\in\mathscr{M}} B^{\triangleleft\triangleright} = G^{\wedge}$. Thus $\tau_{\mathscr{M}}$ is compatible and $(G, \tau_{\mathscr{M}})$ is a Mackey group, in this case it is a $(\mathscr{M})$-Mackey group. □

As a corollary of the above theorem, taking into account Remark 7.2, we obtain the following important result:

**Theorem 7.10** *Let $(G, \tau)$ be a LQC topological group. If every $B \in \mathscr{M}$ is equicontinuous (i.e. $\mathscr{E}(\tau) = \mathscr{M}$), then $(G, \tau)$ is a $(\mathscr{M})$-Mackey group.*

In Theorems 7.6 and 7.7 we proved that the properties of being $(\mathscr{K})$-Mackey ($g$-barrelled) or $(\mathscr{Q})$-Mackey (pre-Mackey) are essentially preserved by quotients. We will prove next (Corollary 7.4) that the same pattern is followed by the property of being $(\mathscr{M})$-Mackey. More precisely, if $G$ is a topological group such that $G = (G, \tau_{\mathscr{M}})$, i.e. $G$ is a $(\mathscr{M})$-Mackey group, then for any subgroup $H \leq G$, $(G/H)_{lqc} = (G/H, \tau_{\mathscr{M}'})$, where the family $\mathscr{M}'$ defined for the duality $< G/H, H^{\perp} >$ plays the same role as the family $\mathscr{M}$ in the duality $< G, G^{\wedge} >$. In other words, $(G/H)_{lqc}$ is also a $(\mathscr{M})$-Mackey group.

**Theorem 7.11** *Let $G$ be a topological group, $H$ a subgroup of $G$ and $A \leq H^{\perp}$. Let $\mathcal{M}' := \bigcup_{\rho,(G/H,\rho)^{\wedge}\leq A} \mathcal{E}(\rho)$. Then, the natural mappings*

$$(G, \tau_{\mathcal{M}}) \longrightarrow ((G, \tau_{\mathcal{M}})/H)_{lqc} \longrightarrow (G/H, \tau_{\mathcal{M}'}) \longrightarrow (G/H, \tau_g(G/H, A))$$

*are continuous homomorphisms.*

*Proof* Set $(G/H, \lambda) := ((G, \tau_{\mathcal{M}})/H)_{lqc}$. By Lemma 7.5 and Proposition 7.1 $\mathcal{E}(\lambda) = \{B \cap H^{\perp} : B \in \mathcal{E}(\tau_{\mathcal{M}})\}$, where $H^{\perp}$ is the annihilator of $H$ in $(G, \tau_{\mathcal{M}})^{\wedge}$. Now by Theorem 7.8 1′) $\mathcal{M}' \subset \mathcal{M}$ and since $B \in \mathcal{M}'$ implies $B \subset A \subset H^{\perp}$ and $\mathcal{M} \subset \mathcal{E}(\tau_{\mathcal{M}})$, we deduce $\mathcal{M}' \subset \mathcal{E}(\lambda)$, concluding that $((G, \tau_{\mathcal{M}})/H)_{lqc} = (G/H, \tau_{\mathcal{E}(\lambda)}) \to (G/H, \tau_{\mathcal{M}'})$ is a continuous homomorphism.

By the definition of $\mathcal{M}'$ it is clear that $\mathcal{M}' \supset \bigcup_{\nu} \mathcal{E}(\nu)$ where $(G/H, \nu)^{\wedge} = A$ and then $(G/H, \tau_{\mathcal{M}'}) \longrightarrow (G/H, \tau_g(G/H, A))$ is a continuous homomorphism. □

**Corollary 7.3** *Let $G$ be a $(\mathcal{M})$-Mackey group, so $G = (G, \tau_{\mathcal{M}})$. Let $H \leq G$, $A \leq H^{\perp}$, and let $\mathcal{M}' = \bigcup_{\rho,(G/H,\rho)^{\wedge}\leq A} \mathcal{E}(\rho)$. Then $(G/H, \tau_{\mathcal{M}'})^{\wedge} \leq H^{\perp}$.*

*Proof* As $(G, \tau_{\mathcal{M}})^{\wedge} = G^{\wedge}$, $((G, \tau_{\mathcal{M}})/H)^{\wedge} = H^{\perp}$. In Theorem 7.11 we proved that the natural mapping

$$(G, \tau_{\mathcal{M}})/H \longrightarrow (G/H, \tau_{\mathcal{M}'})$$

is a continuous homomorphism. Dualizing this expression, we obtain:

$$(G/H, \tau_{\mathcal{M}'})^{\wedge} \leq ((G, \tau_{\mathcal{M}})/H)^{\wedge} = H^{\perp}.$$ □

**Corollary 7.4** *Let $G$ be a $(\mathcal{M})$-Mackey group, so $G = (G, \tau_{\mathcal{M}})$. Let $H$ be a subgroup of $G$. Then the LQC modification of $G/H$ is a $(\mathcal{M})$-Mackey group such that $(G/H)_{lqc} = (G/H, \tau_{\mathcal{M}'})$, with $\mathcal{M}' = \bigcup_{\rho,(G/H,\rho)^{\wedge}\leq H^{\perp}} \mathcal{E}(\rho)$, i.e. $(G/H, \tau_{\mathcal{M}'})^{\wedge} = H^{\perp}$.*

*Proof* If $(G/H, \lambda) := G/H$, it holds: $(G/H)_{lqc} = (G/H, \tau_{\mathcal{E}(\lambda)})$ and $(G/H)^{\wedge} = H^{\perp}$. Now we apply Corollary 7.3 with $A = H^{\perp}$, and we obtain that $(G/H, \tau_{\mathcal{M}'})^{\wedge} = (G/H)^{\wedge} = H^{\perp}$. By Theorem 7.11 $\mathcal{M}' \subset \mathcal{E}(\lambda)$ and by the definition of $\mathcal{M}'$ also $\mathcal{E}(\lambda) \subset \mathcal{M}'$, thus $\mathcal{M}' = \mathcal{E}(\lambda)$ and we have that $(G/H)_{lqc} = (G/H, \tau_{\mathcal{M}'})$. □

**Theorem 7.12** *Let $(G, \tau)$ be a $(\mathcal{M})$-Mackey group, so that $\tau = \tau_{\mathcal{M}}$ and $\mathcal{E}(\tau) = \mathcal{M}$. The following assertions hold:*

(1) *If $\nu$ denotes a new LQC topology on $G$, then $(G, \nu)^{\wedge} \leq G^{\wedge} \Leftrightarrow \nu \subset \tau$.*
(2) *If $H$ is a subgroup of $(G, \tau)$ and $\lambda$ a LQC topology on $G/H$, then the canonical projection $(G, \tau) \longrightarrow (G/H, \lambda)$ is continuous if and only if $(G/H, \lambda)^{\wedge} \leq H^{\perp} \leq G^{\wedge}$.*

*Proof* In order to prove (1), assume that $\mathcal{E}(\tau) = \mathcal{M}$. From the definition of the family $\mathcal{M}$, it is clear that $\nu \subset \tau$ if and only if $\mathcal{E}(\nu) \subset \mathcal{E}(\tau) = \mathcal{M}$ if and only if $(G, \nu)^{\wedge} \leq G^{\wedge}$.

The proof of (2) follows from the fact that $(G, \tau) \longrightarrow (G/H, \lambda)$ is a continuous homomorphism if and only if $\mathscr{E}(\lambda) \subset \mathscr{E}(\tau) = \mathscr{M}$ and by Theorem 7.8. (1′) this last condition is equivalent to $(G/H, \lambda)^\wedge \leq G^\wedge$. □

## 7.7 The Family $\mathscr{S}$

The family $\mathscr{S}$ defined in Sect. 7.3 turns out to be optimal in the following sense: the topology $\tau_{\mathscr{S}}$ of uniform convergence on $\mathscr{S}$ coincides with $\tau_g$, the least upper bound of all LQC topologies compatible with $\tau$. Therefore a ($\mathscr{S}$)-Mackey group defined in the above spirit, is simply a Mackey group.

We begin this section with new representations for the family $\mathscr{S}$. From these representations we characterize when $\tau_g(G, G^\wedge)$ is a compatible topology and thus the duality $< G, G^\wedge >$ admits a Mackey topology.

**Theorem 7.13** *Let G be a topological group. For the duality* $<G, G^\wedge>$, *the family* $\mathscr{S} := \bigcup_{\lambda,(G,\lambda)^\wedge = G^\wedge} \mathscr{E}(\lambda)$ *above defined admits the following representations:*

(1) $\mathscr{S}_1 = \bigcup_{H,\nu} \mathscr{E}(\nu)$ *where* $H \leq G$ *and* $\nu$ *is a topology on* $G/H$ *such that* $(G/H, \nu)^\wedge = H^\perp$.
(2) $\mathscr{S}_2 = \{B \subset G^\wedge : (G, \tau_{\{B\}} \vee \sigma(G, G^\wedge))^\wedge = G^\wedge\}$.
*It also holds:* $\mathscr{S}_2 = \{B \subset G^\wedge : (G/B^\perp, \tau_{\{B\}} \vee \sigma(G/B^\perp, B^{\perp\perp}))^\wedge = B^{\perp\perp}\}$.
(3) $\mathscr{S}_3 = \{B \in \mathscr{M} : \forall F \subset G^\wedge, F$ *finite* $, B \cup F \in \mathscr{M}\}$.

*Proof* (1) In order to prove $\mathscr{S} \subset \mathscr{S}_1$, just take $H = G^{\wedge\perp}$ and apply Lemma 7.1.

Now, by Theorem 7.5 applied to the particular case $A = H^\perp$, if $(G/H, \nu)^\wedge = H^\perp$ then $(G, h^{-1}(\nu) \vee \sigma(G, G^\wedge))^\wedge = G^\wedge$. So $B \in \mathscr{E}(\nu)$ implies
$B \in \mathscr{E}(h^{-1}(\nu)) \subset \mathscr{E}(h^{-1}(\nu) \vee \sigma(G, G^\wedge)) \subset \mathscr{S}$, concluding $\mathscr{S}_1 \subset \mathscr{S}$.

(2) $\mathscr{S}_2 \subset \mathscr{S}$ is trivial. Conversely, if $B \in \mathscr{S}$ there exists then a LQC topology $\lambda$ so that $B \in \mathscr{E}(\lambda)$ and $(G, \lambda)^\wedge = G^\wedge$. Therefore $\tau_{\{B\}} \vee \sigma(G, G^\wedge) \leq \lambda$ and we have:

$$G^\wedge = (G, \sigma(G, G^\wedge))^\wedge \leq (G, \tau_{\{B\}} \vee \sigma(G, G^\wedge))^\wedge \leq (G, \lambda)^\wedge = G^\wedge.$$

Hence $\mathscr{S} \subset \mathscr{S}_2$.

For the second part of (2) apply Theorem 7.5.

(3) We check that $\mathscr{S}_2 = \mathscr{S}_3$. Let $B \in \mathscr{S}_2$, and therefore $(G, \tau_{\{B\}} \vee \sigma(G, G^\wedge))^\wedge = G^\wedge$. It is easy to deduce that

$$\tau_{\{B\}} \vee \sigma(G, G^\wedge) = \tau_{\{B\} \cup \mathscr{F}}, \text{ where } \mathscr{F} = \{F \subset G^\wedge : F \text{ finite }\}.$$

Now, applying Theorem 7.1 (3) we obtain:

$$(G, \tau_{\{B\}} \vee \sigma(G, G^\wedge))^\wedge = (G, \tau_{\{B\} \cup \mathscr{F}})^\wedge = \bigcup_{F \in \mathscr{F}} \bigcup_{n \in \mathbb{N}} (B \cup F, \mathbb{T}_n)^\blacktriangleright.$$

Since $B \in \mathscr{S}_2$, $B$ is already in $\mathscr{M}$ and $(B \cup F, \mathbb{T}_n)^{\blacktriangleright} = (B \cup F, \mathbb{T}_n)^{\triangleright}$ for every $F \in \mathscr{F}$ and $n \in \mathbb{N}$. By Theorem 7.8. (3) this assertion is equivalent to $B \cup F \in \mathscr{M}$ for each $F \in \mathscr{F}$. Thus it follows $\mathscr{S}_2 = \mathscr{S}_3$. □

**Theorem 7.14** *Let G be a topological group. Then, the Mackey topology exists for the duality $< G, G^\wedge >$ (i.e. $(G, \tau_g(G, G^\wedge))^\wedge = (G, \tau_{\mathscr{S}})^\wedge = G^\wedge$) if and only if the family $\mathscr{S}$ is directed (i.e. $\mathscr{S} = \tilde{\mathscr{S}}$).*

*Proof* ⇒): From $(G, \tau_{\mathscr{S}})^\wedge = G^\wedge$ and the definition of the family $\mathscr{S}$, we obtain $\mathscr{E}(\tau_{\mathscr{S}}) \subset \mathscr{S}$. On the other hand $\mathscr{S} \subset \mathscr{E}(\tau_{\mathscr{S}})$ and again by Remark 7.2 the family $\mathscr{S} = \mathscr{E}(\tau_{\mathscr{S}})$ must be directed.

⇐): By Theorem 7.13 (2), if $B \in \mathscr{S}$ then:

$$\mathscr{E}(\tau_{\{B\}}) = \bigcup_{n \in \mathbb{N}} \mathscr{P}((B, \mathbb{T}_n)^{\triangleright}) = \bigcup_{n \in \mathbb{N}} \mathscr{P}((\bigcup_{m=1}^{n} mB)^{\triangleleft\triangleright}) \subset \mathscr{S},$$

so $B^{\triangleleft\triangleright} \in \mathscr{S}$ and $nB \in \mathscr{S}$ for every $n \in \mathbb{N}$.

Since $\mathscr{S} = \tilde{\mathscr{S}}$ and $nB \in \mathscr{S}$ for every $B \in \mathscr{S}$ and $n \in \mathbb{N}$, we have that $\mathscr{S} = \tilde{\mathscr{S}} = \overline{\mathscr{S}}$ is a well directed family. Applying then Theorem 7.1, we obtain the following neighborhood basis of 0 for $\tau_{\mathscr{S}}$, $\mathscr{B}(\tau_{\mathscr{S}}) = \{B^{\triangleleft} : B \in \mathscr{S}\}$. With respect to this basis, we can set $(G, \tau_{\mathscr{S}})^\wedge = \bigcup_{B \in \mathscr{S}} B^{\triangleleft\blacktriangleright}$. Since $B \in \mathscr{S} \subset \mathscr{M}$, $B^{\triangleleft\blacktriangleright} = B^{\triangleleft\triangleright}$ and hence $(G, \tau_{\mathscr{S}})^\wedge = \bigcup_{B \in \mathscr{S}} B^{\triangleleft\triangleright} = G^\wedge$.

We now definitely prove (Theorem 7.16) the stability of the Mackey property through quotients. Explicitly, if $G$ is a Mackey group, for any $H \leq G$, the LQC modification of the quotient group $(G/H)_{lqc}$ is also a Mackey group.

**Theorem 7.15** *Let G be a topological group. Then $\tau_g(G, G^\wedge) = \tau_{\mathscr{S}}$ and for every subgroup $H \leq G$, the following natural mappings*

$$(G, \tau_g(G, G^\wedge)) \longrightarrow ((G, \tau_g(G, G^\wedge))/H)_{lqc} \longrightarrow (G/H, \tau_g(G/H, H^\perp))$$

*are continuous homomorphisms.*

*Proof* The first homomorphism is continuous just by the definition of the LQC-modification. In order to obtain that the second one is also continuous, set $(G/H, \lambda) := ((G, \tau_{\mathscr{S}})/H)_{lqc}$. By Lemma 7.5, $\mathscr{E}(\lambda) = \{B \cap H^\perp : B \in \mathscr{E}(\tau_{\mathscr{S}})\}$, where $H^\perp$ is now the annihilator of $H$ in $(G, \tau_{\mathscr{S}})^\wedge$. Let $\mathscr{S}' = \bigcup_\nu \mathscr{E}(\nu)$ where $\nu$ runs over the LQC topologies with $(G/H, \nu)^\wedge = H^\perp$. The family $\mathscr{S}'$ in $(G/H)^\wedge$ has the same role as the family $\mathscr{S}$ in $G^\wedge$.

Clearly $(G/H, \tau_g(G/H, H^\perp)) = (G/H, \tau_{\mathscr{S}'})$. By Theorem 7.13 (1), $\mathscr{S}' \subset \mathscr{S}$ and if $B \in \mathscr{S}'$ then $B \subset H^\perp$ and we deduce that $\mathscr{S}' \subset \mathscr{E}(\lambda)$. Therefore the mapping

$$((G, \tau_{\mathscr{S}})/H)_{lqc} = (G/H, \tau_{\mathscr{E}(\lambda)}) \longrightarrow (G/H, \tau_{\mathscr{S}'})$$

is a continuous homomorphism. □

**Theorem 7.16** *Let $G$ be a Mackey group and let $H \leq G$. Then the LQC modification of $G/H$ is a Mackey group, so $(G/H)_{lqc} = (G/H, \tau_g(G/H, H^{\perp}))$.*

*Proof* As $G$ is a Mackey group, $G = (G, \tau_{\mathscr{S}})$. From the equality $(G, \tau_{\mathscr{S}})^{\wedge} = G^{\wedge}$ we obtain $((G, \tau_{\mathscr{S}})/H)^{\wedge} = H^{\perp}$, and now dualizing Theorem 7.15 we have

$$(G/H, \tau_g(G/H, H^{\perp}))^{\wedge} = ((G, \tau_{\mathscr{S}})/H)^{\wedge} = H^{\perp}.$$

By the definition of the topology $\tau_g(G/H, H^{\perp})$, we conclude that
$((G, \tau_{\mathscr{S}})/H)_{lqc} = (G/H, \tau_g(G/H, H^{\perp}))$ is a Mackey group.

**Corollary 7.5** *Let $(G, \tau)$ be a LQC topological group. The following assertions are equivalent:*

(1) *$(G, \tau)$ is Mackey.*
(2) *Every $B \in \mathscr{S}$ is equicontinuous, i.e. $\mathscr{E}(\tau) = \mathscr{S}$.*
(3) *For every subgroup $H \leq G$, $(G/H)_{lqc}$ is Mackey.*

## 7.8 Two Consequences

The following theorem was proved in [4]. Here we use an observation of [6] (see the claim) to give a more direct proof.

**Theorem 7.17** *Let $G$ be a countable, Pontryagin reflexive group of bounded torsion. Then $G$ is discrete.*

*Proof* **Claim** [6]. Let $H$ be a MAP group of bounded torsion such that $|H^{\wedge}| < \mathfrak{c}$. Then $H$ carries the Bohr topology and it is a Mackey group. Furthermore, $\mathscr{M} = \mathscr{F}$.

Since $G$ is reflexive, then the dual group with the compact-open topology $H := G_c^{\wedge}$ is also reflexive and of bounded torsion. It is clear that $H^{\wedge} = G^{\wedge\wedge} = G$ is countable. Now we apply the claim to $H$, deducing that $H$ must be precompact. As $H^{\wedge}$ is countable we obtain that $H$ is precompact and metrizable. By [2] or [8], $H_c^{\wedge} = G$ must be discrete. □

**Theorem 7.18** *The topological group $\mathbb{Q} \leq \mathbb{R}$ is not a Mackey group.*

*Proof* Assume $\mathbb{Q}$ is Mackey and consider the subgroup $\mathbb{Z} \leq \mathbb{Q}$. By Theorem 7.16 the group $\mathbb{Q}/\mathbb{Z} \leq \mathbb{T}$ is a Mackey group. On the other hand $\mathbb{Q}/\mathbb{Z}$ is the torsion group of $\mathbb{T}$, i.e. $\mathbb{Q}/\mathbb{Z} = tor\mathbb{T} = \bigoplus \mathbb{Z}(p^{\infty})$, where $\mathbb{Z}(p^{\infty})$ is the Prüfer group, and $p$ runs over the prime numbers.

As $\mathbb{T}$ is a compact group, $\mathbb{Q}/\mathbb{Z}$ must be a precompact group (with the induced quotient topology).

In order to obtain a contradiction we have to prove that the precompact group $\mathbb{Q}/\mathbb{Z} = (tor\mathbb{T}, \sigma(tor\mathbb{T}, \mathbb{Z}))$ is not Mackey. This will follow from the following facts.

**Fact 1** Let $(a_n)_{n\in\mathbb{N}}$ be a sequence in $\mathbb{N}$ such that $\sum_{n\in\mathbb{N}}\frac{1}{a_n}<\infty$. Define the sequence $(\gamma_n)_{n\in\mathbb{N}}$ by $\gamma_1=1$ and $\gamma_n=\prod_{i=1}^{i=n-1}a_i$, for $n>1$. Following [21], we denote by $\mathbb{Z}\{\gamma_n\}$ the group $\mathbb{Z}$ endowed with the finest group topology in which the sequence $\gamma_n$ converges to zero. Then, the topological group $\mathbb{Z}\{\gamma_n\}$ is Pontryagin reflexive (See [12, Theorems 2, 3. p. 2797]) and $\mathbb{Z}\{\gamma_n\}^{\wedge}_c$ is a Polish group.

**Fact 2** [12] Consider the group $\mathbb{Z}\{\gamma_n\}^{\wedge}_c$ of the previous fact, it is clear that algebraically $\mathbb{Z}\{\gamma_n\}^{\wedge}\leq\mathbb{T}$. Define the group $G=\{z\in\mathbb{T}:\gamma_n z=0 \text{ for some } n\in\mathbb{N}\}$. Then the group $G$ is dense in $\mathbb{Z}\{\gamma_n\}^{\wedge}_c$.

Let $\tau$ be the topology of $G\leq\mathbb{Z}\{\gamma_n\}^{\wedge}_c$. As $G$ is dense in $\mathbb{Z}\{\gamma_n\}^{\wedge}_c$, $G^{\wedge}=(\mathbb{Z}\{\gamma_n\}^{\wedge}_c)^{\wedge}=\mathbb{Z}$ and the family of equicontinuous $\mathscr{E}(\tau)$ coincides with that of $(\mathbb{Z}\{\gamma_n\}^{\wedge}_c)^{\wedge}=\mathbb{Z}$.

It is clear that $\mathscr{E}(\tau)\neq\mathscr{F}$, therefore $G$ cannot be precompact, i.e. $\sigma(G,G^{\wedge})=\sigma(G,\mathbb{Z})<\tau$. In other words, the Bohr topology is strictly coarser than the original.

Observe that for the particular case of $a_n=n!$, the group $G$ coincides with the torsion of $\mathbb{T}$, $tor\mathbb{T}$. So we obtain that there is a finer compatible topology on the quotient $\mathbb{Q}/\mathbb{Z}$, the searched contradiction. □

**Acknowledgments** We thank Professors V. Tarieladze and M.J. Chasco for enlightening discussions about the topic. We also thank the Editors, for the enormous work they have taken over them to make this project successful. The second author was partially supported by Ministerio de Educación y Ciencia, grant: MTM2009-14409-C02-01.

## References

1. Arhangel'skii, A.V., Tkachenko, M.G.: Topological Groups and Related Structures. Atlantis Series in Mathematics. **I**. Atlantis Press, Paris (2008)
2. Außenhofer, L.: Contributions to the duality theory of abelian topological groups and to the theory of nuclear groups. Dissertationes Math. (Rozprawy Mat.) **384** (1999)
3. Außenhofer, L., de la Barrera, D.: Linear topologies on $\mathbb{Z}$ are not Mackey topologies. J. Pure Appl. Algebra **216**(6), 1340–1347 (2012)
4. Außenhofer, L., Gabriyelyan, S.S.: On reflexive group topologies on abelian groups of finite exponent. Arch. Math. **99**, 583–588 (2012)
5. Banaszczyk, W.: Additive subgroups of topological vector spaces. Lecture Notes in Mathematics, vol. 1466. Springer, Berlin (1991)
6. Bonales, G., Trigos-Arrieta, F.J., Vera Mendoza, R.: A Mackey-Arens theorem for topological Abelian groups. Bol. Soc. Mat. Mex. III Ser. **9**(1), 79–88 (2003)
7. Bruguera, M.M.: Some properties of locally quasi-convex groups. Topol. Appl. **77**, 87–94 (1997)
8. Chasco, M.J.: Pontryagin duality for metrizable groups. Arch. Math. **70**, 22–28 (1998)
9. Chasco, M.J., Martín-Peinador, E., Tarieladze, V.: On Mackey topology for groups. Stud. Math. **132**(3), 257–284 (1999)
10. Dikranjan, D., Martín-Peinador, E., Tarieladze, V.: Group valued null sequences and metrizable non-Mackey groups. Forum Math. **26**, 723–757 (2014)
11. Dikranjan, D., Lukács, G.: Compact-like abelian groups without nontrivial quasi-convex null sequences. J. Pure Appl. Algebra **218**, 135–147 (2014)
12. Gabriyelyan, S.S.: Groups of quasi-invariance and the Pontriagin duality. Topol. Appl. **157**, 2786–2802 (2010)

13. Kalton, N.J., Peck, N.T., Roberts, J.W.: An F-space sampler. Lecture Notes series, vol. 89. Cambridge University Press, Cambridge (1984)
14. Kąkol, J.: The Mackey-Arens theorem for non-locally convex spaces. Collect. Math. **41**(2), 129–132 (1990)
15. Kelley, J.L., Namioka, I.: Linear Topological Spaces. Van Nostrand, Princeton (1963)
16. Noble, N.: K-groups and duality. Trans. Amer. Math. Soc. **151**, 551–561 (1970)
17. Schaefer, H.H.: Topological Vector Spaces. The Macmillan Co., New York (1966)
18. Varopoulos, N.Th: Studies in harmonic analysis. Proc. Camb. Philos. Soc. **60**, 465–516 (1964)
19. Vilenkin, N.Ya.: The theory of characters of topological Abelian groups with boundedness given. Izvestiya Akad. Nauk SSSR. Ser. Mat. **15**, 439–162 (1951)
20. Warner, S.: Topological Fields. Mathematics Studies, vol. 157. North Holland, Amsterdam (1989)
21. Zelenyuk, E.G., Protasov, I.V.: Topologies on abelian groups. Math. USSR Izv. **37**, 445–460 (1991). Russian original: Izv. Akad Nauk SSSR **54**, 1090–1107 (1990)

# Chapter 8
# Bowen's Entropy for Endomorphisms of Totally Bounded Abelian Groups

**Domingo Alcaraz, Dikran Dikranjan and Manuel Sanchis**

**Abstract** We say that the completion theorem holds for a uniform space $(X, \mathscr{U})$ if, for every uniformly continuous function $\alpha\colon (X, \mathscr{U}) \to (X, \mathscr{U})$, the Bowen entropy of $\alpha$ coincides with the Bowen entropy of $\widetilde{\alpha}$, the extension of $\alpha$ to the completion $(\widetilde{X}, \widetilde{\mathscr{U}})$ of $(X, \mathscr{U})$. We study the completion theorem in the realm of abelian topological groups. Namely, we prove that it fails to be true in a drastic way by showing that every (abstract) abelian group $G$ can be endowed with a totally bounded group topology $\tau$ such that the topological group $(G, \tau)$ has endomorphisms of zero entropy whose extension to the Raĭkov completion of $(G, \tau)$ has infinite entropy. Our proof uses the structure theorems for abelian groups, the properties of the Bohr topology and Pontryagin duality. The case of metrizable groups is also analyzed in the case of Bernoulli shifts of finite groups, dense subgroups of the circle $\mathbb{T}$ and the reals $\mathbb{R}$. The key tool is the so-called $e$-supporting family of an endomorphism $\alpha$ of an abelian metrizable group $G$ with respect to a neighborhood $U$ of the neutral element of $G$. Several open questions are proposed.

**Keywords** Bowen's entropy · Topological entropy · Bohr compactification · Totally bounded group · Metrizable group

**MSC 2000 Classification:** 22A05, 37B40, 54C70

---

D. Alcaraz
Departamento de Matemática Aplicada y Estadística,
Universidad Politécnica de Cartagena, C/Dr. Fleming s/n, 30202 Cartagena, Spain
e-mail: mingo.alcaraz@upct.es

D. Dikranjan
Dipartimento de Matematica e Informatica, Università di Udine, via delle Scienze 206, 33100 Udine, Italy
e-mail: dikran.dikranjan@uniud.it

M. Sanchis (✉)
Institut de Matemàtiques i Aplicacions de Castelló (IMAC), Universitat Jaume I de Castelló, Campus Riu Sec s/n, 12071 Castelló, Spain
e-mail: sanchis@mat.uji.es

J. C. Ferrando and M. López-Pellicer (eds.), *Descriptive Topology and Functional Analysis*, Springer Proceedings in Mathematics & Statistics 80, DOI: 10.1007/978-3-319-05224-3_8, 
© Springer International Publishing Switzerland 2014

## 8.1 Introduction

Adler et al. [1] first introduced the concept of topological entropy for continuous self-maps of compact spaces as an analogue of the measure theoretic entropy. Topological entropy and many of its generalizations became a useful tool in the field of discrete dynamical systems (see, for instance, the survey [12]). One of the most valuable extension of the concept of topological entropy is the notion of Bowen [4] entropy for uniformly continuous self-maps of metric spaces. It is worth noting that Bowen's definition can be reformulated in the slightly more general context of uniform spaces which has the advantage that we need not restrict ourselves to metric spaces. In this framework, the study of Bowen's entropy in the realm of topological groups is widely motivated. For instance, when applied to topological automorphisms of locally compact abelian groups, the invariance properties present in the compact case are retained, and for compact abelian groups the Kolmogorov-Sinai entropy can be expressed in terms of the Bowen's one (see [25] for details). The Kolmogorov-Sinai entropy appears in several interesting situations; for example, under certain conditions, it is the rate of increase of the statical entropy of a simple, fully chaotic conservative system [21]. For its relation with the concept of algebraic entropy, the interested reader can consult among others [7, 8, 12, 28].

Now we recall the definition of Bowen's entropy. Let $(X, \mathscr{U})$ be a uniform space and let $\alpha\colon (X, \mathscr{U}) \to (X, \mathscr{U})$ be a uniformly continuous map. For $U \in \mathscr{U}$, a subset $F$ *$(n, U)$-spans* a compact subset $C$ if for every $x \in C$ there is $y \in F$ such that $\big(\alpha^j(x), \alpha^j(y)\big) \in U$ for each $0 \leq j < n$. A subset $F$ of $C$ is said to be an $(n, U)$-*separated* set with respect to $\alpha$, if for each pair of distinct points $x, y \in F$ there exists $j$ such that $0 \leq j < n$ and $(\alpha^j(x), \alpha^j(y)) \notin U$.

For a compact set $C \subset X$, let $r_n(\alpha, U, C)$ be the smallest cardinality of any set $F$ which $(n, U)$-spans $K$ and let $s_n(\alpha, U, C)$ be the biggest cardinality of an $(n, U)$-separated subset of $K$ (obviously, $s_n(\alpha, U, C) < \infty$). Let

$$r(\alpha, U, C) = \limsup_{n\to\infty} \frac{1}{n} \log r_n(\alpha, U, C)$$

and

$$s(\alpha, U, C) = \limsup_{n\to\infty} \frac{1}{n} \log s_n(\alpha, U, C).$$

Further, let

$$h_r(\alpha, C) = \sup\{r(\alpha, U, C) : U \in \mathscr{U}\} \text{ and } h_s(\alpha, C) = \sup\{s(\alpha, U, C) : U \in \mathscr{U}\}.$$

It turns out, that $h_r(\alpha, C) = h_s(\alpha, C)$ for every compact set $C \subset X$ [9, Lemma 1]. We simply write $h_B(\alpha, C)$ for this common value. The Bowen entropy $h_B(\alpha)$ of $\alpha$ with respect to the uniformity $\mathscr{U}$ is defined as

$$h_B(\alpha) = \sup\{h_B(\alpha, C) : C \subseteq X, C \text{ compact}\}.$$

It is well-known that $h_B(\alpha)$ coincides with the topological entropy whenever the uniform space $(X, \mathscr{U})$ is compact (a proof is available in [12, Corollary 2.14]), and with the entropy associated to a metric $d$, whenever the uniformity $\mathscr{U}$ is the uniformity induced by the metric $d$ [9].

Modifying Bowen's entropy, Kimura [20] introduced a different uniform entropy $h_K$ using *totally bounded* sets $C$ instead of compact ones as in the definition of $h_B$:

$$h_K(\alpha) = \sup\{h_B(\alpha, C) : C \subseteq X, C \text{ totally bounded}\}$$

where, as usual, a subset $A$ of $(X, \mathscr{U})$ is said to be totally bounded if, for all $V \in \mathscr{U}$, there exists a finite subset of $A$ which is $V$-dense in $A$.

Let $(\widetilde{X}, \widetilde{\mathscr{U}})$ denote the completion of the uniform space $(X, \mathscr{U})$ and let $\widetilde{\alpha}: (\widetilde{X}, \widetilde{\mathscr{U}}) \to (\widetilde{X}, \widetilde{\mathscr{U}})$ denote the extension of $\alpha$ to $(\widetilde{X}, \widetilde{\mathscr{U}})$. In the spirit of Kimura's paper [20], we say that *the completion theorem holds* for the entropy $h_B$ of a uniform space $(X, \mathscr{U})$ if the equality $h_B(\alpha) = h_B(\widetilde{\alpha})$ holds for every uniformly continuous map $\alpha: (X, \mathscr{U}) \to (X, \mathscr{U})$. Kimura proves that the completion theorem holds for the entropy $h_K$ in the case of *metric spaces* [20, Theorem 5.4] and in the case of *totally bounded* uniform spaces [20, Theorem 5.3], but may fail for non-metric spaces [20, Example 5.5]. Kimura shows that the completion theorem for Bowen's entropy does not hold even in the class of all totally bounded metric spaces [20, Example 5.6].

Our goal is to present several results concerning the relationship between the Bowen entropy of a compact group endomorphism $\widetilde{\alpha}: \widetilde{G} \to \widetilde{G}$ and the entropy of its restriction to a dense $\widetilde{\alpha}$-invariant subgroup $G$ where, as usual, $\widetilde{\alpha}$-invariant means that $\widetilde{\alpha}(G) \subseteq G$. Since dense subgroups of compact groups are totally bounded groups, we treat this question in the realm of endomorphisms of totally bounded (abelian) groups. Let us note, that for every endomorphism $\alpha : G \to G$ of a totally bounded (abelian) group, one has $h_B(\alpha) \leq h_B(\widetilde{\alpha})$ (see Lemma 8.5) where $\widetilde{\alpha}$ stands for the extension of $\alpha$ to the Raĭkov completion of $G$. Notice that, being $G$ totally bounded, its Raĭkov completion is a compact group. Thus, in equivalent terms, we study the *completion theorem for endomorphisms* and we show that it strongly fails even for endomorphisms of totally bounded abelian groups. To this end we build zero entropy endomorphisms whose extension to the (Raĭkov) completion of the group has infinite entropy. To emphasize better this phenomenon, we shall call these endomorphisms shortly *h-jump endomorphisms*. Obviously, they best witness the failure of the completion theorem for Bowen's entropy. A key role in our example is played by the property of having no infinite compact subsets which ensures zero entropy for all continuous endomorphisms (see Corollary 8.1). Many totally bounded groups have this property, in particular those equipped with the finest totally bounded topology (see Corollary 8.4). Remark that the respective examples from [20, Example 5.5] are very far from being topological groups (actually, the example is a discrete totally bounded space). It is worth noting that when the group in question is not bounded torsion, then our endomorphism $\alpha$ can be chosen to be the multiplication by an integer $k > 1$.

The paper is organized as follows. Section 8.2 is devoted to some general results on entropy of group endomorphisms of topological groups. Here we introduce the class $\mathfrak{E}_0$ of topological groups $G$, such that every continuous endomorphism of $G$ has entropy 0. This class plays an important role in this paper and will be further studied in the forthcoming paper [11], which studies also the larger class $\mathfrak{E}$ of topological groups that have no endomorphisms of infinite entropy.

In Sect. 8.3 we give examples of $h$-jump endomorphisms. Among other things it is showed that if $G^\#$ stands for an abstract abelian group $G$ equipped with its finest totally bounded topology (the so-called the Bohr topology on $G$), then $G^\# \in \mathfrak{E}_0$ always holds true. On the other hand, if $b(\alpha)$ denotes the extension of $\alpha$ to the Raĭkov completion of $G^\#$, we show that it is always possible to find an $\alpha$ such that $h_B(b(\alpha)) = \infty$. Moreover, as we said above, if $G$ is not bounded torsion, then such an $\alpha$ is simply the multiplication $m_k^G$ by any fixed integer $k > 1$ in $G$. Notice that $G^\#$ being totally bounded, its Raĭkov completion is a compact group.

The Bowen entropy is mostly studied in (locally) compact spaces. In Sect. 8.4 we give a specific technique for the computation of the Bowen entropy in metrizable abelian groups that are not necessarily locally compact. We show, among other results, that in various cases one can replace the compact sets in the definition of the Bowen entropy by convergent sequences. Using this idea we obtain positive results concerning Kimura's completion theorem in the realm of totally bounded metrizable groups.

### 8.1.1 Notation and Terminology

Our terminology and notation are standard. For instance, $\mathbb{P}$, $\mathbb{T}$ and $\mathbb{R}$ stand for the prime numbers, the circle and the additive group of the real numbers, respectively. Finally, $\mathbb{Z}(n)$ denotes the cyclic group of order $n$ and $\mathfrak{c}$ the continuum. As usual, $e_G$ stands for the neutral element of a group $G$, in case of additive notation we use simply 0.

A topological group $G$ is said to be *totally bounded* if for each neighborhood $U$ of the identity there exists a finite set $F \subset G$ such that $G = UF$ (equivalently, it is totally bounded for the natural uniformities of the group), and with $H \leq G$ we denote a subgroup $H$ of $G$. The completion with respect to the two-sided uniformity of a topological group $G$ is called the *Raĭkov completion* of $G$ and will be denoted by $\widetilde{G}$. Obviously, the left uniformity, the right uniformity and the two-sided uniformity coincide for abelian groups. By an early result of Weil the same result is valid for totally bounded groups [26, 27].

In the sequel we denote by $\widehat{G}$ the Pontryagin dual of a locally compact abelian group $G$. We feel free of using, without special mention, the basic fact that the Pontryagin dual of a compact abelian group (respectively, a discrete abelian group) is a discrete group (respectively, a compact group). Actually for either a compact or a discrete abelian group $G$, the equality $\widehat{\widehat{G}} = G$ holds. We denote by $\dim G$ the dimension of $G$. According to Pasynkov's celebrated theorem all dimensions coincide for locally compact topological groups [22, 23]. When $G$ is compact abelian, $\dim G = r(\widehat{G})$, where $r(X)$ denotes the free rank of the abelian group $X$. For $p \in \mathbb{P}$ we denote by $\mathbb{Z}_p$ the group of $p$-adic integers.

Let $G$ be an abstract abelian group. The topology of $G^\#$, the finest totally bounded topology on $G$, can be defined as the weak topology associated to the family of all homomorphisms from $G$ into the circle $\mathbb{T}$. Its Raĭkov completion is denoted by $bG$. Since $G^\#$ is totally bounded, $bG$ is a compact abelian group which is called *the Bohr compactification of* $G$. For notions and terminology not defined here on topological groups the reader can consult [10, 18], and on abstract abelian groups [14, 15].

## 8.2 Entropy of Endomorphisms of Topological Groups

In the sequel, given a(n) (abelian) topological group $G$, the uniformity $\mathcal{U}$ is the usual left (right, two-sided) uniformity on $G$. It is well known that a continuous homomorphism from a topological group $G_1$ into a topological group $G_2$ is uniformly continuous respect to the usual uniformities on $G_1$ and $G_2$. It is routine to check that the Bowen entropy with respect to the left uniformity of an endomorphism $\alpha$ of a (not necessarily abelian) topological group $G$ coincides with the Bowen entropy of $\alpha$ with respect to the right uniformity on $G$.

If $f : G \to G$ is an endomorphism of a topological group $G$ and $N$ is a closed normal subgroup of $G$, we say that $N$ is *$f$-invariant* if $f(N) \leq N$. In such a case $f$ induces an endomorphism of the quotient group $G/N$ that we shall denote by $f/N$. The restriction of $f$ to $N$ will be denoted by $f\restriction_N$. We shall often make use of the so called *addition theorem for entropy*:

**Theorem 8.1** ([4, Theorem 19]) *Let $f : G \to G$ be a continuous endomorphism of a metrizable compact group $G$ and let $N$ be a closed normal $f$-invariant subgroup of $G$. Then*

$$h_B(f) = h_B(f\restriction_N) + h_B(f/N). \tag{8.1}$$

This theorem was proved by Bowen in the case of compact metrizable groups. It is commented in [12] (without a proof) that the general case can be obtained from the metric case using the fact that the entropy function on compact groups is continuous on inverse limits (see [12, Theorem 3.12]). However, to the best of authors knowledge, no proof is available in the literature. We give a proof of this fact in Theorem 8.2.

We shall say that a topological group has *small subgroups* if each of its identity neighborhoods contains a nonsingleton subgroup. Compact topological groups which has no small subgroups are nothing else but compact Lie groups. Equivalently, a compact topological group $G$ is a Lie group if $G$ is isomorphic as a topological group to a unitary group of matrices [19, Corollary 2.40]. It is a well-known fact that a compact abelian group is a Lie group if, and only if, it is isomorphic to $\mathbb{T}^n \times F$ for some natural number $n$ and some finite abelian group $F$, that is, if and only if it is the character group of a finitely generated abelian group [19, Proposition 2.42]. Consequently, a connected compact abelian group is a Lie group if and only if it is isomorphic to $\mathbb{T}^n$ for some natural number $n$, i.e, it is an $n$-torus. In the next theorem we use the fact that every compact group is an inverse limit of Lie groups.

For the proof of Theorem 8.2 we need a lemma. Recall first that a subset $A$ of a topological space $X$ is called a $G_\delta$-set if $A$ is an intersection of countably many open sets. It is a well-known fact that if the neutral element $e_G$ of a topological group $G$ is a $G_\delta$-set, then there exists a weaker metrizable topology on $G$ which respect to which $G$ is topologically homogeneous. Therefore, since every one-to-one continuous map of a compact space onto a metrizable space is a homeomorphism, compact subsets of a topological group $G$ are metrizable in the case that $e_G$ is a $G_\delta$-set (see, for instance, [2, Theorem 3.3.16, Corollary 3.3.17]). Thus, we have

**Lemma 8.1** *Let G be a compact group and let N be a closed normal subgroup of G.Then the quotient group $G/N$ is metrizable if and only if N is a $G_\delta$-set.*

In the sequel we refer to such a normal subgroup $N$ of $G$ by briefly saying $N$ is a $G_\delta$-subgroup.

**Theorem 8.2** *Let $f: G \to G$ be a continuous endomorphism of a compact group G. Then for every neighborhood U of the identity there exists a f-invariant closed normal subgroup $N \subseteq U$ such that $G/N$ is metrizable. Moreover, if f is an automorphism of G, then N can be chosen so that the induced endomorphism $f/N: G/N \to G/N$ is an automorphism as well.*

*Proof* By [19, Corollary 2.36], there is a filter base $\mathscr{N}$ of compact normal subgroups converging to the neutral element of $G$ (that is, given a neighborhood $U$ of $e_G$, there is $N \in \mathscr{N}$ with $N \subseteq U$) such that $G/N$ is isomorphic to a closed subgroup of a unitary group of matrices. In particular, $G/N$ is metrizable, i.e., $N$ is a $G_\delta$-subgroup by Lemma 8.1.

Fix a neighborhood $U$ of $e_G$ and pick $N \in \mathscr{N}$ with $N \subseteq U$. As $N$ is a $G_\delta$-subgroup of $G$, also $f^{-n}(N)$ is a $G_\delta$-subgroup of $G$ for every $n$. Then $M = \bigcap_{n\in\mathbb{Z}} f^{-n}(N)$ is still a $G_\delta$-subgroup of $G$. Moreover, the obvious equality

$$f^{-1}(M) = M \tag{8.2}$$

implies that $M$ is $f$-invariant. This proves the first assertion of the theorem, since $G/M$ is metrizable by Lemma 8.1.

To prove the second assertion assume that $f$ is surjective. Since $M$ is $f$-invariant, the following diagram commutes

$$\begin{array}{ccc} G & \xrightarrow{f} & G \\ {\scriptstyle q}\downarrow & & \downarrow{\scriptstyle q} \\ G/M & \xrightarrow{f/M} & G/M \end{array}$$

where $q$ is the quotient map. Since $f$ is surjective, $f/M$ is surjective as well. Moreover, $f/M$ is injective by (8.2). Hence, $f/M$ is an isomorphism. By the compactness of $G/M$, we conclude that $f/M$ is a homeomorphism too. □

Now we can show the addition theorem for entropy in the realm of compact groups.

**Theorem 8.3** *Let $f: G \to G$ be an endomorphism of a compact group $G$ and let $N$ be a closed normal $f$-invariant subgroup of $G$. Then* (8.1) *holds.*

*Proof* Making use of Theorem 8.2 we can fix a filter base $\mathscr{N}$ of $f$-invariant closed normal subgroups $N_\alpha$ of $G$ such that the quotient $K_\alpha := G/N_\alpha$ is metrizable for all $N_\alpha \in \mathscr{N}$. Denote by $g_\alpha : K_\alpha \to K_\alpha$ the induced by $f$ map on the metrizable quotient group $K_\alpha = G/N_\alpha$. The closed normal subgroup $L_\alpha = NN_\alpha/N_\alpha$ of $K_\alpha$ is $g_\alpha$-invariant. Hence, Theorem 8.1, applied to the map $g_\alpha$ gives

$$h_B(g_\alpha) = h_B(g_\alpha/L_\alpha) + h_B(g_\alpha \restriction L_\alpha). \tag{8.3}$$

By the third isomorphism theorem,

$$K_\alpha/L_\alpha = (G/N_\alpha)/(NN_\alpha/N_\alpha) \cong G/NN_\alpha.$$

This isomorphism witnesses the fact that the map $g_\alpha/L_\alpha$ is conjugated to the map $f/NN_\alpha$. Hence,

$$h_B(g_\alpha/L_\alpha) = h_B(f/NN_\alpha). \tag{8.4}$$

Analogously,

$$L_\alpha = NN_\alpha/N_\alpha \cong N/(N \cap N_\alpha),$$

deducing from this isomorphism that the map $g_\alpha \restriction L_\alpha$ is conjugated to the map $(f\restriction_N)/(N \cap N_\alpha)$. This implies the equality

$$h_B(g_\alpha \restriction L_\alpha) = h_B((f\restriction_N)/(N \cap N_\alpha)). \tag{8.5}$$

Combining (8.3), (8.4) and (8.5) we conclude that

$$h_B(g_\alpha) = h_B(f/NN_\alpha) + h_B((f\restriction_N)/(N \cap N_\alpha)) \tag{8.6}$$

for every $N_\alpha \in \mathscr{N}$.

By [19, Proposition 1.33(ii)], the group $G$ is the inverse limit of the groups $G/N_\alpha$, the group $G/N$ is the inverse limit of the groups $G/NN_\alpha$ and the group $N$ is the inverse limit of the groups $N/(N \cap N_\alpha)$, when the subgroup $N_\alpha$ runs over $\mathcal{N}$. More precisely, due to the additional property that the subgroups $N_\alpha$ and $N \cap N_\alpha$ are invariant:

(1) $f$ is the inverse limit of the maps $g_\alpha$.
(2) $f/N$ is the inverse limit of the maps $f/NN_\alpha$.
(3) $f \restriction_N$ is the inverse limit of the maps $(f \restriction_N)/(N \cap N_\alpha)$.

The well-known fact that the entropy function is continuous under taking inverse limits (see, for example, [12, Theorem 3.12]) along with (8.1)–(8.3) imply that

$$\begin{aligned} &h_B(f) = \lim h_B(g_\alpha),\ h_B(f/N) = \lim h_B(f/NN_\alpha) \text{ and} \\ &h_B(f \restriction_N) = h_B((f \restriction_N)/(N \cap N_\alpha)). \end{aligned} \tag{8.7}$$

Now (8.1) follows from (8.6) and (8.7). □

It is not known whether the additivity theorem holds true for locally compact groups (see [12]).

For a topological abelian group $G$ and for every integer $k$ we denote by $m_k^G$ the continuous endomorphism of $G$ defined by the multiplication by $k$, i.e., $m_k^G(x) = kx$ for every $x \in G$. We explicitly formulate the next lemma for the sake of easy reference. Here $\mathbb{K}$ denotes the compact dual group $\widehat{\mathbb{Q}}$, where $\mathbb{Q}$ is the discrete group of the rational numbers.

**Lemma 8.2** *Let $k$ be an integer, let $G$ be a topological abelian group and let $H$ be a subgroup of $G$. Then:*

(a) *$H$ is $m_k^G$-invariant, $m_k^H = m_k^G \restriction_H$ and $m_k^G/H = m_k^{G/H}$.*
(b) *If $G$ is compact and $H$ is closed, then $h_B(m_k^G) = h_B(m_k^H) + h_B(m_k^{G/H})$.*
(c) *(*[4, Theorem 15], [29]*) $h_B(m_k^{\mathbb{R}^n}) = h_B(m_k^{\mathbb{T}^n}) = h_B(m_k^{\mathbb{K}^n}) = n \log k$ for every $k \in \mathbb{N}$.*

Item (b) immediately follows from (a) and Theorem 8.3, while (c) is well-known.

**Lemma 8.3** *If $G$ is a compact abelian group with* $\dim G$ *infinite, then $h_B(m_k^G) = \infty$ for every integer $k > 1$.*

*Proof* Our assumption that $\dim G$ is infinite implies that there exists a continuous surjective homomorphism $f : G \to \mathbb{T}^\omega$. In particular, for every $n$ there exists a continuous surjective homomorphism $f_n : G \to \mathbb{T}^n$. By Lemma 8.2(b) applied to $G$ and $G/H = \mathbb{T}^n$, along with Lemma 8.2(c), $h_B(m_k^G) \geq n \log k$ for every $n$. Therefore, $h_B(m_k^G) = \infty$. This completes the proof. □

**Lemma 8.4** *If every compact subset of a uniform space $(X, \mathcal{U})$ is finite, then $h_B(\alpha) = 0$ for each uniformly continuous function $\alpha$ on $(X, \mathcal{U})$.*

*Proof* Let $|K|$ be the cardinal of a compact subset of $(X, \mathscr{U})$. For each uniformly continuous function $\alpha$ on $(X, \mathscr{U})$, it is clear that $r_n(U, K) \leq |K|$ for each $U \in \mathscr{U}$ and, consequently,

$$\limsup_{n\to\infty} \frac{1}{n} \log r_n(U, K) \leq \limsup_{n\to\infty} \frac{1}{n} \log |K| = 0.$$

Thus, $h_B(\alpha) = 0$. □

If we denote by $\mathfrak{E}_0$ the class of topological groups $G$ such that every endomorphism of $G$ has zero entropy, we have

**Corollary 8.1** *If $G$ is a topological group without infinite compact subsets, then $G \in \mathfrak{E}_0$.*

**Corollary 8.2** *Every countable topological group $G$ without non-trivial convergent sequences belongs to $\mathfrak{E}_0$.*

*Proof* It suffices to note that $G$ cannot have infinite compact sets so the previous corollary applies. □

**Corollary 8.3** *Let $f$ be a continuous endomorphism of a compact topological group $G$. If $N$ is an open invariant normal subgroup of $G$, then $h_B(f) = h(f \restriction N)$.*

*Proof* Since $N$ is an open subgroup, $N$ is clopen. Thus, $G/N$ is a discrete group. The result now follows from Corollary 8.1 and Theorem 8.3. □

It is a well-known fact that every endomorphism $\alpha : G \to G$ is continuous when considered as an endomorphism of $G^\#$. One can deduce from Corollary 8.1 that every endomorphism of $G^\#$ has zero entropy:

**Corollary 8.4** $G^\# \in \mathfrak{E}_0$ *for every abelian group $G$.*

*Proof* As a consequence of the well-known Glicksberg's Theorem [16] which characterizes the compact subsets of the Bohr topology in the larger class of locally compact abelian groups, every compact subset of $G^\#$ is finite. Now Lemma 8.4 applies. □

**Lemma 8.5** *If $\alpha\colon (X, \mathscr{U}) \to (X, \mathscr{U})$ is a uniformly continuous function and $Y$ is an $\alpha$-invariant subspace of $X$, then $h_B(\alpha \restriction_Y) \leq h_B(\alpha)$. In particular, if $G$ is a topological group and $\alpha : G \to G$ is a continuous endomorphism, then $h_B(\alpha) \leq h_B(\widetilde{\alpha})$.*

**Corollary 8.5** *Let $G$ be a topological group with $\widetilde{G} \in \mathfrak{E}_0$, then also $G \in \mathfrak{E}_0$.*

*Proof* Let $f : G \to G$ be a continuous endomorphism. Then its extension $\widetilde{f} : \widetilde{G} \to \widetilde{G}$ has zero entropy by our assumption $\widetilde{G} \in \mathfrak{E}_0$. Hence its restriction $f$ to $G$ has zero entropy as well by Lemma 8.4. Therefore, $G \in \mathfrak{E}_0$. □

## 8.3 $h$-Jump Endomorphisms

As we say above, given an abstract abelian group $G$, we will consider $G$ equipped with the weak topology associated to the family of all homomorphisms from $G$ into the circle $\mathbb{T}$, the so-called *Bohr topology* of $G$ (this is the finest totally bounded topology on $G$). As usual, we will denote the group $G$ endowed with the Bohr topology by $G^\#$ and its Raĭkov completion by $bG$. Since $G^\#$ is totally bounded, $bG$ is a compact abelian group which is called *the Bohr compactification of* $G$.

The so-called *left Bernoulli shift* plays a central role in this section. Let $F$ be a set. We denote by $\sigma_F : F^{\mathbb{N}} \to F^{\mathbb{N}}$ the left Bernoulli shift defined by $(x_n)_n \to (x_{n+1})_n$, i.e.,

$$(x_0, x_1, \ldots, x_n, \ldots) \to (x_1, x_2, \ldots, x_{n+1}, \ldots)$$

for all $(x_n)_n \in F^{\mathbb{N}}$. If $F$ is a group whose cardinality is bigger than one, then $\sigma_F$ is a surjective group endomorphism. In case $F$ is a topological space (or just a finite set considered as a discrete compact space), $F^{\mathbb{N}}$ will carry the product topology. Then the left Bernoulli shift is continuous and $h_B(\sigma_F) = \log|F|$, to be intended as $h_B(\sigma_F) = \infty$ when $F$ is infinite. We start by the main result of this section.

**Theorem 8.4** *For every infinite abelian group $G$ there exists an $h$-jump endomorphism $\alpha : G^\# \to G^\#$, i.e., such that $h_B(b\alpha) = \infty$ for the extension $b\alpha : bG \to bG$ of $\alpha$.*

*Proof* We produce an endomorphism $\alpha : G^\# \to G^\#$ with $h_B(b\alpha) = \infty$ for the extension $b\alpha : bG \to bG$ of $\alpha$. To see that this is an $h$-jump endomorphism it remains to apply Corollary 8.4. To build $\alpha$ we consider three cases depending on the algebraic structure of $G$.

**Case 1** The abelian group $G$ is not bounded torsion. We show that the endomorphism $\alpha = m_k^G$ works in this case for any $k > 1$. Now the compact group $\widehat{G}$ is not torsion. Indeed, torsion compact groups are bounded torsion [18], so in case $\widehat{G}$ were torsion one would have $n\widehat{G} = 0$. But in such a case also $nG = 0$, a contradiction.

Since the compact group $\widehat{G}$ is not torsion, its free rank $r(\widehat{G})$ is infinite (actually, $r(\widehat{G}) \geq \mathfrak{c}$). This essentially follows from the structure theorem for compact abelian groups [18] (for pseudocompact groups this was announced by Comfort and van Mill [5] and proved in [13]). Now consider the discrete group $(\widehat{G})_d$. Since the dimension of its dual $bG = \widehat{(\widehat{G})_d}$ coincides with $r(\widehat{G})$, we can claim that the compact group $bG$ is infinite dimensional. Therefore, by Lemma 8.3, the multiplication $m_k^{bG}$ by $k$ in $bG$ has infinite entropy.

**Case 2** Let $n > 1$ and let $G = \bigoplus_\kappa \mathbb{Z}(n)$, where $\kappa$ is an infinite cardinal. To prove that $G$ admits an endomorphism $\alpha : G \to G$ such that $h_B(b\alpha) = \infty$ we consider the following presentation of $G$:

$$G = A_1 \oplus A_2 \oplus \ldots \oplus A_i \oplus \ldots,$$

where $A_i = \bigoplus_\kappa \mathbb{Z}(n)$ for every $i$. The group $K = \mathbb{Z}(n)^\kappa$ is compact when equipped with the Tichonov topology and all component groups $\mathbb{Z}(n)$ are taken discrete. Obviously, $K \cong K^{\mathbb{N}}$. Let $f : K \to K$ be the map conjugated by this isomorphism to the left Bernoulli shift $\sigma_K$ of $K^{\mathbb{N}}$. As $K$ is infinite, $h(f) = h(\sigma_K) = \infty$.

As the map $f : K \to K$ satisfies $f(G) = G$, we can consider the restriction $\alpha = f \restriction_G$. It remains to see that $h_B(b\alpha) = \infty$. To this end denote by $\tau$ the topology induced on $G$. As $G$ is a dense subgroup of $K$, $K$ is the completion of $(G, \tau)$. Since $\tau$ is totally bounded, the identity map $1_G : G^\# \to (G, \tau)$ is continuous, so can be extended to a continuous homomorphism $q : bG \to K$ of the compact completions of $G^\#$ and $(G, \tau)$. Moreover, by the density of $G$ in $K$, it follows that $q(bG)$ is both dense and closed in $K$, so $q(bG) = K$. Finally, the obvious equality $\alpha \circ 1_G = 1_G \circ \alpha$ extends by continuity to $b\alpha \circ q = q \circ f$, witnessing that $b\alpha$ dominates $f$. Therefore, $h(b\alpha) \geq h(f) = \infty$.

**Case 3** There exits a natural $n > 1$ such that $nG = 0$. By Prüfer's theorem $G$ is a direct sum of cyclic groups. Hence, $G = \bigoplus_{i=1}^m G_i$, where each $G_i$ has the form $G_i = \bigoplus_{\kappa_i} \mathbb{Z}(n_i)$ and $n_i$ is a divisor of $n$ for $i = 1, 2, \ldots, m$ as $nG_i = 0$. Suppose that $G_i$ is infinite for some $i = 1, 2, \ldots, m$. Let $H = \bigoplus_{j \neq i}^m G_j$, so that $G = G_i \oplus H$. By Case 2 there exists an endomorphism $\alpha : G_i \to G_i$ such that $h_B(b\alpha) = \infty$. Let $\gamma : G \to G$ be defined by $\gamma = \alpha \oplus id_H$. Since the functor of Bohr compactification commutes with (finite) products, one has $bG = bG_i \oplus bH$ and $b(\gamma) = b\alpha \oplus id_{bH}$. Then $h_B(b\gamma) = h_B(b\alpha) \oplus h_B(id_{bH}) = h_B(b\alpha) = \infty$. □

The above theorem shows that every infinite abelian group $G$ admits at least one totally bounded group topology $\tau$ (namely, the Bohr topology) and a continuous $h$-jump endomorphism $\alpha : (G, \tau) \to (G, \tau)$. Now, making use of a result of Comfort, Razckowski and Trigos-Arrieta [6] (see also [3]) we obtain the following optimal result in the case of countable groups (for the optimality part recall that every countable *set* admits at most $2^\mathfrak{c}$ many topologies).

**Theorem 8.5** *Every countably infinite abelian group $G$ that is not bounded torsion admits $2^\mathfrak{c}$ many pairwise non-homeomorphic totally bounded group topologies $\{\tau_i : i < 2^\mathfrak{c}\}$ of weight $\mathfrak{c}$ such that:*

(a) $(G, \tau_i) \in \mathfrak{C}_0$ *for each* $i < 2^\mathfrak{c}$.
(b) *For each $i < 2^\mathfrak{c}$ and for each $k > 1$ the group endomorphism $\widetilde{m_k^G}$ is an h-jump endomorphism of $\widetilde{(G, \tau_i)}$; in particular, $\widetilde{(G, \tau_i)}$ has endomorphisms of infinite entropy.*

*Proof* Let $K$ be the compact Pontryagin dual of $G$. Then $K$ is a compact metrizable group. Moreover, $K$ is non-torsion, so according to [6] one can find $2^\mathfrak{c}$ many pairwise non-homeomorphic totally bounded group topologies $\tau_i$ on $G$ without non-trivial convergent sequences such that the group $H_i$ of continuous characters of $(G, \tau_i)$ is a free subgroup of rank $\mathfrak{c}$ of $K$. Since $(G, \tau_i)$ is countable and without non-trivial convergent sequences, Corollary 8.2 yields that $(G, \tau_i) \in \mathfrak{C}_0$. This proves (a).

On the other hand, for every $i < 2^\mathfrak{c}$ the completion $K_i = \widetilde{(G, \tau_i)}$ has the same continuous characters as the group $(G, \tau_i)$. In other words, the underlying group of the (discrete) Pontryagin dual of $K_i$ is the (discrete) group $H_i$. So $K_i \cong \widehat{H_i} \cong \mathbb{T}^{\mathfrak{c}}$ is a connected compact group of weight $\mathfrak{c}$. In particular, $K_i$ is infinite-dimensional. Hence by Lemma 8.4 the endomorphism $m_k^{K_i}$ of $K_i$ has infinite Bowen entropy, i.e., $m_k^G = m_k^{K_i} \restriction_G$ is an $h$-jump endomorphism. □

## 8.4 The e-Support of an Endomorphism and Kimura's Completion Theorem

We have no examples of $h$-jump endomorphisms of *metrizable* abelian groups. Related to this question, we study in this section several versions of Kimura's completion theorem for the Bowen entropy in the case of metrizable abelian groups. This is why, from now on $G$ will be a metrizable abelian group and $(U_n)$ a decreasing sequence of symmetric neighborhoods of $e_G$ in $G$ that form a local base at $e_G$.

**Definition 8.1** Let $G$ be a topological group and let $\alpha : G \to G$ be a continuous endomorphism. For every neighborhood $U$ of $e_G$ and every positive integer $n$ define the *n-th cotrajectory* of $U$ (under $\alpha$) by $C_n(U, \alpha) = U \cap \alpha^{-1}(U) \cap \ldots \cap \alpha^{-n+1}(U)$.

Obviously, $\{C_n(U, \alpha) : n \in \mathbb{N}\}$ is a decreasing chain of open sets and the entropy is intended to measure the speed of decrease of $C_n(U, \alpha)$. In order to do this one can use the growth of the sequences of numbers $\{r_n(\alpha, U, K) : n \in \mathbb{N}\}$ and $\{s_n(\alpha, U, K) : n \in \mathbb{N}\}$, defined with respect to a pair $U$, $K$ of a neighborhood of $e_G$ and a compact set.

**Lemma 8.6** *Let $G$ be an abelian topological group and let $\alpha : G \to G$ be a continuous endomorphism. In the above notation, for a finite subset $F$ of $G$ and a symmetric open neighborhood $U$ of 0, the following assertions hold:*

(a) *$F$ $(n, U)$-spans $K$ if and only if $K \subseteq F + C_n(U, \alpha)$.*
(b) *$F \subset K$ is $(n, U)$-separated if and only if $(F - F) \cap C_n(U, \alpha) = \{0\}$.*

*Moreover, if $F$ is a subgroup of $G$, then $F$ is $(n, U)$-separated if and only if $F \cap C_n(U, \alpha) = \{0\}$.*

*Proof* (a) A subset $F$ $(n, U)$-*spans* a compact subset $K$ if for every $x \in K$ there is $y \in F$ such that $\alpha^j(x) - \alpha^j(y) \in U$ for each $0 \leq j < n$. In other words, $\alpha^j(x - y) \in U$ for each $0 \leq j < n$. This means that

$$x - y \in \bigcap_{j=0}^{n-1} \alpha^{-j}(U) = C_n(U, \alpha).$$

(b) The set $F$ is $(n, U)$-separated with respect to $\alpha$, if and only if for each pair of distinct points $x, y \in F$ there exists $j$ such that $0 \leq j < n$ and $\alpha^j(x) - \alpha^j(y) \notin U$. In other words, $\alpha^j(x - y) \notin U$ for some $0 \leq j < n$. This means that $F$ is $(n, U)$-separated if and only if $x - y \notin C_n(U, \alpha)$ for each pair of distinct points $x, y \in F$, i.e., $(F - F) \cap C_n(U, \alpha) = \{0\}$. □

It is clear from the definitions, that if $\alpha : G \to G$ is a continuous endomorphism, then there exists a family $(K_n)$ of compact sets such that $h_B(\alpha) = \lim_n s(\alpha, K_n, U_n)$. Since $s(\alpha, -, -)$ is monotone with respect to both arguments, we can assume with no loss of generality that the sequence $(K_n)$ is increasing. We refer to the family $(K_n)$ as e-*supporting family* of $\alpha$. In this context, the following question arises in a natural way:

*Question 8.1* Let $G$ be a metrizable abelian group as above and let $\alpha : G \to G$ be a continuous endomorphism. When does $\alpha$ admit an $e$-supporting family $(K_n)$ such that all compacts $K_n$ are converging (to 0) sequences ?

We conjecture, that under mild conditions the answer to this question is: *almost always*.

In the sequel we show that the $e$-supporting family $(K_n)$ may witness $h_B(\alpha) = \lim_n s(\alpha, K_n, U_n)$ having all neighborhoods $U_n$ equal to a fixed neighborhood $U$. Motivated by this fact we propose also the following:

**Definition 8.2** Let $G$ be a metrizable abelian group, let $U$ be a fixed neighborhood of 0 and let $\alpha : G \to G$ be a continuous endomorphism.

(a) A family of compact subsets $(K_n)$ of $G$ such that $h_B(\alpha) = \lim_n s(\alpha, K_n, U)$ is called an *e-supporting family* of $\alpha$ with respect to $U$.
(b) If in the $e$-supporting family $(K_n)$ with respect to $U$, as in (a), each $K_n$ has the form
$$K_n = \{0\} \cup \{x_m^{(n)} : m \in \mathbb{N}\},$$
where $(x_m^{(n)})$ is a faithfully enumerated sequence with $\lim_m x_m^{(n)} = 0$, we say that $(K_n)$ is an *e-supporting family of sequences* of $\alpha$ with respect to $U$.

In the next theorem we see that one may have an e-supporting family of sequences of $\alpha$, with respect to an appropriately chosen $U$, that consists of a *single* sequence.

**Theorem 8.6** *Let $F$ be a finite abelian group and let $G = \bigoplus_{j=0}^{\infty} G_j$, where each $G_j \cong F$, equipped with the product topology induced from the Cartesian power $F^{\mathbb{N}}$. Let $\alpha : G \to G$ be the left Bernoulli shift. Then:*

(a) *There exists a convergent to 0 sequence $K$ and a neighborhood $U_0$ of 0 in $G$, such that $h_B(\alpha) = s(\alpha, U_0, K)$.*
(b) $h_B(\alpha) = h_B(\widetilde{\alpha}) = \log|F|$, *i.e., $\alpha$ satisfies Kimura's theorem for $h_B$.*

*Proof* Since the extension $\tilde{\alpha} : F^{\mathbb{N}} \to F^{\mathbb{N}}$ of $\alpha$ is the left Bernoulli shift of $F^{\mathbb{N}}$ having entropy $\log|F|$, we obviously have

$$h_B(\alpha) \le h_B(\widetilde{\alpha}) = \log|F|. \tag{8.8}$$

For every $m \in \mathbb{N}$ let

$$H_m = \bigoplus_{j=0}^{m} G_j \text{ and } U_m = \bigoplus_{j=m+1}^{\infty} G_j,$$

so that

$$H_m \cap U_m = 0 \text{ and } G = H_m \oplus U_m. \tag{8.9}$$

The subgroups $U_m$ form a base of neighborhoods of 0 in $G$.

We first prove (a), so our aim is to define a convergent to 0 sequence $K$, such that $s(\alpha, U_0, K) \geq \log |F|$. Then (8.8) will yield

$$\log |F| \geq h_B(\alpha) \geq s(\alpha, U_0, K) \geq \log |F| = h_B(\widetilde{\alpha}),$$

proving (b).

In oder to define $K$ let

$$X_m := \bigoplus_{j=m+1}^{m^2} G_j \text{ and } K = \{0\} \cup \bigcup_{m=1}^{\infty} X_m.$$

For an arbitrarily fixed $m$ one has that $X_s \subseteq U_m$ for all $s \geq m$, so $U_m$ contains all but a finitely many elements of $K$. Therefore, any enumeration of $K$ gives a sequence converging to 0. In particular, $K$ is compact.

On the other hand, $U_m = \alpha^{-m} U_0$, so $C_m(U_0, \alpha) = U_m$. Therefore, (8.9) yields that $H_m$ is $(m, U_0)$-separated according to Lemma 8.6. As $X_m \subseteq H_{m^2}$, we deduce that $X_m$ is $(m^2, U_0)$-separated. So, the conjunction of $X_m \subseteq K$ and $|X_m| = |F|^{m^2-m}$ entails that

$$s_{m^2}(\alpha, U_0, K) \geq |F|^{m^2-m}.$$

Thus, $\log s_{m^2}(\alpha, U_0, K) \geq (m^2 - m) \log |F|$. Therefore,

$$\begin{aligned} s(\alpha, U_0, K) = \limsup_n \frac{\log s_n(\alpha, U_0, K)}{n} &\geq \limsup_m \frac{\log s_{m^2}(\alpha, U_0, K)}{m^2} \\ \geq \limsup_m \frac{(m^2 - m) \log |F|}{m^2} &= \lim_m \frac{(m^2 - m) \log |F|}{m^2} = \log |F|, \end{aligned}$$

i.e., the compact set $K$ has the desired properties. □

In the sequel $A$ will denote the algebraic closure of $\mathbb{Q}$ in $\mathbb{R}$. In other words, $A$ is the subfield of $\mathbb{R}$ consisting of all real algebraic numbers. In the sequel we shall be interested only of the underlying (additive) group structure of $A$. Obviously, $A$ is countable and contains $\mathbb{Q}$. So, $A$ is dense in $\mathbb{R}$ and $A/\mathbb{Z}$ is a countable dense subgroup of $\mathbb{T}$ containing $\mathbb{Q}/\mathbb{Z}$.

**Theorem 8.7** *Let $G$ be a subgroup of $\mathbb{T}$ containing $A/\mathbb{Z}$ and let $\alpha : G \to G$ be a continuous endomorphism. Then:*

(a) *$\alpha$ has an e-supporting family of convergent sequences $K_t = (x_m^{(t)})_m$ ($t = 2, 3, \dots$) with respect to an appropriate neighborhood $U$ of 0 in $G$.*
(b) *$h_B(\alpha) = h_B(\tilde{\alpha})$, i.e., $\alpha$ satisfies Kimura's theorem for $h_B$.*

*Proof* Being an infinite subgroup of $\mathbb{T}$, $G$ is dense in $\mathbb{T}$. Hence, $\widetilde{G} = \mathbb{T}$ and for the extension $\widetilde{\alpha} : \mathbb{T} \to \mathbb{T}$ of $\alpha$ there exists $k \in \mathbb{Z}$ such that $\widetilde{\alpha}(x) = kx$ for all $x \in \mathbb{T}$. Since both assertions are trivially true for $k = 0, \pm 1$ and since $h_B(-\alpha) = h_B(\alpha)$, we can assume without any loss of generality that $k > 1$ in the sequel. Then

$$h_B(\alpha) \leq h_B(\widetilde{\alpha}) = \log k. \tag{8.10}$$

In order to prove (a) we define a neighborhood $U$ of 0 in $G$ and a converging to 0 sequence $K_t = (x_n^{(t)})_n$ for every natural $t > 1$, so that

$$s(\alpha, U, K_t) \geq \frac{t}{t+1} \log k. \tag{8.11}$$

Then,

$$h_B(\alpha) \geq \sup_t s(\alpha, K_t, U) \geq \sup_t \frac{t}{t+1} \log k = \log k = h_B(\widetilde{\alpha}),$$

so (8.10) yields (a) and (b).

For every natural $s > 0$ let $T_s = \{x \in \mathbb{T} : \|x\| < 1/4s\}$. We will be mainly concerned with $T_k$ and $U = G \cap T_k$. Note that $C_n(T_k, \widetilde{\alpha}) = T_{k^n}$ and $C_n(U, \alpha) = G \cap T_{k^n}$. Therefore,

(S) *any finite set $F$ in $G$ with $(F - F) \cap T_{k^n} = \{0\}$ is $(n, U)$-separated.*

For every integer $t > 1$ define a converging to 0 sequence $K_t$ in $G$ (actually, in $A/\mathbb{Z}$) by letting

$$x_0^{(t)} = 0,\ x_m^{(t)} = \frac{1}{\sqrt[t]{m}} \text{ for } m > 0, \text{ and } K_t = \{0\} \cup \{x_m^{(t)} : m \in \mathbb{N}\}.$$

This gives

$$\|x_{m-1}^{(t)} - x_m^{(t)}\| = \left| \frac{1}{\sqrt[t]{m-1}} - \frac{1}{\sqrt[t]{m}} \right| \geq \frac{1}{t\sqrt[t]{m^{t+1}}} \tag{8.12}$$

after an application of the mean value theorem to the function $x^{-\frac{1}{t}}$. Since this function is convex, we deduce also that

$$\|x_{j-1}^{(t)} - x_j^{(t)}\| \geq \|x_j^{(t)} - x_{j+1}^{(t)}\| \text{ for every } j > 1. \tag{8.13}$$

Let $K_m = \{x_1, x_2, \dots, x_m\}$. So (8.12) and (8.13) imply that $\|x - y\| \geq \frac{1}{t\sqrt[t]{m^{t+1}}}$ whenever $x, y \in K_m$ are distinct. Therefore, if we choose $m$ such that

$$\frac{1}{t\sqrt[t]{m^{t+1}}} \geq 1/4k^n, \tag{8.14}$$

then $(K_m - K_m) \cap T_{k^n} = \{0\}$, so $K_m$ is $(n, U)$-separated by Lemma 8.6.

To ensure (8.14), it is enough to pick $m$ with

$$m \leq C.k^{\frac{tn}{t+1}}, \tag{8.15}$$

where $C = (4/t)^{t/(t+1)}$. Since $|K_m| = m$, choosing $m := \lfloor \sqrt{C}.k^{\frac{tn}{t+1}} \rfloor$, (8.15) gives $s_n(\alpha, U, K_t) \geq C.k^{\frac{tn}{t+1}}$, Hence,

$$\begin{aligned} s(\alpha, U, K_t) &= \limsup_n \frac{\log s_n(\alpha, U, K_t)}{n} \\ &\geq \limsup_n \frac{\log C + \frac{tn}{t+1} \log k}{n} \\ &= \lim_n \frac{\log C + \frac{tn}{t+1} \log k}{n} = \frac{t}{t+1} \log k. \end{aligned}$$

This proves (8.11) and the theorem. □

Let us give a brief comment of the above constructions.

*Remark 8.1* (a) The main idea of the construction is to find a neighborhood $U$ of 0 such that $C_n(U, \alpha)$ forms a local base at 0 and using this fact build for every integer $t > 1$ a sequence $(x_m)$ (depending on $t$) that *slowly* converges to 0, so that for the compact set $K_t = (x_m)$ the values of $s_n(\alpha, K_t, U)$ are as big as possible (asymptotically, $\geq C \cdot h_B(\alpha)^{\frac{tn}{t+1}}$).

(b) The family $K_t = (x_m)$ of compact set built in the above theorem *does not depend on the endomorphism* $\alpha : G \to G$. Indeed, the family $K_t = (x_m)$ is contained in the subgroup $A/\mathbb{Z}$, so for every subgroup $G$ of $\mathbb{T}$ containing $A/\mathbb{Z}$ and every continuous endomorphism $\alpha : G \to G$ there exists a neighborhood $U$ of 0 in $G$ such that the assertions (a) and (b) from the theorem hold true for this *fixed* family $K_t = (x_m)$. Finally, a more careful analysis of the proof shows that also the neighborhood $U$ can be chosen *independently on the endomorphism* $\alpha : G \to G$ (e.g., $U = T_1 \cap G$ will work; this choice introduces only some additional small technical complications, the main line of the proof remains the same).

(c) Using the same argument one can prove the same properties for continuous endomorphisms $\alpha : G \to G$ of an arbitrary subgroup $G$ of $\mathbb{R}$ containing $A$.

It is reasonable to expect that the construction from Theorem 8.7 can be appropriately modified to produce an e-supporting family of sequences contained in an *arbitrary* infinite (hence, dense) subgroup $H$ of $\mathbb{T}$.

**Conjecture 8.1** *Theorem* 8.7 *can be extended to arbitrary infinite subgroups of* $\mathbb{T}$.

One can consider also the following obvious counterpart for arbitrary finite-dimensional tori:

*Question 8.2* (a) Can Theorem 8.7 can be extended to arbitrary subgroups $H$ of $\mathbb{T}^d$? (b) What about some special endomorphisms like $\alpha = m_k^H$ of $H$?

The next question is motivated by Theorem 8.6:

*Question 8.3* Let $G$ be a metrizable abelian group and let $\alpha : G \to G$ be a continuous endomorphism. When $\alpha$ has an $e$-supporting family formed by a *single* sequence $K$?

Let us discuss situations in metrizable topological groups when Kimura's completion theorem may hold true for the Bowen entropy. So far, according to Theorem 8.7b, every subgroup of $\mathbb{T}$ containing the countable subgroup $A/\mathbb{Z}$ satisfies Kimura's completion theorem for $h_B$. According to Remark 8.1 (c), also all subgroups of $\mathbb{R}$ containing $A$ satisfy Kimura's completion theorem for $h_B$.

Similarly, one obtains:

**Theorem 8.8** *If Conjecture* 8.1 *holds true, then every subgroup* $H$ *of* $\mathbb{T}$ *satisfies Kimura's completion theorem.*

Indeed, in case $H$ is finite, then $H$ coincides with its completion so there is nothing to prove. If $H$ is infinite, then $H$ is dense in $\mathbb{T}$. Then to every continuous endomorphism $\alpha : H \to H$ one applies the extended form of Theorem 8.7 (b) for *all* infinite subgroups of $\mathbb{T}$ granted by Conjecture 8.1.

Analogously, a positive answer to Question 8.2 implies that Kimura's completion theorem holds true for all subgroups of $\mathbb{T}^d$.

Other instances when Kimura's completion theorem for continuous endomorphism of topological groups holds true are provided by Corollary 8.5. Namely, if $K \in \mathfrak{C}_0$ is a compact (or just complete) group, then every dense subgroup $G$ of $K$ satisfies Kimura's completion theorem for $h_B$ by Corollary 8.5. The class $\mathfrak{C}_0$ is studied in detail in the forthcoming paper [11]. In the next example we anticipate a prominent source of such groups.

*Example 8.1* Let $G$ be a subgroup of a compact group of the form $K = \prod_p \mathbb{Z}_p^{k_p} \times F_p$, where $k_p \in \mathbb{N}$ and $F_p$ is a finite abelian $p$-group for every prime $p$. Then $G$ satisfies Kimura's completion theorem. Indeed, if $G$ is dense in $K$, this follows from Corollary 8.5, as mentioned above. Otherwise, it suffices to observe that every closed subgroup $N$ of $K$ has the same form as $K$, so $N \in \mathfrak{C}_0$ and the assertion follows as before.

Finally, let us comment some possible consequences of a negative solution of Conjecture 8.1.

Actually, we do not know the answer even for the case of the subgroup $\mathbb{Q}/\mathbb{Z}$ of $\mathbb{T}$. Nevertheless, one can prove the following inspiring partial result in this case.

**Proposition 8.1** *There is a convergent to 0 sequence $K$ in $\mathbb{Q}/\mathbb{Z}$ and a neighborhood $U$ of 0 in $\mathbb{Q}/\mathbb{Z}$, such that for every continuous endomorphism $\alpha : \mathbb{Q}/\mathbb{Z} \to \mathbb{Q}/\mathbb{Z}$ one has*

$$s(\alpha, U, K) \geq \frac{1}{2} \cdot h_B(\widetilde{\alpha}). \tag{8.16}$$

*Proof* The subgroup $H = \mathbb{Q}/\mathbb{Z}$ is dense in $\mathbb{T}$ and for the extension $\widetilde{\alpha} : \mathbb{T} \to \mathbb{T}$ of $\alpha$ there exists $k \in \mathbb{Z}$ such that $\widetilde{\alpha}(x) = kx$ for all $x \in \mathbb{T}$. Since (8.16) is trivially true for $k = 0, \pm 1$ and since $h_B(-\alpha) = h_B(\alpha)$, we can assume without any loss of generality that $k > 1$ in the sequel. Then $h_B(\widetilde{\alpha}) = \log k$, so our aim will be to prove

$$s(\alpha, U, K) \geq \frac{\log k}{2}. \tag{8.17}$$

We intend to take as $K$ the converging to 0 sequence $x_n = \frac{1}{n}$ in $H$, i.e., $K = \{0\} \cup \{x_n : n > 0\}$. The neighborhood $U = H \cap T_k$ of 0 in $H$ satisfies $C_n(U, \alpha) = H \cap T_{k^n}$, so (S) from the proof of Theorem 8.7 remains true in $H$ as well. Now one can use the fact that

$$\|x_{m-1} - x_m\| = \left|\frac{1}{m-1} - \frac{1}{m}\right| = \frac{1}{m(m-1)} < \frac{1}{m^2}$$

so that with $K_m = \{x_1, x_2, \ldots x_m\}$, $(K_m - K_m) \cap T_{k^n} = \{0\}$ provided $m^2 \leq 4k^n$. This proves, as in the proof of Theorem 8.7, that $K_m$ is $(n, U)$-separated, so $s_n(\alpha, U, K) \geq 2k^{\frac{n}{2}}$. Consequently, (8.17) is obtained arguing as in the proof of Theorem 8.7. □

Obviously, (8.16) implies the following weaker form of Kimura's completion theorem for $h_B$

$$h_B(\alpha) \geq \frac{1}{2} \cdot h_B(\widetilde{\alpha}). \tag{8.18}$$

Hence, in the case of failure of Conjecture 8.1 for arbitrary infinite subgroups $H$ of $\mathbb{T}$, one may try to replace the Kimura's completion theorem for $h_B$ by a suitable weaker form as in (8.18), replacing $\frac{1}{2}$ by some positive constant $C_H$ eventually depending on $H$. Of course, this can be extended to arbitrary metrizable totally bounded groups at three levels:

*Question 8.4* Let $G$ be a metrizable totally bounded (abelian) group.

(a) Does there exist a positive constant $C_G$ such that for every continuous endomorphism $\alpha : G \to G$ one has

$$h_B(\alpha) \geq C_G \cdot h_B(\widetilde{\alpha})?$$

(b) Does there exist a positive constant $C_\alpha$ for every continuous endomorphism $\alpha : G \to G$ such that

$$h_B(\alpha) \geq C_\alpha \cdot h_B(\widetilde{\alpha})?$$

(c) Is it true that for every continuous endomorphism $\alpha : G \to G$ with $h_B(\alpha) = 0$ also $h_B(\widetilde{\alpha}) = 0$?

A positive answer to the weakest property (c) will already imply that the metrizable totally bounded abelian groups admit no $h$-jump endomorphisms.

**Acknowledgments** The authors wish to thank the Editors for their kind invitation to participate in this Special Issue. Our contribution is dedicated to Prof. J. Kakol on occasion of his 60th birthday. The first authors was partially supported by "Progetti di Eccellenza 2011/12" of Fondazione CARIPARO. The third author was supported by the Spanish Ministry of Science and Education (Grant number MTM2011-23118), and by Bancaixa (Projecte P1-1B2011-30).

## References

1. Adler, R.L., Konheim, A.G., McAndrew, M.H.: Topological entropy. Trans. Am. Math. Soc. **114**, 309–319 (1965)
2. Arhangel'skii, A., Tkachenko, M.: Topological Groups and Related Structures. Atlantis Studies in Mathematics, vol. 1. Atlantis Press, Paris, World Scientific, Hackensack (2008)
3. Barbieri, G., Dikranjan, D., Milan, C., Weber, H.: Convergent sequences in precompact group topologies. Appl. Gen. Topology **6**(2), 149–169 (2005)
4. Bowen, R.: Entropy for group endomorphisms and homogeneous spaces. Trans. Am. Math. Soc. **153**, 401–414 (1971)
5. Comfort, W.W., van Mill, J.: Concerning connected, pseudocompact abelian groups. Topology Appl. **33**(1), 21–45 (1989)
6. Comfort, W.W., Raczkowski, S.U., Trigos-Arrieta, F.J.: Making group topologies with, and without, convergent sequences. Appl. Gen. Topology **7**(1), 109–124 (2006)
7. Dikranjan, D., Giordano Bruno, A.: The connection between topological and algebraic entropy. Topology Appl. **159**(13), 2980–2989 (2012)
8. Dikranjan, D., Goldsmith, B., Salce, L., Zanardo, P.: Algebraic entropy for Abelian groups. Trans. Am. Math. Soc. **361**, 3401–3434 (2009)
9. Hood, B.M.: Topological entropy and uniform spaces. J. Lond. Math. Soc. **8**, 633–641 (1974)
10. Dikranjan, D., Prodanov, I.R., Stoyanov, L.N.: Topological groups (characters, dualities and minimal group topologies). Monographs and Textbooks in Pure and Applied Mathematics, vol. 130. Marcel Dekker Inc., New York (1990)
11. Dikranjan, D., Sanchis, M.: On the e-spectrum of a topological group (work in progress)
12. Dikranjan, D., Sanchis, M., Virili, S.: New and old facts about entropy in uniform spaces and topological groups. Topology Appl. **159**, 1916–1942 (2012)
13. Dikranjan, D., Shakhmatov, D.: Algebraic structure of pseudocompact groups. Mem. Am. Math. Soc. **133**(633), (1998)
14. Fuchs, L.: Infinite abelian groups, I. Pure and Applied Mathematics, vol. 36. Academic Press, New York (1970)
15. Fuchs, L.: Infinite abelian groups, II. Pure and Applied Mathematics, vol. 36-II. Academic Press, New York (1973)
16. Glicksberg, I.: Uniform boundedness for groups. Can. J. Math. **14**, 269–276 (1962)
17. Goodwyn, L.W.: The product theorem for topological entropy. Trans. Am. Math. Soc. **158**, 445–452 (1971)
18. Hewitt, E., Ross, K.A.: Abstract harmonic analysis, I: Structure of topological groups. Integration theory, group representations. Die Grundlehren der mathematischen Wissenschaften, Bd. 115, Academic Press, New York (1963)
19. Hofmann, K.K., Morris, S.A.: The structure of compact groups. A primer for the studentÃŚa handbook for the expert. De Gruyter Studies in Mathematics, vol. 25. De Gruyter, Berlin (2013)

20. Kimura, T.: Completion theorem for uniform entropy. Comm. Math. Univ. Carolin. **39**(2), 389–399 (1998)
21. Latora, V., Baranger, M., Rapisarda, A., Tsallis, C.: The rate of entropy increase at the edge of chaos. Phys. Lett. A **273**(1–2), 97–103 (2000)
22. Pasynkov, B.A.: The coincidence of various definitions of dimensionality for locally bicompact groups. Dokl. Akad. Nauk SSSR **132**, 1035–1037 (1960)
23. Pasynkov, B.A.: Almost-metrizable topological groups. Dokl. Akad. Nauk SSSR **161**, 281–284 (1965). (in Russian)
24. Peters, J.: Entropy of discrete abelian groups. Adv. Math. **33**(1), 1–13 (1979)
25. Peters, J.: Entropy of automorphisms on L.C.A. groups. Pac. J. Math. **96**(2), 475–488 (1981)
26. Weil, A.: Sur les Spaces a Structure Uniforme and sur la Topologie Generale. Publ. Math. Univ. Strasbourg, Hermann (1937)
27. Weil, A.: L'intégration sur les groupes topologique et ses applications. Actualités Sci. Ind., vol. 869. Hermann, Paris (1950)
28. Weiss, M.D.: Algebraic and other entropies of group endomorphisms. Math. Syst. Theory **8**(3), 243–248 (1974/75)
29. Yuzvinskiĭ, S.A.: Calculation of the entropy of a group-endomorphism. Sibirsk. Mat. Ž. **8**, 230–239 (1967)

# Chapter 9
# On Preserved and Unpreserved Extreme Points

**Antonio José Guirao, Vicente Montesinos and Václav Zizler**

**Abstract** An extreme point of the closed unit ball of a Banach space is said to be *preserved* if it is extreme of the closed unit ball of the bidual space; otherwise it is called *unpreserved*. The beginning of the present work takes the form of a survey on this topic, presenting some elementary facta about those concepts—usually with new proofs—and discussing in particular Katznelson's solution to a Phelps' question on preserved extreme points, not available, to our knowledge, in the literature. In a second part, some new results are presented. Since some of them depend on the concept of polyhedrality, we first review several results on this topic. Then we present Godun renorming theorem for the class of nonreflexive Banach spaces, and Morris renorming result—with a new proof—on separable Banach spaces containing a copy of $c_0$. We show that, under some extra conditions—polyhedrality—a similar renorming, this time adding smoothness, can be defined ensuring strict convexity with all points in the unit sphere unpreserved extreme. We finalize this work by presenting what—to our knowledge—is the first nonseparable result of this kind for the natural class of the weakly compactly generated Banach spaces.

**Keywords** Unpreserved extreme points · Extreme points · Renorming · Strictly convex norm · Disk algebra · Krein–Milman property · Radon-Nikodým property · Banach space containing $c_0$ · Weakly Lindelöf determined spaces

**MSC 2010 Classification:** Primary 46B03, Secondary 46B22, 46B04, 46B10, 46B20

---

A. J. Guirao · V. Montesinos (✉)
Instituto de Matemática Pura y Aplicada,
Universitat Politècnica de València, C/ Vera s/n, 46022 Valencia, Spain
e-mail: vmontesinos@mat.upv.es

A. J. Guirao
e-mail: anguisa2@mat.upv.es

V. Zizler
Department of Mathematics, University of Alberta, Edmonton, ABT6G 2G1, Canada
e-mail: vasekzizler@gmail.com

J. C. Ferrando and M. López-Pellicer (eds.), *Descriptive Topology and Functional Analysis*, 163
Springer Proceedings in Mathematics & Statistics 80, DOI: 10.1007/978-3-319-05224-3_9,
© Springer International Publishing Switzerland 2014

## 9.1 Some Distinguished Points of the Unit Sphere of a Banach Space

One of the most useful ways for describing the geometrical structure of a set in a vector space is by exhibiting its "corners", as it is plain by looking at a "polygon" in $\mathbb{R}^n$. The technical word for a "corner" is "extreme point", a concept that can be defined for any nonempty subset $M$ of a vector space as a point $e \in M$ that belongs to no open interval $(x, y) \subset M$. Since all Hausdorff vector topologies on a finite-dimensional vector space—in particular, on a line—agree, this concept is—despite the appearance of the term "open" in the definition—a truly algebraic one. The importance of such a concept is mainly due to the famous Krein–Milman theorem, stating that *any nonempty convex compact subset $K$ of a locally convex space is the closed convex hull* $\overline{\text{conv}}(\text{Ext}(K))$ *of the set* $\text{Ext}(K)$ *of its extreme points*; in particular, such a set $K$ *has* at least an extreme point. A more general statement reads: *A compact subset $M$ of a locally convex space has the same closed convex hull as the set* $\text{Ext}(M)$ *of its extreme points.* A useful companion to the Krein–Milman theorem is Milman's theorem, stating that *if $M$ is a subset of a locally convex space such that* $\overline{\text{conv}}(M)$ *is compact, then* $\text{Ext}(\overline{\text{conv}}(M)) \subset \overline{M}$, and $\text{Ext}(\overline{\text{conv}}(M)) \subset \text{Ext}(\overline{M})$.

In a vector space $E$, a *polytope* is, by definition, the convex hull of a finite set $A := \{x_1, x_2, \ldots, x_n\}$ of points in $E$. Note, then, that a polytope $P$ is a subset of a finite-dimensional subspace $F$ of $E$, and that $P$ has a finite number of extreme points. Indeed, if $S$ is the convex and compact set $\{(\lambda_i)_{i=1}^n \in \mathbb{R}^n : \lambda_i \geq 0,\ i = 1, 2, \ldots, n,\ \sum_{i=1}^n \lambda_i = 1\}$, then $P$ is the image of the linear mapping

$$T : S \to F$$

given by

$$T(\lambda_1, \lambda_2, \ldots, \lambda_n) = \sum_{i=1}^{n} \lambda_i x_i, \quad \text{for } (\lambda_i)_{i=1}^n \in S,$$

and, if $F$ is endowed with the unique Hausdorff vector topology on it, the mapping $T$ is continuous. This shows that $P$ is a—convex—compact subset of $F$. Since $P = \text{conv}(A)$, by using Milman's theorem we get that $\text{Ext}(P) \subset \overline{A}$ $(= A)$, hence $\text{Ext}(P)$ is finite.

On the other hand, every compact convex subset $P$ of a locally convex space $E$ having a finite number of extreme points is a polytope, since the Krein–Milman theorem ensures that $P = \overline{\text{conv}}(\text{Ext}(P))$. By the argument above, the set $\text{conv}(\text{Ext}(P))$ is already a—convex—compact set, hence $P = \text{conv}(\text{Ext}(P))$, and $P$ is a polytope.

A panoply of another distinguished points of $B_X$ (from a geometrical point of view), and closely related to the notion of extreme point, have been introduced in the literature. The most significant are listed below. We need the following two definitions: A subset of $B_X$ of the form $\{x \in B_X : x^*(x) > 1 - \alpha\}$, (of the form $\{x \in B_X : x^*(x) \geq 1 - \alpha\}$) where $x^* \in S_{X^*}$ and $\alpha > 0$, is said to be a *w-slice of* $B_X$ (a *closed* $w$-slice of $B_X$, respectively). A subset of $B_{X^*}$ of the form

$\{x^* \in B_{X^*} : x^*(x) > 1 - \alpha\}$ (of the form $\{x^* \in B_{X^*} : x^*(x) \geq 1 - \alpha\}$), where $x \in S_X$ and $\alpha > 0$, is said to be a *$w^*$-slice of $B_{X^*}$* (a *closed* $w^*$-slice of $B_{X^*}$, respectively).

A point $x_0$ in $S_X$ is said to be:

- an *exposed point* of $B_X$ whenever there exists $x^* \in S_{X^*}$ such that

$$\{x \in B_X : x^*(x) = 1\} = \{x_0\};$$

- a *strongly exposed point* of $B_X$ whenever $x_0$ is exposed (by, say, $x^* \in S_{X^*}$), and $\operatorname{diam}\{x \in B_X : x^*(x) > 1 - \varepsilon\} \to 0$ as $\varepsilon \downarrow 0$;
- a *denting point* of $B_X$ if for every $\varepsilon > 0$ there exists a $w$-slice $S$ of $B_X$ such that $x_0 \in S$ and $\operatorname{diam}(S) < \varepsilon$;
- a (($w$)-) *strongly extreme point* of $B_X$ if given two sequences $\{y_n\}$ and $\{z_n\}$ in $B_X$ such that $(y_n + z_n) \to 2x_0$, then $y_n \to x_0$—respectively, $y_n \xrightarrow{w} x_0$;
- a *point of continuity* of $S_X$ if weak and norm topologies coincide at $x_0$.

It is easy to prove the following chain of implications: strongly exposed $\Rightarrow$ denting $\Rightarrow$ strongly extreme $\Rightarrow$ $w$–strongly extreme $\Rightarrow$ extreme. It is well known that an extreme point which is a point of continuity is also a denting point (and conversely) [32].

It is also easy to prove that if $X$ is locally uniformly rotund, then every point in $S_X$ is strongly exposed. A Banach space $X$ has the Radon–Nikodým property (RNP for short) if, and only if, every nonempty closed convex subset of $X$ is the closed convex hull of the set of its strongly exposed points; see [7, Theorem 11.3] and the references therein. A Banach space $X$ is said to be *midpoint locally uniformly rotund* (MLUR for short) if every point in $S_X$ is strongly extreme. For a good account of the literature concerning MLUR Banach spaces see, e.g., [31] and references therein.

## 9.2 Preserved and Unpreserved Extreme Points

Any Banach space $X$ can be canonically embedded in its bidual space $X^{**}$, something that we shall implicitly assume without warning. The closed unit ball $B_X$ of $X$ is just $B_{X^{**}} \cap X$, in the same way that $S_X = S_{X^{**}} \cap X$. Since $B_{X^{**}}$ is $w^*$-compact, the Krein–Milman theorem readily implies that $B_{X^{**}}$ *has* extreme points (in fact, that $B_{X^{**}} = \overline{\operatorname{conv}}^{w^*}(\operatorname{Ext}(B_{X^{**}}))$), in contrast with the situation in $B_X$; it may happen that $\operatorname{Ext}(B_X) = \emptyset$, as in the case of the space $(c_0, \|\cdot\|_\infty)$.

A point $x_0^{**} \in \operatorname{Ext}(B_{X^{**}}) \cap X$ is clearly in $\operatorname{Ext}(B_X)$. Points $x_0 \in \operatorname{Ext}(B_X)$ for which the reverse implication holds—i.e., such that they are also extreme points of $B_{X^{**}}$—are called *preserved extreme points*. Extreme points of $B_X$ that are not extreme in $B_{X^{**}}$ are called, naturally, *unpreserved*. Those are the basic objects we are presently interested in, so we repeat the definition below.

**Definition 9.1** Let $(X, \|\cdot\|)$ be a Banach space. A point $x \in S_X$ is said to be a *preserved extreme point of* $B_X$ whenever it is an extreme point of $B_{X^{**}}$ (in other words, whenever it is simultaneously an extreme point of $B_X$ and an extreme point of $B_{X^{**}}$).

An extreme point $x$ of $B_X$ that is not an extreme point of $B_{X^{**}}$ is said to be an *unpreserved extreme point of* $B_X$.

It is worth to mention that $\text{Ext}(B_X) = \text{Ext}(B_{X^{**}})$ *if, and only if, the space* $X$ *is reflexive*. One direction in this double implication is a triviality. The other is a consequence of James' deep characterization of reflexivity. Indeed, assume that $\text{Ext}(B_X) = \text{Ext}(B_{X^{**}})$. Fix $x^* \in S_{X^*}$. The $w^*$-compactness of $B_{X^{**}}$ implies that $x^*$ attains its supremum on $B_{X^{**}}$. Now, the nonempty set $\{x^{**} \in B_{X^{**}} : x^{**}(x^*) = 1\}$ is a convex $w^*$-compact subset of $X^{**}$, hence, by the Krein–Milman theorem, it has extreme points. It is easy to see that each of those is an extreme point of $B_{X^{**}}$. By the assumption, it belongs to $X$, and we conclude that every $x^* \in S_{X^*}$ attains its supremum on $B_{X^{**}}$ (and this is also the supremum of $x^*$ on $B_X$) at a point in $B_X$. James' theorem concludes then that $X$ is reflexive. In other words, if $X$ is a nonreflexive Banach space, there are surely extreme points of $B_{X^{**}}$ that are not in $X$. Note, too, that if $(X^{**}, \|\cdot\|)$ is *strictly convex* (R for short), then every point in $\text{Ext}(B_X)(= S_X)$ is preserved; indeed, every point in $S_{X^{**}}$ is already an extreme point, being this in fact the definition of strict convexity.

Let us concentrate on $\text{Ext}(B_X)$. Two mutually exclusive classes of extreme points appear, since an extreme point is either preserved or unpreserved.

The following result gives two equivalent formulations for a point in $B_X$ to be preserved extreme. It combines a characterization due to H. P. Rosenthal—the equivalence (i) $\Leftrightarrow$ (ii)—and a result implicitly in [32]—the equivalence (i) $\Leftrightarrow$ (iii). The proof of some of the equivalences is ours.

**Proposition 9.1** *Let $X$ be a Banach space, and let $x_0 \in S_X$. Then, the following are equivalent:*

(i) *$x_0$ is a preserved extreme point of $B_X$.*
(ii) *The family of $w$-slices containing $x_0$ forms a base of neighborhoods of $x_0$ in $(B_X, w)$.*
(iii) *$x_0$ is $w$-strongly extreme.*

*Proof* (i) $\Rightarrow$ (ii) If $x_0$ is a preserved extreme point of $B_X$, then $x_0$ is an extreme point of $B_{X^{**}}$. Let $U$ be a $w$-neighborhood of $x_0$ relative to $B_X$. We can find then $\{x_1^*, x_2^*, \dots, x_n^*\} \subset S_{X^*}$ and $\varepsilon > 0$ such that $W := \{x \in B_X : |x_i^*(x - x_0)| \le \varepsilon,\ i = 1, 2, \dots, n\} \subset U$. Then $W^{**} := \{x^{**} \in B_{X^{**}} : |x_i^*(x^{**} - x_0)| \le \varepsilon\}$ is a $w^*$-neighborhood of $x_0$ in $B_{X^{**}}$. Apply now Choquet's Lemma (see, e.g., [7, Lemma 3.69]) to conclude that there exists a $w^*$-slice $S^{**}$ of $B_{X^{**}}$ containing $x_0$ such that $S^{**} \subset W^{**}$. This implies that $S^{**} \cap X \subset W^{**} \cap X\ (= W)$. Since $S^{**} \cap X$ is a $w$-slice of $B_X$ that contains $x_0$ the conclusion follows.

(ii) $\Rightarrow$ (iii) Assume that $x_0$ has the property that the family of $w$-slices of $B_X$ containing $x_0$ is a base of $w$-neighborhoods of $x_0$ in $B_X$. Let $\{x_n\}_{n=1}^\infty$ and $\{y_n\}_{n=1}^\infty$ be two sequences in $B_X$ such that $(1/2)(x_n + y_n) \to x_0$. The assumed property says that given an arbitrary $w$-neighborhood $U$ of $x_0$ we can find $x^* \in S_{X^*}$ and $\varepsilon > 0$ such that $S(x^*, \varepsilon) := \{x \in B_X : x^*(x) > 1 - \varepsilon\} \subset U$. Thus, there exists $n_0 \in \mathbb{N}$ such that $(1/2)(x_n + y_n) \in S(x^*, \varepsilon)$ for every $n \geq n_0$. This implies that, for $n \geq n_0$, either $x_n$ or $y_n$—or both—belongs to $S(x^*, \varepsilon)$, hence to $U$. This also implies that $x_0 \in \mathrm{Ext}(B_X)$. Indeed, if $x_0 = (1/2)(x + y)$, where $x, y \in B_X$ and $x \neq x_0$ (hence $y \neq x_0$), the existence of a neighborhood $U$ of $x_0$ such that $U$ misses both $x$ and $y$ leads to a contradiction.

We shall prove that the set $M := \{x_n : n = 0, 1, 2, \ldots\}$ is $w$-compact. Observe first that it is $w$-closed. Indeed, if a sequence $\{x_{n_k}\}_{k=1}^{\infty}$ in $M$ is $w$-convergent, and it converges to some $x \in X \setminus \{x_n,\ n = 0, 1, 2, \ldots\}$, then, due to the fact that $(1/2)(x_n + y_n) \to x_0$, the corresponding sequence $\{y_{n_k}\}_{k=1}^{\infty}$ is also $w$-convergent, say to some $y \in X$, and $y \neq x_0$. It follows that $x_0 = (1/2)(x + y)$, contradicting the fact that $x_0$ is an extreme point of $B_X$.

Let us prove now that $M$ is $w$-relatively countably compact. This, by the Eberlein-Šmulyan Theorem, will conclude that $M$ is $w$-relatively compact, hence $w$-compact by the previous observation, and so $w$-sequentially compact. Fix an arbitrary sequence in $M$. We claim that the sequence has a $w$-adherent point in $X$. This is easy if the sequence has finite rank; otherwise, it has a subsequence that, in turn, is a subsequence of $\{x_n\}_{n=1}^{\infty}$. Call this subsequence $\{x_{n_k}\}_{k=1}^{\infty}$. We have two possibilities:

(a) There exists a $w$-neighborhood $U$ of $x_0$ and some $k_0 \in \mathbb{N}$ such that $x_{k_n} \notin U$ for each $k \geq k_0$. Let $V$ be an arbitrary $w$-neighborhood of $x_0$. There exists $k_1 \geq k_0$ such that $(1/2)(x_n + y_n) \in V \cap U$ for all $k \geq k_1$. Since $x_{n_k} \notin V \cap U$ for all $k \geq k_1$, the argument above shows that $y_{n_k} \in V \cap U$ for all $k \geq k_1$. This proves that $y_{n_k} \overset{w}{\to} x_0$ as $k \to \infty$, hence $x_{n_k} \overset{w}{\to} x_0$ as $k \to \infty$, a contradiction.

(b) For every $w$-neighborhood $U$ of $x_0$ and for every $k_0 \in \mathbb{N}$, there exists $k \geq k_0$ such that $x_{n_k \in U}$. This shows that $x_0$ is a $w$-adherent point of $\{x_{n_k}\}_{k=1}^{\infty}$, as we claimed.

Since $M$ is $w$-sequentially compact, any subsequence $\{x_{n_k}\}_{k=1}^{\infty}$ of the sequence $\{x_n\}_{n=1}^{\infty}$ has a $w$-convergent subsequence, and the corresponding sequence $\{y_{n_k}\}$ is $w$-convergent, too. If the $w$-limit is not $x_0$, again we reach a contradiction with the fact that $x_0$ is an extreme point. All together, $x_n \overset{w}{\longrightarrow} x_0$ as $n \to \infty$, as we wanted to show.

(iii) $\Rightarrow$ (i) Let $x_0$ be a non-preserved extreme point of $B_X$. We can find then $y^{**}$ and $z^{**}$ in $B_{X^{**}}$ such that $2x_0 = y^{**} + z^{**}$ and $y^{**} \neq z^{**}$. Find $x^* \in S_{X^*}$ and $\alpha < \beta$ in $\mathbb{R}$ such that $x^*(z^{**}) < \alpha < \beta < x^*(y^{**})$. We can find two nets, $\{z_i : i \in I,\ \leq\}$ and $\{y_i : i \in I,\ \leq\}$ in $B_X$ such that $z_i \overset{w^*}{\to} z^{**}$ and $y_i \overset{w^*}{\to} y^{**}$, and

$$x^*(z_i) < \alpha < \beta < x^*(y_i) \quad \text{for all } i \in I. \tag{9.1}$$

Observe that $(z_i + y_i) \overset{w^*}{\to} 2x_0$ and, since all elements $z_i + y_i$ and $2x_0$ are in $X$ for all $i \in I$, we get $(z_i + y_i) \overset{w}{\to} 2x_0$. This shows, by Mazur's theorem, that $2x_0 \in \overline{\mathrm{conv}}^{\|\cdot\|}\{z_i + y_i : i \in I\}$. Assume that $x_0$ is $w$-strongly extreme. We get, in particular, that given $\varepsilon > 0$ there exists $\delta > 0$ such that

$$\|(z+y)-2x_0\| < 2\delta \quad \text{implies} \quad |x^*(z-y)| < \varepsilon. \tag{9.2}$$

Put $\varepsilon := \beta - \alpha$ and find $\delta$ accordingly. Since $2x_0 \in \overline{\text{conv}}^{\|\cdot\|}\{z_i + y_i : i \in I\}$, we can find $n \in \mathbb{N}$ and $\{\lambda_1, \lambda_2, \ldots, \lambda_n\}$ in $[0, 1]$ such that $\sum_{j=1}^{n} \lambda_j = 1$ and

$$\left\| \sum_{j=1}^{n} \lambda_j (z_{i_j} + y_{i_j}) - 2x_0 \right\| = \left\| \sum_{j=1}^{n} \lambda_j z_{i_j} + \sum_{j=1}^{n} \lambda_j y_{i_j} - 2x_0 \right\| < 2\delta.$$

Letting $z := \sum_{j=1}^{n} \lambda_j z_{i_j}$ and $y := \sum_{j=1}^{n} \lambda_j y_{i_j}$, we obtain from (9.1) that $x^*(z) < \alpha < \beta < x^*(y)$, and we reach a contradiction with (9.2). This shows that $x_0$ is not $w$-strongly extreme. □

## 9.3 A Brief Introduction to Polyhedrality

### *9.3.1 A Sufficient Condition for Polyhedrality and Some Positive Examples*

In [29], V. Klee extended the concept of a polytope for the case of the closed unit ball $B_X$ of an infinite-dimensional Banach space $X$—the intersection of $B_X$ with any finite-dimensional subspace of $X$ is a (finite-dimensional) polytope, and the extension to any closed convex and bounded subset of a Banach space was considered by M.I. Kadets in 1981, see [17]. Several other definitions of an infinite-dimensional polytope appear in the literature. For a comprehensive account see [6, 17]. In this survey we shall stick to Klee's definition given above, and, accordingly, to the definition of polyhedrality that follows.

**Definition 9.2** A Banach space $(X, \|\cdot\|)$ is said to be *polyhedral*—or that its norm is *polyhedral*—whenever its closed unit ball is a polytope, i.e., the closed unit ball of any of its finite-dimensional subspaces is a (finite-dimensional) polytope.

In his 1959 paper, V. Klee gave the following result which implies in particular that in the previous definition *two-dimensional* subspaces can be used instead of the more general *finite-dimensional* subspaces.

**Theorem 9.1** ([28, Theorem 4.7]) *Let $C \subset \mathbb{R}^n$ be a convex body with the origin as interior point. Then it is a polytope if and only if $C \cap V$ is a polytope whenever $V \subset \mathbb{R}^n$ is 2-dimensional subspace.*

Polyhedrality is an isometric property. Indeed, if $K$ is an infinite compact topological space, the Banach space $X := (C(K), \|\cdot\|_\infty)$ is not polyhedral; see Corollary 9.2. However, if $K$ is countable, $X$ is a Banach space with a countable boundary, namely $B := \{\pm\delta_k : k \in K\}$, and it follows from Theorem 9.3 that the space $X$ has an equivalent polyhedral norm.

Let $(X, \|\cdot\|)$ be a Banach space. A subset $N$ of $B_{X^*}$ is said to be *norming*—for $X$—whenever $\|x\|_N := \sup\{|x^*(x)|: x^* \in N, \ \|x^*\| \leq 1\}$, $x \in X$, is an equivalent norm on $X$, and $N$ is called *1-norming* if $\|\cdot\|_N = \|\cdot\|$. By an abuse of notation, a *subspace* $Y$ of $X^*$ is said to be *norming* if the set $Y \cap B_{X^*}$ is norming. Note that every norming subspace $N$ of $X^*$ is $w^*$-dense in $X^*$, due to the fact that $x^*(x) = 0$ for every $x^* \in N$ implies $x = 0$.

Examples of *closed norming subspaces* of $X^*$ are the kernels of elements in $X^{**} \setminus X$ (for nonreflexive spaces $X$, to be sure), something that can be proved by using the so-called *parallel hyperplane lemma*, see, e.g., [7, Exercise 2.13].

Observe that $\overline{\text{conv}}^{w^*}(N) = B_{X^*}$ for any 1-norming subset $N$ of $B_{X^*}$. Indeed, otherwise there exists, by the separation theorem, an element $x \in S_X$ such that $x^*(x) \leq \alpha < 1$ for all $x^* \in N$, something impossible if $N$ is a 1-norming set. In particular, norming subsets of $B_{X^*}$ are $w^*$-linearly dense in $X^*$.

Examples of 1-norming sets are $S_{X^*}$ and, more generally, any $w^*$-dense subset of $S_{X^*}$. A particular example of this situation is the set $N := F^{-1}(0) \cap B_{X^*}$, where $F$ is any nonzero element in $X^{**}$ such that $\|F + x\| \geq \|x\|$ for all $x \in X$,—see, e.g., [7, Exercise 5.4]. Indeed, we know that, in this case, $N$ is 1-norming. Moreover, by the fact noted above that $N$ is $w^*$-dense in $B_{X^*}$, given $x^* \in S_{X^*}$ there exists a net $\{x_i^*: i \in I, \ \leq\}$ in $N$ that $w^*$-converges to $x^*$. The $w^*$-lower semicontinuity of the norm in $X^*$ shows that, by passing to a subnet if necessary, we may assume that $\{\|x_i^*\|: i \in I, \ \leq\} \to 1$. The net $\{x_i^*/\|x_i^*\|: i \in I, \ \leq\}$ is in $N \cap S_{X^*}$, and it $w^*$-converges to $x^*$. This proves that $N \cap S_{X^*}$ is $w^*$-dense in $S_{X^*}$.

The fact that, in general, $\overline{\text{conv}}^{w^*}(N) = B_{X^*}$ for a 1-norming set $N$ implies, by Milman's converse to the Krein–Milman Theorem, that $\text{Ext}(B_{X^*}) \subset \overline{N}^{w^*}$. The set $\text{Ext}(B_{X^*})$ of all extreme points of $B_{X^*}$ is another example of a 1-norming subset of $B_{X^*}$—a consequence of the Krein–Milman Theorem.

A *boundary* of $X$ is a subset $B$ of $S_{X^*}$ such that every $x \in S_X$ attains its supremum on $B_X$ at some point of $B$. An example of a boundary of $X$ is the set $\text{Ext}(B_{X^*})$, again a consequence of the Krein–Milman Theorem. Indeed, any $x \in S_X$ attains its supremum on $B_{X^*}$ at some point of $S_{X^*}$, and the "face" defined by $x$ on $B_{X^*}$—i.e., the nonempty set $\{x^* \in S_{X^*}: x^*(x) = 1\}$—is a compact convex subset of $(B_{X^*}, w^*)$, hence having extreme points that are, clearly, extreme points of $B_{X^*}$.

Trivially, every boundary $B$ of $X^*$ is 1-norming. The converse is false (the open unit ball of $X^*$ is a 1-norming subset of $B_{X^*}$ that is not a boundary), and this can be even the case for 1-norming subsets of $S_{X^*}$. For example, if $(X, \|\cdot\|) := (C[0,1], \|\cdot\|_\infty)$, and $\delta_x$ denotes the Dirac delta function at $x \in [0,1]$, the subset $\{\delta_q : q \in \mathbb{Q}\}$ of $S_{X^*}$ is clearly a 1-norming subset of $B_{X^*}$. However, it is not a boundary of $X$, since there are real-valued continuous functions on $[0,1]$ that attain their supremum on points in $[0,1] \setminus \mathbb{Q}$.

The following definition was introduced by A. Gleit and R. McGuigan in [18]. See also [11] and [13].

**Definition 9.3** Let $X$ be a Banach space. A subset $B$ of $B_{X^*}$ is said to have *property* $(*)$ if given any $w^*$-limit point $x_0^*$ of $B$ (i.e., any $w^*$-neighborhood of $x_0^*$ contains infinitely many points of $B$), we have $x_0^*(x) < 1$ whenever $x \in S_X$. If a Banach space $(X, \|\cdot\|)$ has a 1-norming subset of $B_{X^*}$ having $(*)$, we say that *the norm has* $(*)$.

For an example of a Banach space having $(*)$ see Corollary 9.1. Observe that *every* 1*-norming subset* $N$ *of* $B_{X^*}$ *having property* $(*)$ *must be a boundary*. Indeed, let $x \in S_X$. Find a sequence $\{x_n^*\}_{n=1}^\infty$ in $N$ such that $x_n^*(x) \to 1$. The sequence $\{x_n^*\}_{n=1}^\infty$ has a $w^*$-cluster point $x_0^*$. Observe that $x_0^*(x) = 1$. Assume that $x_0^*$ is a limit point of the set $\{x_n^* : n \in \mathbb{N}\}$. Then $x_0^*(x) < 1$, and we reach a contradiction. It follows then that there is a $w^*$-neighborhood $U$ of $x_0^*$ such that $U \cap N$ is finite. This surely implies $x_0^* \in N$, and so $N$ is a boundary, as claimed.

For separable Banach spaces $X$, there is a condition on the dual space $X^*$ that implies $(*)$, and that has been used in the study of quotients of polyhedral spaces [12]. If $x^* \in X^*$, put $J_{X^{**}}(x^*) := \{x^{**} \in S_{X^{**}} : x^{**}(x^*) = \|x^*\|\}$, and $J_X(x^*) := J_{X^{**}}(x^*) \cap X$. Denote by $NA(X)$ the set of all the norm-attaining functionals in $X^*$.

**Proposition 9.2** ([12]) *Let* $X$ *be a separable Banach space such that for all* $x^* \in S_{X^*} \cap NA(X)$ *there exists a* $w^*$*-open neighborhood* $V(x^*)$ *of* $x^*$ *such that* $y^* \in V(x^*) \cap S_{X^*}$ *implies* $J_X(y^*) \subset J_X(x^*)$. *Then there exists a boundary* $B$ *of* $X$ *such that no* $w^*$*-limit point of* $B$ *belongs to* $S_{X^*} \cap NA(X)$.

For a proof see, e.g., [20, Proposition III.9], where it is mentioned that the converse of Proposition 9.2 holds true [12]. Lemma III.7 in [20] states that every subspace $X$ of $c_0$ satisfies a strengthening of the hypothesis of Proposition 9.2: If $x^* \in S_{X^*} \cap NA(X)$, there exists a $w^*$-neighborhood $V(x^*)$ of $x^*$ such that if $y^* \in S_{X^*} \cap V$ then $J_{X^{**}}(y^*) \subset J_{X^{**}}(x^*)$.

Observe that $B$ in Proposition 9.2 has property $(*)$. Indeed, if $x_0^* \in S_{X^*}$ is a $w^*$-limit point of $B$, it may belong to the open unit ball of $X^*$—and then $x_0^*(x) < 1$ for all $x \in S_X$, or to the unit sphere. In this case, according to Proposition 9.2 it does not belong to $NA(X)$, hence $x_0^*(x) < 1$ for every $x \in S_X$.

The following is a well-known sufficient condition for polyhedrality.

**Lemma 9.1** (Gleit-McGuigan, [18]) *Let* $(X, \|\cdot\|)$ *be a Banach space, and let* $N$ *be a subset of* $S_{X^*}$ *that is* 1*-norming and has property* $(*)$. *Then* $X$ *is polyhedral.*

*Proof* By a previous observation, $N$ is a boundary of $X$. Fix a finite-dimensional subspace $F \subset X$, and let $q : X^* \to F^*$ be the restriction mapping. The set $N_F := q(\{x^* \in N : \exists x \in S_F,\ x^*(x) = 1\})$ is a boundary of $F$. Compactness of $S_F$ implies that $w^*$-accumulation points of $N_F$ are norm-attaining functionals. Hence $N_F$ needs to be finite which, in turn, implies that $B_F$ is a polytope. Since $F$ is arbitrary, the space $X$ is polyhedral. □

Condition $(*)$ is not necessary for polyhedrality. Indeed, V. Fonf and L. Veselý survey in [17] eight different definitions of polihedrality—introducing accordingly eight pairwise different classes of Banach spaces—and being the most general the one we use here. In their survey the definition "Ext($B_{X^*}$) has property $(*)$" is called "definition (IV)". See also R. Durier and P. L. Papini work in [6]. By the way, condition $(*)$ is necessary for polyhedrality if $X^*$ is an $L$-space; see again [18].

**Corollary 9.1** *The Banach space $(c_0, \|\cdot\|_\infty)$ is polyhedral.*

*Proof* The norm $\|\cdot\|_\infty$ on $c_0$ has property $(*)$. Indeed, the set $\mathrm{Ext}(B_{\ell_1})$ $(=\{\pm e_n^*\colon n \in \mathbb{N}\})$ is a boundary of $(c_0, \|\cdot\|_\infty)$ and it has zero as its unique $w^*$-accumulation point. □

*Remark 9.1* Note that the Banach space $(c_0, \|\cdot\|_\infty)$ *has no proper* 1*-norming closed subspace*. Indeed, let $N$ be a 1-norming closed subspace of $\ell_1$, for $c_0$. Fix $j \in \mathbb{N}$. Since $N$ is 1-norming, there exists a sequence $\{x_n^*\}_{n=1}^\infty$ in $B_N$ such that $x_n^*(j) = \langle e_j, x_n^*\rangle \to \|e_j\|_\infty = 1$. For a fixed $\varepsilon \in (0,1)$, chose $n \in \mathbb{N}$ such that $0 \le 1 - x_n^*(j) < \varepsilon/2$. Then, if we denote by $e_n^*$ the $n$-th unit vector of the canonical basis of $\ell_1$,

$$\begin{aligned}\|e_j^* - x_n^*\|_1 &= \sum_{k\neq j} |x_n^*(k)| + (1 - x_n^*(j))\\ &= (\|x_n^*\|_1 - x_n^*(j)) + 1 - x_n^*(j) \le 1 - x_n^*(j) + 1 - x_n^*(j) < \varepsilon,\end{aligned}$$

which implies that $\mathrm{dist}_1(e_j^*, N) = 0$, where $\mathrm{dist}_1$ denotes the metric defined by $\|\cdot\|_1$ in $\ell_1$. Since $N$ is closed, we get $e_j^* \in N$. This happens for all $j \in \mathbb{N}$. The $\|\cdot\|_1$-linear denseness of $\{e_j^*\colon j \in \mathbb{N}\}$ in $\ell_1$ implies that $N = \ell_1$.

An alternative proof uses the fact that every proper 1-norming closed subspace of $X^*$ is contained in a 1-norming proper closed hyperplane $H$ of $X^*$. In $\ell_1$, $H$ is precisely $\ker x_0^{**}$ for some $x_0^{**} \in \ell_\infty \setminus c_0$, and it can be proved easily that there is always $x \in c_0$ such that $\|x_0^{**} - x\|_\infty < \|x\|_\infty$. We may use then the aforesaid characterization of elements in $X^{**} \setminus X$ whose kernel is a 1-norming—closed—hyperplane of $X^*$; see, e.g., [7, Exercise 5.4].

We give here another proof of the polyhedral character of $(c_0, \|\cdot\|_\infty)$; see [20]. Let $E$ be a finite-dimensional subspace of $c_0$. For each $x \in S_E$ there exists $\varepsilon(x) > 0$ and a finite subset $F(x)$ of $X^*$ such that $\|y\|_\infty = \sup\{|\langle y, x^*\rangle|\colon x^* \in F(x)\}$ for all $y \in B_E(x; \varepsilon(x))$. Since $S_E$ is compact, there are $x_1, \ldots, x_n$ in $S_E$ such that

$$S_E \subset \bigcup_{i=1}^n B_E(x_i, \varepsilon(x_i)).$$

Put $F := \bigcup_{i=1}^n F(x_i)$. Then $F$ is a finite subset of $X^*$ such that

$$\|x\|_\infty = \sup\{|\langle x, x^*\rangle|\colon x^* \in F\}$$

for all $x \in E$, hence $E$ is isometric to a subspace of $(\mathbb{R}^{|F|}, \|\cdot\|_\infty)$, a polyhedral space.

The proof in [20], recorded above, relies on a particular property of the norm $\|\cdot\|_\infty$ of $c_0$ that, when isolated, gives a nowadays well-studied concept: The norm $\|\cdot\|$ of a Banach space is said *to depend locally on a finite number of coordinates* if given any $x_0 \in S_X$ there exists $\delta > 0$, continuous linear functionals $\{\psi_1, \psi_2, \ldots, \psi_n\} \subset X^*$, and a continuous function $f\colon \mathbb{R}^n \to \mathbb{R}$ such that, for every $x \in B(x_0; \delta)$ we have

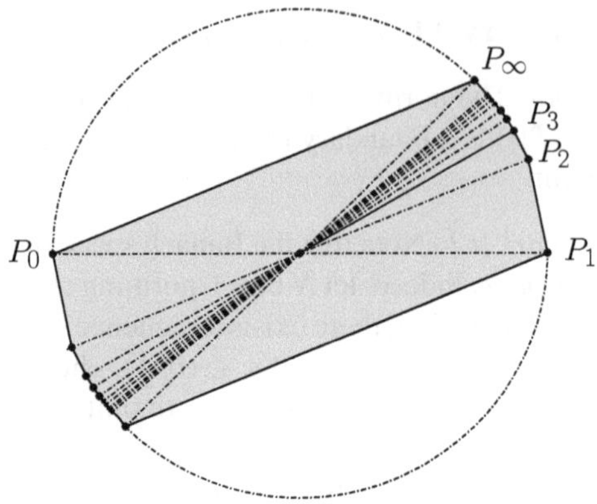


**Fig. 9.1** The construction to prove that $c$ is not polyhedral

$\|x\| = f(\psi_1(x), \psi_1(x), \ldots, \psi_1(x))$. The norm $\|\cdot\|_\infty$ of $c_0$ depends locally on a finite number of coordinates (see the argument above): if $E$ is a finite-dimensional subspace of $c_0$, and $x \in S_E$, then the function $f$ is the maximum function, and $\psi_i(x)$ is the $i$-th coordinate for $i = 1, 2, \ldots, |F(x)|$, where $F(x)$ is the finite set associated to $x$ as above.

We record here another result on polyhedrality; see also [7, Theorem 10.9].

**Theorem 9.2** ([10, 36]) *Banach spaces with a countable James boundary are $c_0$-saturated, i.e., each closed subspace contains an isomorphic copy of $c_0$.*

## *9.3.2 Some Typical Examples of Nonpolyhedral Spaces*

1. The space $(c, \|\cdot\|_\infty)$ is not polyhedral. The following argument was kindly provided by L. Veselý (personal communication): Consider the points $P_n := \exp\{i(1-1/n)\pi/4\}$ in the plane, for all $n \in \mathbb{N}$; see Fig. 9.1. Let $a_n x + b_n y = 1$ be the equation of the line through $P_n$ and $P_{n+1}$ for all $n \in \mathbb{N}$, and $a_0 x + b_0 y = 1$ the equation of the line through $P_\infty := \exp(\pi/4)$ and $P_0 := (-1, 0)$. Then $a := (a_n)_{n\geq 0}$ and $b := (b_n)_{n\geq 0}$ are elements in $c$, and their linear span $L$ is isometric to a plane equipped with the norm whose closed unit ball is the set $\overline{\text{conv}}\{\pm P_1, \pm P_2, \ldots, \pm P_\infty\}$.
2. The space $(C(K), \|\cdot\|_\infty)$, for $K$ an infinite compact topological space, is not polyhedral; see Corollary 9.2. This is a well known result. A direct argument based on the technique used in the proof for the space $(c, \|\cdot\|_\infty)$—see this section (1) above—shall be provided later. Both proofs we present here, using again the non-polyhedrality character of $(c, \|\cdot\|_\infty)$, needs first the following preliminary result.

**Lemma 9.2** *If $K$ is a infinite compact topological space, then there is an infinite sequence of nonempty pairwise disjoint open subsets of $K$.*

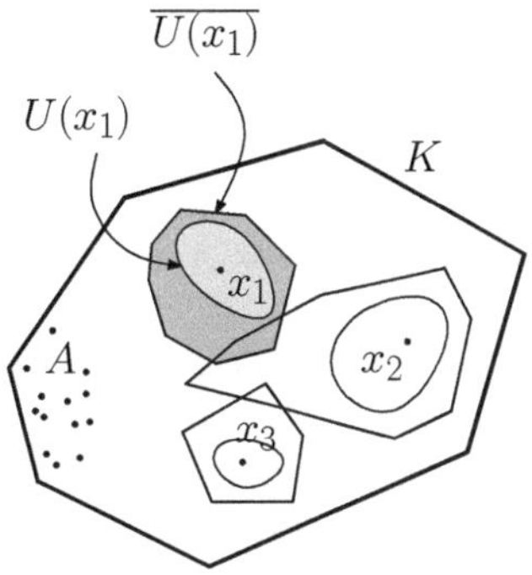

**Fig. 9.2** The first steps of the construction in Lemma 9.2

*Proof* (See Fig. 9.2) Let $A$ be the set of all isolated points in $K$. Since for each $x \in A$ the set $\{x\}$ is open, $A$ is an open subset of $K$. Assume first that $A$ is infinite. Choose a countable subset $\{x_n : n \in \mathbb{N}\}$ of $A$. Then $\{\{x_n\} : n \in \mathbb{N}\}$ is an infinite sequence of pairwise disjoint open subsets of $K$. Assume that, on the contrary, $A$ is finite. Put $K_1 := K \setminus A$, an infinite compact subset of $K$ consisting only of accumulation points. All the concepts in the rest of the argument refer to the compact space $K_1$. Take $x_1$ and $x_2$ in $K_1$, $x_1 \neq x_2$ and find an open neighborhood $U(x_1)$ of $x_1$ such that $x_2 \notin \overline{U(x_1)}$. Since $K_1 \setminus \overline{U(x_1)}$ is open and nonempty, it contains infinitely many points—recall that $x_2$ is an accumulation point. Choose $x_3 \in K_1 \setminus \overline{U(x_1)}$ and an open neighborhood $U(x_2)$ of $x_2$ in $K_1 \setminus \overline{U(x_1)}$ such that $x_3 \notin \overline{U(x_2)}$. Again, $K_1 \setminus (\overline{U(x_1)} \cup \overline{U(x_2)})$ is an open neighborhood of $x_3$, hence it contains infinitely many points of $K_1$. Choose $x_4 \in K_1 \setminus (\overline{U(x_1)} \cup \overline{U(x_2)})$, $x_4 \neq x_3$ and an open neighborhood $U(x_3)$ of $x_3$ in $K_1 \setminus (\overline{U(x_1)} \cup \overline{U(x_2)})$ such that $x_4 \notin \overline{U(x_3)}$. Continue in this way. We obtain a pairwise disjoint sequence $\{U(x_n) : n \in \mathbb{N}\}$ of open subsets of $K_1$. To finalize the proof, it is enough to observe that, since $A$ is finite, each $U(x_n)$ may be chosen to be an open subset of $K$ that is contained in $K_1$. □

**Proposition 9.3** *The Banach space* $(C(K), \|\cdot\|_\infty)$ *contains an isometric copy of* $(c, \|\cdot\|_\infty)$, *for any infinite compact topological space* $K$.

*Proof* Lemma 9.2 ensures the existence of an infinite sequence of nonempty pairwise disjoint open subsets of $K$, $\{U_n\}_{n\in\mathbb{N}}$. For every $n \in \mathbb{N}$, take an arbitrary $x_n \in U_n$ and, by using Urysohn's Lemma, a continuous function $f_n \in C(K)$ whose support is included in $U_n$ but $f_n(x_n) = 1$ and its range is $[0, 1]$. Denote by $\mathbb{1}$ the characteristic function of $K$ on $K$. Define $T : c \to C(K)$ by $T(a_n) = (\lim_n a_n)\mathbb{1} + \sum_{n\in\mathbb{N}}(a_n - \lim_n a_n) f_n$ for all $(a_n) \in c$.

First of all, let us show that $T$ is well defined. Take $(a_n) \in c$ and set $f := T(a_n)$, $a := \lim_n a_n$. We shall prove that $f$ is continuous. Consider the open set $A := \cup_n U_n$ and observe that $f(t) = a$ for all $t \notin A$. Therefore, $f$ is continuous on $K \setminus \overline{A}$. Now fix $t \in \overline{A}$. In case there exists $n \in \mathbb{N}$ such that $t \in U_n$, then $f_{|U_n}$ coincides with $a\mathbb{1}_{|U_n} + (a_n - a) f_{n|U_n}$, which implies that $f$ is continuous at $t$. Otherwise, $t \in \partial A$, $f(t) = a$. Assume that there exists a net $\{t_\alpha\}_\alpha$ converging to $t$ such that $\{f(t_\alpha)\}_\alpha$ does not converge to $a$. Then we can assume, taking a subnet if necessary, that $\{t_\alpha\}_\alpha \subset A$.

For each $\alpha$ denote by $n(\alpha)$ the unique natural number such that $t_\alpha \in U_{n(\alpha)}$. Then

$$f(t_\alpha) = a + (a_{n(\alpha)} - a) f_{n(\alpha)}(t_\alpha)$$

and so, since $t \notin A$, the net $\{n(\alpha)\}_\alpha$ repeats terms just finitely, which implies that

$$|f(t_\alpha) - a| \leq |a_{n(\alpha)} - a||f_{n(\alpha)}(t_\alpha)| \leq |a_{n(\alpha)} - a| \to 0.$$

This yields a contradiction that finishes to prove that $f$ is continuous.

Finally, to show that $T$ is an isometry it is enough to observe that given $(a_n) \in c$ the range of $T(a_n)$, as a function on $K$, coincides with $\overline{\text{conv}}\{a_n : n \in \mathbb{N}\}$, which implies that $\|T(a_n)\|_{C(K)} = \|(a_n)\|_c$. □

**Corollary 9.2** *If $K$ is an infinite compact topological space, then $(C(K), \|\cdot\|_\infty)$ is not polyhedral.*

*Proof* According to Proposition 9.3, the space $(C(K), \|\cdot\|_\infty)$ has a subspace isometrically isomorphic to $(c, \|\cdot\|_\infty)$. We have shown in this section (1) that $(c, \|\cdot\|_\infty)$ is not polyhedral, hence the space $(C(K), \|\cdot\|_\infty)$ is not polyhedral either. □

We mentioned above that it is possible to give a direct proof of the non-polyhedrality character of $(C(K), \|\cdot\|_\infty)$ based on ideas in this section (1). This goes as follows.

*Proof* Veselý's example in this section (1) above gives us two convergent sequences $(a_n)$ and $(b_n)$ with limits $a$ and $b$, respectively, and a non-polyhedral norm $\|\cdot\|_V$ on $\mathbb{R}^2$. We can construct two continuous functions as in the proof of Proposition 9.3, namely $f = a + \sum_n (a_n - a) f_n$ and $g = b + \sum_n (b_n - b) f_n$. Thus, a linear isometry can be defined from $(\mathbb{R}^2, \|\cdot\|_V)$ into $(C(K), \|\cdot\|_\infty)$ whose range is $\text{span}\{f, g\}$. This shows that $(C(K), \|\cdot\|_\infty)$ is not polyhedral. □

Theorem 9.3 below is the contribution of V. Fonf and P. Hájek. It will be used in the proof of our Theorem 9.9. Note that spaces with a countable James boundary $B$ are already separable and Asplund, due to the fact that the boundary $B$ is *strong*, i.e., $\overline{\text{conv}}^{\|\cdot\|}(B) = B_{X^*}$ (see [19]).

**Theorem 9.3** ([10, 23]) *For a Banach space $X$, the following are equivalent:*

(i) *$X$ has a countable James boundary.*
(ii) *$X$ has a James boundary that can be covered by a countable number of $\|\cdot\|$-compact subsets of $X^*$.*
(iii) *$X$ is separable and has an equivalent norm that depends locally on a finite number of coordinates.*
(iv) *$X$ is separable and has an equivalent norm that is $C^\infty$-smooth away from the origin and depends locally on a finite number of coordinates.*
(v) *$X$ is separable and isomorphic to a polyhedral Banach space.*

We would like to mention the recent work [15] on isomorphic polyhedral Banach spaces where some sufficient conditions for polyhedrality are provided. Related to this, see also [14] and [16].

## 9.4 Katznelson Example and Related

In his study of non-preservedness of extremality, and in order to guarantee that there are, after all, unpreserved extreme points in the unit sphere of some Banach spaces, P. Morris [33] refers to R. Phelps' paper [38]. Phelps says there (added in proof):

> K. de Leeuw has suggested and Y. Katznelson has proved that the answer to the question raised at he beginning of Sect. 4—i.e., whether there is any Banach space $X$ and an extreme point $x$ of $B_X$ such that $j(x)$ is not an extreme point of $B_{X^{**}}$, where $j\colon X \to X^{**}$ is the canonical embedding—is negative for the Banach space of all complex valued functions which are analytic for $|z| < 1$ and continuous for $|z| \le 1$, with the supremum norm.

Since we haven't found the original—apparently unpublished—example of Katznelson we provide here our approach to this.

We start by some standard definitions. Let $\mathbb{T} = \{z \in \mathbb{C}\colon |z| = 1\}$, $\mathbb{D} = \{z \in \mathbb{C}\colon |z| < 1\}$ and $\overline{\mathbb{D}} = \mathbb{D} \cup \mathbb{T}$. Denote by $\mathbb{A}$ the Banach subalgebra of $(C(\overline{\mathbb{D}}), \|\cdot\|_\infty)$ consisting on those functions that are analytic on $\mathbb{D}$—and continuous on $\overline{\mathbb{D}}$, since we are in $C(\overline{\mathbb{D}})$. The Banach algebra $\mathbb{A}$ is called the *disc algebra*. The following notation is also standard: Given a continuous function $f$ on $\mathbb{D}$, and given $0 \le r < 1$, put $f_r(e^{i\theta}) := f(re^{i\theta})$. Let $\sigma$ be the normalized Lebesgue measure on $\mathbb{T}$. Put then $\|f_r\|_p := (\int_{\mathbb{T}} |f_r|^p\, d\sigma)^{1/p}$ for $0 < p < \infty$, and $\|f_r\|_\infty := \sup_\theta |f_r(e^{i\theta})|$. We also put $\|f_r\|_0 := \exp \int_{\mathbb{T}} \log^+ |f_r|\, d\theta$. If $f \in H(\mathbb{D})$ and $0 \le p \le \infty$, we put $\|f\|_p := \sup\{\|f_r\|_p\colon 0 \le r < 1\}$. If $0 < p \le \infty$, the space $H^p$ consists of all elements $f \in H(\mathbb{D})$ for which $\|f\|_p < \infty$. Functions $f$ in $H^p$ for $0 < p < \infty$ have *nontangential*, i.e., *radial*, *limits* $f^*(e^{i\theta})$ (a.e.) on $\mathbb{T}$, and $f^* \in L^p(\mathbb{T})$; see, e.g., [40, Theorem 17.11]. If there is no possibility of misunderstanding, we shall denote by $f$ the radial limit $f^*$ of a function $f$ on d.

Let us remember the following important fact.

**Theorem 9.4** ([40, Theorem 17.18]) *Let $f$ and $g$ be two functions in $H^p$ for some $0 < p \le \infty$. If the set $\{z \in \mathbb{T}\colon f(z) = g(z)\}$ has positive Lebesgue measure, then $f = g$ on* d.

The following theorem allows to construct certain functions in $\mathbb{A}$.

**Theorem 9.5** ([40, Theorem 17.16] and [26, p. 79]) *Let $f$ be a real-valued Lebesgue integrable function on $\mathbb{T}$ such that* $\exp(f)$ *is also Lebesgue integrable on $\mathbb{T}$. Then, the function*

$$h(z) := \exp\left(\frac{1}{2\pi}\int_{-\pi}^{\pi} \frac{e^{i\theta}+z}{e^{i\theta}-z} f(e^{i\theta})\, d\theta\right), \quad \textit{for } z \in \mathbb{D},$$

*belongs to $H^1$. Moreover, $h(z) := \lim_{r\to 1-} h(rz)$ exists (a.e.) on $\mathbb{T}$ and $|h(z)| =$* $\exp(f(z))$ *(a.e.) on $\mathbb{T}$. Additionally, if $f$ is continuously differentiable on a closed "interval" $\{e^{i\theta} : |\theta-\theta_0| \le \delta\}$ in $\mathbb{T}$, then the previous limit exists and is uniform on the interval. In particular, the function $h$ is continuous on the interval.*

Clearly, if for $f \in S_{\mathbb{A}}$ there exists $g \neq 0$ in $\mathbb{A}$ such that $|f(z)| + |g(z)| \leq 1$ for all $z \in \mathbb{T}$, one has that $\|f \pm g\|_\infty \leq 1$. Hence, $f$ is not an extreme point of $B_{\mathbb{A}}$. It is possible to find a neat characterization of the extreme points of $\mathbb{A}$.

**Proposition 9.4** ([26], pp. 138–139) *A function $f$ in $B_{\mathbb{A}}$ is an extreme point of $B_{\mathbb{A}}$ if and only if*

$$\int_{-\pi}^{\pi} \log(1 - |f(e^{i\theta})|)\, d\theta = -\infty.$$

*Proof* Assume that the integral condition holds. Let $g \in \mathbb{A}$ such that $\|f \pm g\|_\infty \leq 1$. The parallelogram identity in $\mathbb{C}$ applied to every fixed $z \in \overline{\mathbb{D}}$ gives

$$|f(z)|^2 + |g(z)|^2 \leq 1.$$

In particular, $|g(e^{i\theta})|^2 \leq 1 - |f(e^{i\theta})|^2$ $(= (1 - |f(z)|)(1 + |f(z)|))$, and so, since $|f(z)| \leq 1$ for all $z \in \mathbb{T}$,

$$2 \int_{-\pi}^{\pi} \log |g(e^{i\theta})|\, d\theta \leq 2\pi \log(2) + \int_{-\pi}^{\pi} \log(1 - |f(e^{i\theta})|)\, d\theta = -\infty.$$

By [40, Theorem 17.17] the function $g$ is identically 0. We conclude that $f$ is an extreme point.

Now assume that the integral condition of $f$ fails to hold, i.e., that $\log(1-|f(e^{i\theta})|)$ is integrable. We can choose a continuous function $u$ on the unit circle such that $0 \leq u \leq 1 - |f|$, $\log u$ is integrable, and $u$ is continuously differentiable on each open arc of the set where $|f| \neq 1$. Put then

$$g(z) = \exp\left(\frac{1}{2\pi} \int_{-\pi}^{\pi} \frac{e^{i\theta} + z}{e^{i\theta} - z} \log u(e^{i\theta})\, d\theta\right), \quad \text{for } z \in \mathbb{D},$$

and let $g(z) := \lim_{r\to 1-} g(rz)$ for $z \in \mathbb{T}$. By Theorem 9.5, $|g|$ coincides with $u$ on $\mathbb{T}$ and $g \in \mathbb{A}$. Indeed, observe that Theorem 9.5 gives that $g$ is continuous on $\{z \in \mathbb{T} \colon |f| \neq 1\}$; this together with the fact that $|g_{|\mathbb{T}}| = u$ gives the continuity on the whole $\mathbb{T}$. In particular, $|g(e^{i\theta})| \leq 1 - |f(e^{i\theta})|$. Hence, $f$ is not extreme. □

**Corollary 9.3** *A function $f \in S_{\mathbb{A}}$ is an extreme point of $B_{\mathbb{A}}$ if and only if the only function $g \in \mathbb{A}$ satisfying*

$$|f(z)| + |g(z)| \leq 1 \quad \textit{for all } z \in \mathbb{T} \tag{9.3}$$

*is the null function.*

*Proof* Necessity was discussed in the paragraph preceding Theorem 9.4. Sufficiency can be seen as follows. If $f$ is not an extreme point, then Proposition 9.4 tells us that the function $\log(1 - |f(e^{i\theta})|)$ is integrable. It follows from the second part of the proof of Proposition 9.4 itself that there exists a non null function $g \in \mathbb{A}$ satisfying (9.3). □

Corollary 9.3 can be proved without using the integral characterization in Proposition 9.4. This direct proof is due to Phelps [38]. We already mentioned that necessity is almost obvious. So we focuss just on the proof of sufficiency.

*Proof* (Alternative proof of Corollary 9.3) Let us assume that there exists a function $h \neq 0$ in $\mathbb{A}$ such that $\|f \pm h\|_\infty \leq 1$. Then, by the parallelogram identity in $\mathbb{C}$ we have, for each $z \in \overline{\mathbb{D}}$, that $|f(z)|^2 + |h(z)|^2 \leq 1$, and so

$$|h(z)|^2 \leq 1 - |f(z)|^2 = (1 - |f(z)|)(1 + |f(z)|) \leq 2(1 - |f(z)|).$$

This shows that $|f(z)| + \frac{1}{2}|h(z)|^2 \leq 1$ for all $z \in \overline{\mathbb{D}}$, and we can let $g = \frac{1}{2}h^2$ in order to get a non-identically zero function $g$ such that $|f(z)| + |g(z)| \leq 1$ for all $z \in \mathbb{T}$. □

*Remark 9.2* We should point out that the previous proof by Phelps as well as the majority of those in this section where analyticity is not used, can be proved for general uniform algebras.

A consequence of Corollary 9.3 and Theorem 9.4 is that $f \in S_{\mathbb{A}}$ is an extreme point whenever the set $\{z \in \mathbb{T} : |f(z)| = 1\}$ has positive Lebesgue measure. A more precise result, due to Phelps [37], is that functions satisfying this latest condition are precisely the exposed points of $B_{\mathbb{A}}$.

It is known that $B_{\mathbb{A}}$ has no point of continuity in $B_{\mathbb{A}}$ ([35], where it is proved, more generally, for the closed unit ball of any infinite-dimensional uniform algebra). This, in particular, implies that there is no denting or strongly exposed point; that the closed unit ball of any infinite-dimensional uniform algebra has no strongly exposed points was proved in [1]. Additionally, it has been recently shown that there is no point of sequential continuity of $B_{\mathbb{A}}$, see [39]. However, the exposed points are dense in $S_{\mathbb{A}}$, see [42], and the closed convex hull of the finite Blaschke products coincides with $B_{\mathbb{A}}$, see [9]. In view of Proposition 9.5 below, we conclude that $B_{\mathbb{A}}$ is the closed convex hull of its strong extreme points and of its preserved extreme points.

Given $\alpha \in \mathbb{D}$ it is customary to denote by $\varphi_\alpha$ the holomorphic function defined on $\overline{\mathbb{D}}$ by $\varphi_\alpha(z) = (z - \alpha)/(1 - \overline{\alpha}z)$. It is well known that $\varphi_\alpha$ is one-to-one, maps $\mathbb{D}$ onto $\mathbb{D}$, $\mathbb{T}$ onto $\mathbb{T}$ and $\alpha$ to 0. Its inverse is $\varphi_{-\alpha}$. Observe that $\varphi_\alpha \in \mathbb{A}$ for every $\alpha$ and that given a finite sequence $\{\alpha_1, \ldots, \alpha_n\}$ and $c \in \mathbb{C}$ the function $\varphi = c\,\varphi_{\alpha_1} \cdots \varphi_{\alpha_n}$ belongs also to $\mathbb{A}$. When $|c| = 1$ those functions are called *finite Blaschke products.* It is then clear that any finite Blaschke product $\varphi$ satisfies $|\varphi(z)| = 1$ for every $z \in \mathbb{T}$. The functions in $\mathbb{A}$ satisfying this later condition are called *inner functions* of $\mathbb{A}$.

The ideas of the following result comes from the proof in [34] of the fact that (1) $\iff$ (3). The definition of strong extremality given in the aforesaid reference

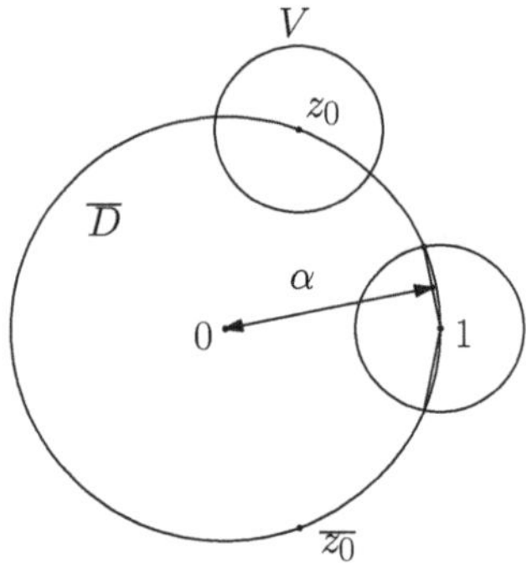


**Fig. 9.3** $|h(z)| \le \alpha$ for $z \in \overline{\mathbb{D}} \setminus V$

is the following: A point $x_0$ in $S_X$ is *strongly extreme* whenever given $\varepsilon > 0$ there exists $\delta > 0$ such that $\|x_0 \pm x\| \le 1 + \delta$ and $x \in B_X$ together imply that $\|x\| < \varepsilon$. It is easy to show that a point $x_0 \in S_X$ is strongly extreme according to this definition if, and only if, it is strongly extreme according to the definition in page x.

**Proposition 9.5** *Let $f$ be a function in $\mathbb{A}$ such that $\|f\| = 1$. The following conditions are equivalent:*

1. *$f$ is a strongly extreme point of $B_{\mathbb{A}}$.*
2. *$f$ is a w-strongly extreme point of $B_{\mathbb{A}}$—i.e., a preserved extreme point.*
3. *$f$ is an inner function of $\mathbb{A}$.*

*Proof* (1) $\Longrightarrow$ (2) is obvious.

(2) $\Longrightarrow$ (3): Assume that (3) is not satisfied. Then there exists $z_0 \in \mathbb{T}$ such that $|f(z_0)| < 1$. Since $f$ is continuous we can find an open set $V \subset \overline{\mathbb{D}}$ containing $z_0$ and a constant $0 < r < 1$ such that $|f(z)| \le r < 1$ for every $z \in V$. We want to show that $f$ is not a weak-strongly extreme of $B_{\mathbb{A}}$. Take $\varepsilon = (1 - r)/2$ and $\delta_{z_0} \in S_{\mathbb{A}^*}$. We can find a function $h \in \mathbb{A}$ such that $h(z_0) = 1 = \|h\|_\infty$ and $|h(z)| < 1$ for every $z \in \overline{\mathbb{D}} \setminus V$ (for example $h(z) = (1 + z\overline{z_0})/2$, see Fig. 9.3). Since $h$ is continuous and $\overline{\mathbb{D}} \setminus V$ compact, we can find $0 < \alpha < 1$ such that $|h(z)| \le \alpha < 1$ for every $z \in \overline{\mathbb{D}} \setminus V$. Fix $n \in \mathbb{N}$. Take $m \in \mathbb{N}$ such that $\alpha^m < 1/n$ and define $g_n := (1 - r)h^m$. Finally take $x_n = (\|f + g_n\|_\infty)^{-1}(f + g_n)$ and $y_n = (\|f - g_n\|_\infty)^{-1}(f - g_n)$. Observe first that

$$|f \pm g_n| \le \begin{cases} |f| + |g| \le r + (1 - r) \le 1 & \text{in } V, \\ |f| + |g| \le 1 + 1/n & \text{in } \overline{\mathbb{D}} \setminus V, \end{cases}$$

which implies that $\|f \pm g_n\|_\infty \le 1 + 1/n$. On the other hand, since $\|f\|_\infty = 1$, there exists $z_1 \in \mathbb{T}$ such that $|f(z_1)| = 1$ (of course $z_1 \in \overline{\mathbb{D}} \setminus V$) then

$$\|f \pm g_n\|_\infty \ge |\delta_{z_1}(f \pm g_n)| \ge |f(z_1)| - |g(z_1)| \ge 1 - 1/n.$$

It follows that

$$\lim_n \left\| \frac{x_n + y_n}{2} - f \right\|_\infty = 0.$$

However,

$$\lim_n \delta_{z_0}(x_n - f) = \lim_n g_n(z_0) = 1 - r = 2\varepsilon = \lim_n \delta_{z_0}(f - y_n).$$

Therefore, $f$ is not a $w$-strongly extreme point of $B_{\mathbb{A}}$.

(3) $\Longrightarrow$ (1): Assume that $f$ is not a strong extreme of $B_{\mathbb{A}}$. Then, there exists $\varepsilon > 0$ such that for every $\delta > 0$ there exits a pair $g$ and $h$ in $S_{\mathbb{A}}$ such that $\|g + h - 2f\|_\infty \leq 2\delta$ but $\|g - h\|_\infty \geq 3\varepsilon$. Fix $\delta > 0$ such that $(1+\delta)^2 - \varepsilon^2 < 1$. Set $2w := g - h$ then

$$\|f - w\|_\infty = \left\| f - \frac{g-h}{2} \right\|_\infty \leq \|h\|_\infty + \left\| f - \frac{g+h}{2} \right\|_\infty \leq 1 + \delta.$$

Analogously, can be seen that $\|f + w\| \leq 1 + \delta$. Let $z_0 \in \mathbb{T}$ such that $\|w\|_\infty = |w(z_0)| > \varepsilon$. Then

$$|f(z_0) \pm w(z_0)| = |(f \pm w)(z_0)| \leq \|f \pm w\|_\infty \leq 1 + \delta.$$

The parallelogram identity on $\mathbb{C}$ implies that $|f(z_0)|^2 + |w(z_0)|^2 \leq (1+\delta)^2$. Hence,

$$|f(z_0)|^2 \leq (1+\delta)^2 - |w(z_0)|^2 < (1+\delta)^2 - \varepsilon^2 < 1.$$

This, in turn, implies that $|f(z_0)| < 1$ so $f$ is not an inner function of $\mathbb{A}$. □

To sum up, any extreme point of $B_{\mathbb{A}}$—for example, any exposed point—which is not an inner function gives an unpreserved extreme point. Let us construct such an element in $\mathbb{A}$. Take a continuously differentiable function $w\colon [-\pi, \pi] \to [0, 1]$ such that $w(-\pi) = w(\pi) = 0$, $w(\theta) = 1$ for $\theta \in [-\pi/2, \pi/2]$ and that $w(\theta) > 0$ for $\theta \in (-\pi, \pi)$. Apply Theorem 9.5 to the function $\log w$ to obtain a function $f\colon \overline{\mathbb{D}} \to \mathbb{C}$ which is analytic on $\mathbb{D}$, continuous on $\mathbb{T}\backslash\{-1\}$ and such that $|f(e^{i\theta})| = 1$ for $\theta \in [-\pi/2, \pi/2]$ and $f(-1) = 0$. Therefore $f \in \mathbb{A}$ (continuity on $z = -1$ follows from continuity of $w$) is an extreme point since the set $\{z \in \mathbb{T}\colon |f(z)| = 1\}$ has positive Lebesgue measure, but it is not an inner function of $\mathbb{A}$ since $f(-1) = 0$. Explicitly,

$$f(z) := \exp\left( \frac{1}{2\pi} \int_{-\pi}^{\pi} \frac{e^{i\theta} + z}{e^{i\theta} - z} \log w(\theta)\, d\theta \right), \quad \text{for } z \in \mathbb{D}.$$

## 9.5 Preserved Extreme Points and the Radon–Nikodým Property

A Banach space $X$ is said to have the *Krein–Milman* property (KMP for short) whenever every closed convex and bounded subset of $X$ is the closed convex hull of the set of its extreme points. A geometric description of a Banach space $X$ with the Radon–Nikodým property is made by requesting that every bounded subset of $X$ has nonempty slices of arbitrary small diameter. Note that every space with the RNP has the KMP, see e.g., [7, Chapt. 11]. It is an open problem whether every space with the KMP has the RNP. This problem has been solved in the positive in dual spaces in [27], and, for the notion of preserved extreme points, in the following result. Here, $\mathrm{dist}(A, B) := \inf\{\|a - b\| : a \in A,\ b \in B\}$ for two non-empty subsets $A$ and $B$ of a Banach space $X$.

**Theorem 9.6** *Let X be a Banach space. Then, the following conditions are equivalent:*

(i) *X fails the RNP.*
(ii) [41] *For every* $\varepsilon > 0$ *there exists an equivalent norm* $\|\cdot\|$ *on* $X$ *such that* $\mathrm{dist}(\mathrm{Ext}(B_{(X^{**},\|\cdot\|)}), X) \geq 1 - \varepsilon$.
(iii) [3, 43] *There exists an equivalent norm* $\|\cdot\|$ *on X such that each extreme point of* $B_{(X,\|\cdot\|)}$ *is unpreserved—and* $\mathrm{dist}(\mathrm{Ext}(B_{(X^{**},\|\cdot\|)}), X) > 0$.

We shall not prove this theorem. We only note that (ii) $\Rightarrow$ (iii) is obvious, and that (iii) $\Rightarrow$ (i) follows from the result that, if $\|\cdot\|$ is an equivalent norm in $X$, and $X$ has the RNP, then $B_{(X,\|\cdot\|)}$ has a strongly exposed point, a consequence of a deep characterization of the RNP mentioned in page x. We also noted there that every strongly exposed point of $B_{(X,\|\cdot\|)}$ is a preserved extreme point, something that has a simple proof.

*Remark 9.3* 1. A simple consequence of (i) $\Leftrightarrow$ (ii) in Theorem 9.6 is the following: *If a Banach space X fails the RNP then, for every* $\varepsilon > 0$, *there exists an equivalent norm* $\|\cdot\|$ *on X such that every slice S of* $B_{(X,\|\cdot\|)}$ *has* $\|\cdot\|$*-diameter greater than* $1 - \varepsilon$. This was proved in [24] by using the Bishop–Phelps Theorem. We give here an alternative proof that does not use this result.
Fix $\varepsilon > 0$, and choose $\varepsilon' \in (0, \varepsilon)$. Let $\|\cdot\|$ be the norm associated to $\varepsilon' > 0$ given by Theorem 9.6. Let $S = S(f, \delta) := \{x \in B_{(X,\|\cdot\|)} : f(x) > 1 - \delta\}$ be an arbitrary slice of $B_{(X,\|\cdot\|)}$, where $f \in S_{(X^*,\|\cdot\|)}$ and $\delta > 0$. Assume that $\mathrm{diam}(S) \leq 1 - \varepsilon$. The function $f$ attains its supremum on $B_{(X^{**},\|\cdot\|)}$. Let $F := \{x^{**} \in S_{(X^{**},\|\cdot\|)} : f(x^{**}) = 1\}$. The set $F$ is $w^*$-compact convex and extremal of $B_{(X^{**},\|\cdot\|)}$. Find, by the Krein–Milman Theorem, an extreme point $x_0^{**}$ of $F$. Then $x_0^{**}$ is an extreme point of $B_{(X^{**},\|\cdot\|)}$. Fix an arbitrary point $x_0 \in S$; we have $\|x - x_0\| \leq 1 - \varepsilon$ for all $x \in S$. This shows that $S \subset B(x_0, 1 - \varepsilon)$, hence $\overline{S}^{w^*} \subset \overline{B(x_0, 1 - \varepsilon)}^{w^*} = x_0 + (1 - \varepsilon)B_{(X^{**},\|\cdot\|)}$. Obviously, $x_0^{**} \in \overline{S}^{w^*}$, so $\|x_0^{**} - x_0\| \leq 1 - \varepsilon$. This means that $x_0^{**}$ is an extreme point of $B_{(X^{**},\|\cdot\|)}$ such

Fig. 9.4 The construction in Remark 9.3 (1)

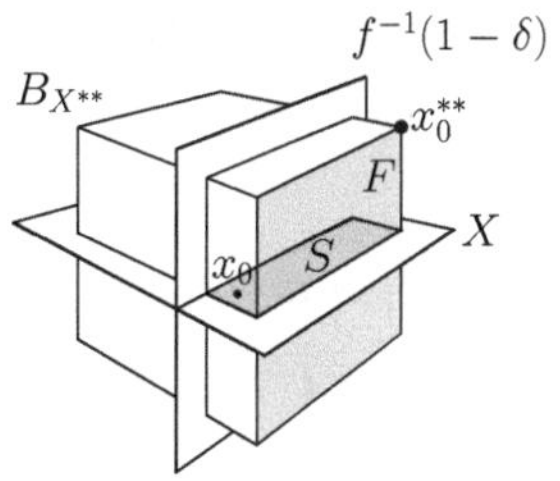


that its distance to $X$ in the norm $\|\cdot\|$ is less than or equal to $1-\varepsilon(<(1-\varepsilon'))$, a contradiction (Fig. 9.4) □

problem.

**Problem 9.1** ([41]) Assume that a Banach space $X$ fails the RNP and $\varepsilon > 0$ is given. Can $X$ be renormed so that all the slices of the new ball have diameter greater than or equal to $2-\varepsilon$?

2. The norm in (ii) (and (iii)) in Theorem 9.6 is not necessarily strictly convex, Theorem 9.8 below shows that *every separable Banach space that contains a copy of $c_0$ has an equivalent norm $\|\cdot\|$ that is strictly convex and such that no extreme point of $B_{(X^{**},\|\cdot\|)}$* is in $X$.

## 9.6 Preserved Extreme Points and Reflexivity

We proved in Sect. 9.2 above that reflexivity of a Banach space can be characterized in terms of the extreme points of the closed unit ball of its bidual space.

Section 9.4 was devoted to prove the existence of—nonreflexive—Banach spaces having at least an unpreserved extreme point, a question raised by Phelps and answered by Katznelson. It is interesting that, after all, such points exist *in every nonreflexive Banach spaces, after a suitable renorming*. This is the content of Theorem 9.7 below, due to B. V. Godun [21]. We slightly modify the original proof in order to use the same technique in proving Theorem 9.8.

The following standard facta will be used.

**Lemma 9.3** *Let $(X,\|\cdot\|)$ and $(Y,|\cdot|)$ be Banach spaces. Let $T\colon X\to Y$ be a continuous linear mapping. Define*

$$|||x||| = \|x\| + |Tx|, \quad \textit{for } x\in X. \tag{9.4}$$

*Then,*

(i) $|||\cdot|||$ *is an equivalent norm on $X$.*

(ii) *The corresponding norm on $X^{**}$ is given by*

$$|||x^{**}||| = \|x^{**}\| + |T^{**}x^{**}|. \tag{9.5}$$

(iii) *If $|\cdot|$ is strictly convex and $T$ is one-to-one, then $|||\cdot|||$ is strictly convex.*

**Lemma 9.4** *Let $(X, \|\cdot\|)$ be a Banach space. Let $N$ be a norming subspace of $X^*$, and let $\|\cdot\|_N$ be the equivalent norm defined by $N$ on $X$. Then its dual norm $\|\cdot\|_N$ on $X^*$ induces on $N$ the original dual norm.*

*Proof* Observe first that $\|x\|_N \leq \|x\|$ for all $x \in X$, so $\|x^*\|_N \geq \|x^*\|$ for all $x^* \in X^*$. It remains then to prove the reverse inequality. For this, fix $x^* \in N$ such that $\|x^*\| \leq 1$. If $x \in X$ satisfies $\|x\|_N \leq 1$, then $|x^*(x)| \leq 1$. This shows that $\|x^*\|_N \leq 1$. For $x^* \neq 0$, apply the argument above to $x^*/\|x^*\|$ to get $\|x^*\|_N \leq \|x^*\|$, as we wanted to show. □

Now we are ready to prove the following result.

**Theorem 9.7** *Let $(X, \|\cdot\|)$ be a Banach space. Then, the following are equivalent:*

(i) *The space $X$ fails to be reflexive.*
(ii) [21] *There exists an equivalent norm $|||\cdot|||$ on $X$ and an extreme point of $B_{(X,|||\cdot|||)}$ that is unpreserved.*

*Proof* Obviously, only (i) ⇒ (ii) needs a proof. Assume first that $X$ is separable—and nonreflexive. Fix $x_0^{**} \in X^{**}\setminus X$, and let $N := \operatorname{Ker} x_0^{**}$ ($\subset X^*$), a norming subspace of $X^*$ (see the introduction of Sect. 9.3.1), so $\|x\|_N := \sup\{|x^*(x)| : x^* \in N,\ \|x^*\| = 1\}$, for $x \in X$, defines an equivalent norm on $X$ whose higher dual norms are denoted, as usual, again by $\|\cdot\|_N$. Let $\{x_n^* : n \in \mathbb{N}\}$ be a subset of $S_{(N,\|\cdot\|_N)}$ such that $x_n^* \xrightarrow{w^*} 0$, and $\overline{\operatorname{span}}^{w^*}\{x_n^* : n \in \mathbb{N}\} = X^*$ (find, for example, a Markushevich basis $\{x_n; x_n^*\}$ in $X \times X^*$ such that $x_n^* \in S_{(N,\|\cdot\|_N)}$ for all $n \in \mathbb{N}$; this is always possible for a separable Banach space and a $w^*$-dense subspace $N$ of $X^*$, see, e.g., [25, Chapt. 1]). We may now define two one-to-one continuous linear operators $S : X \to c_0$ and $T : X \to \ell_2$ by

$$S(x) = \left(x_n^*(x)\right)_{n=1}^{\infty}, \qquad T(x) = \left(\frac{1}{2^n} x_n^*(x)\right)_{n=1}^{\infty}, \qquad \text{for all } x \in X. \tag{9.6}$$

Put

$$A_k = \{x \in X : \|x\|_N \leq k\|S(x)\|_\infty\}, \quad \text{for } k \in \mathbb{N}. \tag{9.7}$$

We obtain an increasing sequence $\{A_k\}_{k=1}^{\infty}$ of—homogeneous—subsets of $X$. Observe that $X = \bigcup_{k=1}^{\infty} A_k$. Indeed, $0 \in A_k$ for all $k \in \mathbb{N}$, and for $x \in X$, $x \neq 0$, we have $S(x) \neq 0$, hence $\|S(x)\|_\infty > 0$; thus we can find $k \in \mathbb{N}$ big enough such that $\|x\|_N \leq k\|S(x)\|_\infty$.[1] Fix $k \in \mathbb{N}$ such that $A_k \neq \{0\}$ and define a new equivalent

[1] We define this, seemingly, artificial sequence $\{A_k\}$ to allow further manipulations in subsequent arguments, although, strictly speaking, we need only here a set $A_k \neq \{0\}$ in the sequence.

norm $|||\cdot|||$ on $X$ by

$$|||x||| = \max\left\{\frac{1}{2k}\|x\|_N, \|S(x)\|_\infty\right\} + \|T(x)\|_2, \quad \text{for all } x \in X. \tag{9.8}$$

By Lemma 9.3, $|||\cdot|||$ is strictly convex in $X$, and its bidual norm $|||\cdot|||$ on $X^{**}$ is given by

$$|||x^{**}||| = \max\left\{\frac{1}{2k}\|x^{**}\|_N, \|S^{**}(x^{**})\|_\infty\right\} + \|T^{**}(x^{**})\|_2, \quad \text{for all } x^{**} \in X^{**}. \tag{9.9}$$

Fix $x \in A_k$ (so $\|x\|_N \leq k\|S(x)\|_\infty$) such that $|||x||| = 1$. This is possible due to the fact that the set $A_k$ is homogeneous and not reduced to $\{0\}$. Since $|||\cdot|||$ is strictly convex, $x$ is an extreme point of $B_{(X,|||\cdot|||)}$. Fix $\delta > 0$ so small that $\|x \pm \delta x_0^{**}\|_N < 2\|x\|_N$. Note that $S^{**}(x \pm \delta x_0^{**}) = S(x)$ and $T^{**}(x \pm \delta x_0^{**}) = T(x)$. Then

$$\begin{aligned} &|||x \pm \delta x_0^{**}||| \\ &= \max\left\{\frac{1}{2k}\|x \pm \delta x_0^{**}\|_N, \|S^{**}(x \pm \delta x_0^{**})\|_\infty\right\} + \|T^{**}(x \pm \delta x_0^{**})\|_2 \\ &\leq \max\left\{\frac{1}{2k}2\|x\|_N, \|S(x)\|_\infty\right\} + \|T(x)\|_2 \\ &= \|S(x)\|_\infty + \|T(x)\|_2 = |||x||| \ (= 1), \end{aligned}$$

hence $x$ is not an extreme point of $B_{(X^{**},|||\cdot|||)}$.

If $X$ is not separable, it is enough to apply a separable-reduction argument by using the next lemma. □

**Lemma 9.5** *Let $(X, \|\cdot\|)$ be a Banach space, and let $Y$ be a closed subspace. Then, every equivalent norm on $Y$ can be extended to an equivalent norm on $X$ in such a way that, if some extreme point $y_0$ of $B_{(Y,|||\cdot|||)}$ is unpreserved, then $y_0$ is an extreme point of $B_{(X,|||\cdot|||)}$ that is unpreserved.*

*Proof* Assume that $\|y\| \leq |||y||| \leq c\|y\|$ for some $c > 0$ and for all $y \in Y$; see Fig. 9.5. Let $|\cdot|$ be the Minkowski functional of the set $\overline{\text{conv}}(B_{(Y,|||\cdot|||)} \cup \frac{1}{c}B_{(X,\|\cdot\|)})$ which is an equivalent norm on $X$ that induces the norm $|||\cdot|||$ on $Y$. Finally, put $|||x||| := |x| + \text{dist}_{|\cdot|}(x, Y)$ for $x \in X$. This is again an equivalent norm in $X$ that induces on $Y$ the norm $|||\cdot|||$. We shall prove first that $y_0$ is an extreme point of $B_{(X,|||\cdot|||)}$. Indeed, assume that $2y_0 = x_1 + x_2$, where $x_1$ and $x_2$ are elements in $X$ such that $|||x_1||| = |||x_2||| = 1$. If $x_1 \in Y$ and $x_2 \in Y$, we get $x_1 = x_2$. Otherwise,

$$\begin{aligned} 1 = |||y_0||| = |y_0| &\leq \frac{1}{2}(|x_1| + |x_2|) \\ &< \frac{1}{2}(|x_1| + \text{dist}_{|\cdot|}(x_1, Y) + |x_2| + \text{dist}_{|\cdot|}(y_2, Y)) = \frac{1}{2}(|||x_1||| + |||x_2|||) = 1, \end{aligned}$$

a contradiction.

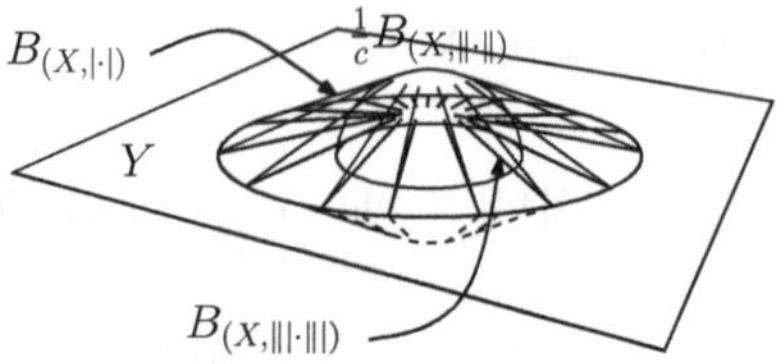


**Fig. 9.5** The three balls in Lemma 9.5

Since $y_0$ is not an extreme point of $B_{(Y^{**},|||\cdot|||)}$, and the space $(Y^{**}, |||\cdot|||)$ is isometrically isomorphic to a subspace of $(X^{**}, |||\cdot|||)$, it follows that $y_0$ is not an extreme point of $B_{(X^{**},|||\cdot|||)}$, as claimed. □

*Remark 9.4*
1. The proof of (i) ⇒ (ii) in Theorem 9.7 shows that, in a *separable* Banach space $X$, the nonreflexivity is equivalent to the existence of a *strictly convex* norm $|||\cdot|||$ on $X$ such that some—extreme—point of $S_{(X,|||\cdot|||)}$ is unpreserved.
2. The Baire Category Theorem ensures that for some $k \in \mathbb{N}$, the set $A_k$ defined in (9.7) has a nonempty interior. For the norm $|||\cdot|||$ defined in (9.8), and for all $x \in A_k$ with $|||x||| = 1$, we conclude that $|||x \pm \delta x_0^{**}||| \leq 1$, for the same $\delta > 0$, i.e., all points $x \in S_{|||\cdot|||} \cap A_k$ are "uniformly" unpreserved extreme of $B_{(X,|||\cdot|||)}$.

## 9.7 Building Norms with all Points in $S_X$ Unpreserved Extreme

The following result, due to P. Morris [33], solves Phelps' problem in a dramatic form. It shows that it is possible to renorm a separable Banach space that contains an isomorphic copy of $c_0$ in such a way that *all* points in its new unit sphere are extreme, and *all of them are unpreserved*. Later on—Theorem 9.9—we shall prove that, adding some extra condition on the space, we may even obtain a new equivalent norm with good smoothness properties and having the same striking property regarding extremality.

We give a name to the property under discussion in the next definition.

**Definition 9.4** A Banach space $(X, \|\cdot\|)$—or its norm $\|\cdot\|$—is said to have *property (M)* if $\|\cdot\|$ is strictly convex and every point $x \in S_{(X,\|\cdot\|)}$ is unpreserved extreme.

The proof we provide here of the next result relies on the technique used for proving Theorem 9.7, and differs from Morris' original proof. We need first the following simple and certainly well-known result, whose proof we sketch below.

**Lemma 9.6** *The $c_0$-sum of countably many copies of the space $(c, \|\cdot\|_\infty)$, denoted $c_0(c, c, \ldots)$, is linearly isomorphic to $(c_0, \|\cdot\|_\infty)$.*

*Proof* As it is well known, there is an isomorphism $T$ from $c$ onto $c_0$—see, e.g., [7, Exercise 5.16]. There exists then $0 < A \leq B$ such that $A\|x\|_\infty \leq \|T(x)\| \leq B\|x\|_\infty$ for every $x \in c$. If $\mathbf{x} =: (x_1, x_2, \ldots) \in c_0(c, c, \ldots)$, for $n \in \mathbb{N}$. Put

$$T(x_n) := (x_{n,1}, x_{n,2}, \ldots) \in c_0,$$

and define

$$S(\mathbf{x}) := (x_{1,1}, x_{2,1}, x_{1,2}, x_{3,1}, x_{2,2}, x_{1,3}, x_{4,1}, \ldots).$$

It is clear that $S(\mathbf{x}) \in c_0$, and that $\|S\| \leq B$. On the other hand, if $0 \neq \mathbf{x} = (x_1, x_2, \ldots) \in c_0(c, c, \ldots)$, we have, due to the fact that $\|\mathbf{x}\|_\infty = \sup\{\|x_n\|_\infty : n \in \mathbb{N}\}$, that there exists $n \in \mathbb{N}$ such that $\|x_n\|_\infty > \|\mathbf{x}\|_\infty/2$. We have then $\|Tx_n\|_\infty \geq A\|x_n\|_\infty > (A/2)\|\mathbf{x}\|_\infty$. We can find then $k \in \mathbb{N}$ such that $|x_{n,k}| = |T(x_n)_k| > (A/2)\|\mathbf{x}\|_\infty$. This shows that $\|S(\mathbf{x})\| > (A/2)\|\mathbf{x}\|_\infty$. It is clear that $S$ maps $c_0(c, c, \ldots)$ onto $c_0$. □

**Theorem 9.8** ([33]) *Assume that $(X, \|\cdot\|)$ is a separable Banach space that contains an isomorphic copy of $c_0$. Then $X$ admits an equivalent norm $|||\cdot|||$ having property* (M).

*Proof* (i) We shall prove it first for the space $(c_0, \|\cdot\|_\infty)$ itself. This follows from the proof of Theorem 9.7 as we shall show presently. Keep the notation there, letting $x_0^{**} = (1, 1, 1, \ldots)$ $(\in \ell_\infty \setminus c_0)$. The subspace $N$ is norming. If, for $n \in \mathbb{N}$, the symbol $e_n^*$ denotes the $n$-th canonical unit vector of $\ell_1$, the countable set $\Gamma := \{(1/2)(e_n^* - e_m^*) : n, m \in \mathbb{N},\ n < m\}$ $(\subset \ell_1)$ is in $S_{(N,\|\cdot\|_N)}$ (see Lemma 9.4) and $\overline{\text{span}}^{w^*}(\Gamma) = \ell_1$. The mapping $S$ defined in the proof of Theorem 9.7 for the set $\Gamma$ is a one-to-one continuous linear operator from $X$ into the $c_0$-sum of countably many copies of $(c, \|\cdot\|_\infty)$, i.e., the space $Z := c_0(c \oplus c \oplus c \oplus \ldots)$ endowed with the supremum norm, a space linearly isomorphic to $(c_0, \|\cdot\|_\infty)$, see Lemma 9.6. Precisely, put

$$S(x) := \left(\left(\left\langle \frac{e_1^* - e_m^*}{2}, x\right\rangle\right)_{m>1}, \left(\left\langle \frac{e_2^* - e_m^*}{2}, x\right\rangle\right)_{m>2}, \ldots\right), \quad \text{for } x \in c_0.$$

Obviously $\|x\|_\infty \geq \|S(x)\|_\infty \geq (1/2)\|x\|_\infty$ and $\|x\|_N \leq \|x\|_\infty$ for every $x \in c_0$, so

$$\frac{1}{2}\|x\|_N \leq \frac{1}{2}\|x\|_\infty \leq \|S(x)\|_\infty, \quad \text{for all } x \in c_0,$$

hence $c_0 = A_2$, see formula (9.7). The—strictly convex—norm $|||\cdot|||$ defined by (9.8) in the proof of Theorem 9.7 for $k = 2$ satisfies the requirements—even more: We found that a single direction in $\ell_\infty$ is enough to check that no element $x \in S_{(c_0,|||\cdot|||)}$ is a preserved extreme point, and every $x \in S_{(c_0,|||\cdot|||)}$ is extreme and "uniformly" non-extreme in $B_{(\ell_\infty,|||\cdot|||)}$, see Remark 9.4 (2).

(ii) Assume now that $X$ contains an isomorphic copy of $c_0$. By Sobczyk's theorem (see, e.g., [7, Theorem 5.11]), this copy is complemented in $X$, i.e., $X$ is isomorphic to $(G \oplus c_0, |\cdot|)$, where $|(g, x)| = \max\{\|g\|, |||x|||\}$, $g \in G$ $(\subset X)$, $x \in c_0$, and $|||\cdot|||$ is the norm on $c_0$ found in (i). Recall that there exists an element $x_0^{**}$ in $\ell_\infty \setminus c_0$ and $\delta > 0$ such that $|||x \pm \delta x_0^{**}||| = |||x|||$ $(= 1)$ for all $x \in S_{(c_0, |||\cdot|||)}$. By using the technique in the proof of Theorem 9.7, we may define a continuous linear and one-to-one operator $T_G\colon G \to \ell_2$—note that $G$ is separable—and a similar operator $T_{c_0}\colon c_0 \to \ell_2$ such that $T_{c_0}^{**} x_0^{**} = 0$. Let $T\colon G \oplus c_0 \to \ell_2 \oplus \ell_2$ be defined by $T(g, x) = (T_G g, T_{c_0} x)$, $g \in G$, $x \in c_0$. Then $T$ is a one-to-one continuous linear operator from $(G \oplus c_0, |\cdot|)$ into $\ell_2 \oplus_2 \ell_2$—a space isometric to $\ell_2$. Put

$$\nu(g, x) := |(g, x)| + \|T(g, x)\|_2, \quad \text{for } (g, x) \in G \oplus c_0. \tag{9.10}$$

As before, the space $(G \oplus c_0, \nu)$ is strictly convex. We shall prove that no element $(g, x) \in S_{(G \oplus c_0, \nu)}$ is an extreme point of the bidual space of $(G \oplus c_0, \nu)$. Assume, without loss of generality, that $|T| = 1$ and $|||x_0^{**}||| = 1$. Observe that

$$1 = |(g, x)| + \|T(g, x)\|_2 \le 2|(g, x)| \tag{9.11}$$

Set $\eta = \min\{\delta/4, 1/4\}$. First note that

$$\begin{aligned} \nu\Big((g, x) \pm (0, \eta x_0^{**})\Big) &= |(g, x \pm \eta x_0^{**})| + \|T^{**}(g, x \pm \eta x_0^{**})\|_2 \\ &= |(g, x \pm \eta x_0^{**})| + \|T(g, x)\|_2 \end{aligned} \tag{9.12}$$

We split our argument into two cases.

1. Assume first that $\|g\| > 2|||x|||$. Then by using (9.11) we obtain $|||\eta x_0^{**}||| \le \frac{1}{4} \le \frac{1}{2}\|g\|$ which implies that

$$|(g, x \pm \eta x_0^{**})| = \max\{\|g\|, |||x \pm \eta x_0^{**}|||\} = \|g\| = |(g, x)|. \tag{9.13}$$

2. Otherwise, $\|g\| \le 2|||x|||$. In this case it can be easily seen, using again (9.11), that $|||x||| \ge \frac{1}{4}$; this implies, in turn, that $|||x \pm \eta x_0^{**}||| = |||x|||$. It follows that, again,

$$|(g, x \pm \eta x_0^{**})| = \max\{\|g\|, |||x \pm \eta x_0^{**}|||\} = \max\{\|g\|, |||x|||\} = |(g, x)|. \tag{9.14}$$

It follows from (9.12), (9.13) in case (i) and (9.14) in case (ii), that $(g, x)$ is an unpreserved extreme point of $B_{(X,\nu)}$. □

*Remark 9.5* For $X$ as in Theorem 9.8, as it was already mentioned for the Banach space $c_0$, a single point in $X^{**} \setminus X$ allows for checking that every point in $S_{(X,\nu)}$ is unpreserved extreme, and that, in fact, all of them are "uniformly" unpreserved extreme.

*Remark 9.6* The first part of the proof of Theorem 9.7 is based on the existence of a $w^*$-linearly dense sequence $\{x_n^*\}$ in a norming subspace $N$ of $X^*$ such that $w^*$-$\lim_n x_n^* = 0$ and that for some $k \in \mathbb{N}$ and $x \neq 0$ in $X$, we have $\|x\|_N <$

$k \sup\{|x_n^*(x)|: n \in \mathbb{N}\}$. Those vectors $x$ form the set denoted $A_k$. In some cases, and for some $k \in \mathbb{N}$, we have $A_k = X$, as in the case $X = c_0$; see our proof of Theorem 9.8. Up to isomorphisms and subspaces, this is the only case, since the following result holds:

**Proposition 9.6** *Let $(X, \|\cdot\|)$ be a Banach space. The following assertions are equivalent:*

(i) *$X$ is linearly isomorphic to a subspace of $c_0$.*
(ii) *There exists $K > 0$ and a sequence $\{x_n^*\}_{n=1}^{\infty}$ in $X^*$ such that $w^*\text{-}\lim_n x_n^* = 0$ and for all $x \in X$,*

$$\|x\| \leq K \sup\{|x_n^*(x)|: n \in \mathbb{N}\}.$$

*Proof* Assume first that there exists a linear isomorphism $T$ from $(X, \|\cdot\|)$ into $(c_0, \|\cdot\|_\infty)$. There exists then $K > 0$ such that $\|x\| \leq K\|Tx\|_\infty$ for all $x \in X$. For $n \in \mathbb{N}$ put $x_n^* := T^*e_n^*$, where $e_n^*$ is the $n$-th vector of the canonical basis of $\ell_1$. Since $e_n^* \overset{w^*}{\to} 0$, we have $x_n^* \overset{w^*}{\to} 0$. Moreover

$$\|x\| \leq K\|Tx\|_\infty = K \sup\{|e_n^*(Tx)|: n \in \mathbb{N}\} = K \sup\{|x_n^*(x)|: n \in \mathbb{N}\},$$

for all $x \in X$, and this is (ii).

Assume now that (ii) holds. Define $T: X \to c_0$ by $Tx(n) := x_n^*(x)$ for all $n \in \mathbb{N}$ and $x \in X$. From the Uniform Boundedness Principle, the set $\{x_n^*: n \in \mathbb{N}\}$ is bounded in $X^*$, say $\|x_n^*\| \leq B$ for all $n \in \mathbb{N}$ and some $B > 0$. Then

$$\begin{aligned}\|Tx\|_\infty &= \sup\{|Tx(n)|: n \in \mathbb{N}\} = \sup\{|x_n^*(x)|: n \in \mathbb{N}\}\\ &\leq B\|x\| \leq BK \sup\{|x_n^*(x)|: n \in \mathbb{N}\} = BK\|Tx\|_\infty,\end{aligned}$$

hence $T$ is a—linear—isomorphism from $(X, \|\cdot\|)$ into $(c_0, \|\cdot\|_\infty)$. □

Of course, there is no improvement in Morris' result by requesting that $X$ should contain a linearly isomorphic copy of a subspace of $c_0$, since such a subspace contains in turn a subspace isomorphic to $c_0$.

**Problem 9.2** ([33]) Which spaces can be renormed to have property *(M)*? *Morris observes that the class of separable spaces having property (M) must be in between the separable spaces that contain $c_0$ and those that do not have RNP. For the second inclusion see Theorem 9.6. Consider, alternatively, that a closed convex and bounded set $C$ with the RNP in a Banach space is the closed convex hull of its strongly exposed points; see, e.g.,* [2, Theorem 5.17]. *This shows that no Banach space with the RNP has an equivalent norm $|||\cdot|||$ such that all points in $S_{(X,|||\cdot|||)}$ are unpreserved extreme. This has some interesting implications on the density character of $X^{**}/X$ and implies, in particular, that quasireflexive Banach spaces can not share property (M). In fact we have the following two results (Proposition* 9.7 *and Corollary* 9.4*).*

**Proposition 9.7** ([30]) *Let $X$ be a Banach space such that $X^{**}/X$ is separable. Then $X$ has the RNP.*

Before presenting the proof of Proposition 9.7, let us mention that there is a well-known class of Banach spaces that share some of the properties of separable Banach spaces. It consists of the weakly compactly generated ones. A Banach space $X$ is said to be *weakly compactly generated* (WCG for short) if there exists a weakly compact subset $K$ of $X$ such that $\overline{\text{span}}\,(K) = X$. Separable Banach spaces are WCG, as well as reflexive Banach spaces. The space $c_0(\Gamma)$ is also WCG for every set $\Gamma$, and so it is $L_1(\mu)$ for any $\sigma$-finite measure $\mu$. For this and other properties of the class of WCG spaces we refer, e.g., to [7] and [25].

*Proof* (Proof of Proposition 9.7) Assume that $X^{**}/X$ is separable. By a result of M. Valdivia in [44], the space $X$ can be written as $X = R \oplus S$, where $R$ is reflexive and $S$ is separable. Observe that $X^{**}/X$ is isomorphic to $S^{**}/S$, hence $S^{**}$ itself is separable. This shows that $X^{**}$ is the direct sum of a reflexive and a separable space, so $X^{**}$ is weakly compactly generated. Let $K$ be a $w$-compact linearly dense subset of $X^{**}$. Let $Y$ be a separable subspace of $X^*$, and let $q : X^{**} \to Y^*$ be the canonical quotient mapping. Since the $w$ and the $w^*$-topologies on $q(K)$ agree, and $q(K)$ is $w^*$-separable, it follows that $q(K)$ is $w$- (and so $\|\cdot\|$-) separable, and so it is $Y^*$. This shows that $X^*$ is Asplund, hence $X^{**}$ (and so $X$) has the RNP. □

*Remark 9.7* The argument in the proof of Proposition 9.7 can be refined to show Kuo's result that every Banach space such that its dual is a subspace of a weakly compactly generated Banach space is Asplund.

**Corollary 9.4** *A necessary condition for a Banach space to have property* (M) *is that the space $X^{**}/X$ would be nonseparable.*

Theorem 9.9 below shows that, by adding an extra property on the space, we may obtain equivalent norms having property (M) and enjoying very good smoothness properties. Observe that the spaces satisfying the hypothesis of Theorem 9.9 are, precisely, the spaces satisfying any of the five equivalent conditions in Theorem 9.3 (in particular, they are Asplund and separable, since any countable James boundary $B$ is a strong boundary, meaning that $\overline{\text{conv}}^{\|\cdot\|}(B) = B_{X^*}$; see [19]), and they contain isomorphic copies of $c_0$. We need a lemma that was already used by Morris in [33] in his proof of Theorem 9.8. Our proof is different.

**Lemma 9.7** *There exists an infinite-dimensional $w^*$-closed subspace $M_0$ of $\ell_\infty$ such that $M_0 \cap c_0 = \{0\}$.*

*Proof* Note that every separable Banach space is isometric to a subspace of $\ell_\infty$, thus in particular $\ell_\infty$ contains an isometric copy $Z$ of a given infinite-dimensional separable reflexive space. By a result of Rosenthal (see, e.g., [7, Lemma 4.62]), $Z$ is $w^*$-closed. Observe that $Z \cap c_0$ must be finite-dimensional, as any infinite-dimensional subspace of $c_0$ contains a copy of $c_0$. Then, a finite-codimensional subspace $M_0$ of $Z$ is what we need to finish the proof. □

**Theorem 9.9** ([22]) *Let $(X, \|\cdot\|_0)$ be a Banach space having a countable James boundary. Then there exists an equivalent norm $|||\cdot|||$ on $X$ with property* (M) *and $C^\infty$-smooth away from the origin.*

*Proof* By Theorem 9.3 above, the space $X$ has an equivalent $C^\infty$-smooth norm $\|\cdot\|$ that depends locally on a finite number of coordinates. Moreover, it contains an isomorphic copy $Z$ of $c_0$; see Theorem 9.2 above. The space $Z^{**}$ can be canonically identified to a closed subspace of $X^{**}$. Let $M$ be a $w^*$-closed infinite-dimensional subspace of $Z^{**}$ such that $M \cap Z = \{0\}$; it exists thanks to Lemma 9.7 above. It is clear, too, that $M \cap X = \{0\}$.

Let $N := M_\perp \subset X^*$ (the orthogonal is taken with respect to the duality $\langle X^{**}, X^*\rangle$). Find a sequence $\{\phi_n\}$ in $N$ such that $\overline{\text{span}}\ \{\phi_n : n \in \mathbb{N}\} = N$ and $\sum_{n=1}^\infty \|\phi_n\|^2 < +\infty$. Define a linear operator $T: X \to \ell_2$ by $Tx := (\langle x, \phi_n\rangle)_{n=1}^\infty$ for $x \in X$; then $T$ is clearly bounded and one-to-one, and the mapping $x \to \|Tx\|_2$ from $X$ into $\mathbb{R}$ is certainly $C^\infty$-smooth away from the origin.

Define a norm $|||\cdot|||$ on $X$ by

$$|||x||| := \|x\| + \|Tx\|_2 \quad \text{for all } x \in X. \tag{9.15}$$

Clearly $|||\cdot|||$ is strictly convex—see e.g., [5, Chapt. II]—and $C^\infty$-smooth away from the origin. Let us show that every point $x_0$ in $S_{|||\cdot|||}$ is unpreserved extreme. Find $\delta > 0$ such that $\|\cdot\|$ depends on $B_{\|\cdot\|}(x_0;\delta)$ on finitely many coordinates $\{\psi_1, \psi_2, \ldots, \psi_n\}$ $(\subset X^*)$, i.e., $\|x\| = f(\psi_1(x), \psi_2(x), \ldots, \psi_n(x))$ for $x \in B_{\|\cdot\|}(x_0;\delta)$, where $f: \mathbb{R}^n \to \mathbb{R}$ is a continuous function. Due to the fact that $M$ is infinite-dimensional, we can find $h^{**} \in M \cap \bigcap_{k=1}^n \ker \psi_k$ with $0 < \|h^{**}\| \le \delta$.

Find a net $\{h_i : i \in I, \le\}$ in $B_{\|\cdot\|}(0;\delta)$ that $w^*$-converges to $h^{**}$. Observe that $x_0 + h_i \in B_{\|\cdot\|}(x_0;\delta)$, hence

$$\|x_0+h_i\| = f(\psi_1(x_0+h_i), \psi_2(x_0+h_i), \ldots, \psi_n(x_0+h_i)), \quad \text{for all } i \in I. \tag{9.16}$$

Note that $\psi_k(x_0 + h_i) \to \psi_k(x_0 + h^{**})$ for all $k = 1, 2, \ldots, n$, and so, by (9.16),

$$\begin{aligned} \|x_0 + h_i\| &= f(\psi_1(x_0 + h_i), \psi_2(x_0 + h_i), \ldots, \psi_n(x_0 + h_i)) \\ &\to f(\psi_1(x_0 + h^{**}), \psi_2(x_0 + h^{**}), \ldots, \psi_n(x_0 + h^{**})) \\ &= f(\psi_1(x_0), \psi_2(x_0), \ldots, \psi_n(x_0)) = \|x_0\|. \end{aligned} \tag{9.17}$$

Since

$$x_0 + h_i \overset{w^*}{\to} x_0 + h^{**}, \tag{9.18}$$

we get from (9.17) and (9.18) that $\|x_0 + h^{**}\| \le \|x_0\|$. In the same way we get $\|x_0 - h^{**}\| \le \|x_0\|$, so finally by a standard convexity argument,

$$\|x_0\| = \|x_0 + h^{**}\| = \|x_0 - h^{**}\|.$$

Regarding the norm $|||\cdot|||$, we have then

$$|||x_0 + h^{**}||| = \|x_0 + h^{**}\| + \|T^{**}(x_0 + h^{**})\|,$$

as it is easy to show, hence, since $T^{**}(h^{**}) = 0$,

$$|||x_0 + h^{**}||| = \|x_0\| + \|Tx_0\| = |||x_0||| = 1. \tag{9.19}$$

Analogously,

$$|||x_0 - h^{**}||| = |||x_0||| = 1. \tag{9.20}$$

Equations (9.19) and (9.20) together show that $x_0$ is an unpreserved extreme point of $B_{|||\cdot|||}$. □

## 9.8 A Nonseparable Result

In the previous section, all results deal with separable Asplund spaces. It is natural to ask whether something can be said in a nonseparable context. The space $c_0(\Gamma)$, where $\Gamma$ is uncountable, is the first case to be considered. In fact, it happens that Banach spaces in a quite wide class have an equivalent renorming with property (M) as soon as they contain a copy of $c_0$ (see Theorem 9.10). This extends significatively Theorem 9.8, and includes the case of the Banach space $c_0(\gamma)$ for an infinite set $\gamma$; see Corollary 9.5. A still more general result pertaining to the class of Banach spaces that can be decomposed as the direct sum of a strictly convex space and some $c_0(\gamma)$ holds, and will appear elsewhere.

We considered in page 9.7 the class consisting of the weakly compactly generated Banach spaces. A larger class consists of the so-called weakly Lindelöf determined. A Banach space $X$ is said to be *weakly Lindelöf determined* (WLD for short) if $(B_{X^*}, w^*)$ is a Corson compact, i.e., it is homeomorphic to a subset $K$ of $[-1, 1]^\gamma$, for some set $\gamma$, such that each point in $K$ has only a countable number of nonzero coordinates. For this and other properties of the class of WLD spaces we refer, e.g., to [7] and [25].

**Theorem 9.10** *Let* $(X, \|\cdot\|)$ *be a weakly Lindelöf determined Banach space (in particular, let* $X$ *be a weakly compactly generated Banach space) that contains an isomorphic copy of* $c_0$*. Then* $X$ *has an equivalent norm with property* (M).

*Proof* The space $X$ has the separable complementation property, see, e.g., [8] and [7, Theorem 13.11]. In particular, there exists a separable complemented subspace $Y$ of $X$ that contains $c_0$. Put $X = G \oplus Y$. The space $G$, as a subspace of a weakly Lindelöf determined, is weakly Lindelöf determined itself; in particular, it has an M-basis (see, e.g., [25, Theorem 5.2]) and so there exists a linear, continuous, and one-to-one operator $S$ from $G$ into some $c_0(\gamma)$; see, e.g., [25, Theorem 5.3]. The space $Y$ is separable and contains a copy of $c_0$; it follows from Theorem 9.8 that there exists $\delta > 0$, an element $y_0^{**} \in Y^{**}$ and an equivalent norm $|||\cdot|||$ on $Y$ such that $|||y \pm \delta y_0^{**}||| = 1$ for all $y \in S_{(Y,|||\cdot|||)}$. Moreover, there exists a linear continuous and one-to-one operator $T : Y \to \ell_2$, as it was defined in (ii) in the proof of Theorem 9.8, such that $T^{**}y_0^{**} = 0$.

Define on $X$ an equivalent norm $|\cdot|$ by

$$|x| = |(g, y)| := \max\{\|g\|, |||y|||\}, \quad \text{where } x = g + y,\ g \in G,\ y \in Y,$$

and $\|\cdot\|$ on $G$ is the norm on $G$ induced by $\|\cdot\|$ on $X$. Let $\|\cdot\|_D$ be Day's norm on $c_0(\gamma)$. This norm is strictly convex [4]. Finally, define

$$\nu(g, y) := |(g, y)| + \left(\|Sg\|_D^2 + \|Ty\|_2^2\right)^{1/2}, \quad \text{for } (g, y) \in G \oplus Y.$$

It follows easily that $\nu$ is an equivalent norm on $X$. Observe that the norm $(g, y) \to \left(\|Sg\|_D^2 + \|Ty\|_2^2\right)^{1/2}$ defined on $X = G \oplus Y$ is strictly convex. As a consequence, the space $(G \oplus Y, \nu)$ is strictly convex. We shall prove that no element $(g, y) \in S_{(G\oplus Y,\nu)}$ is an extreme point of its bidual space.

Observe that there exists $K > 0$ (independent of the election of $(g, y)$) such that

$$1 = |(g, y)| + \left(\|Sg\|_D^2 + \|Ty\|_2^2\right)^{1/2} \leq K|(g, y)| \tag{9.21}$$

Without loss of generality, assume that $K = 2$ and that $|||y_0^{**}||| = 1$. Set $\eta = \min\{\delta/4, 1/4\}$. First note that

$$\begin{aligned} \nu\Big((g, y) \pm (0, \eta y_0^{**})\Big) &= |(g, y \pm \eta y_0^{**})| + \left(\|Sg\|_D^2 + \|T^{**}(y \pm y_0^{**})\|_2^2\right)^{1/2} \\ &= |(g, x \pm \eta y_0^{**})| + \left(\|Sg\|_D^2 + \|Ty\|_2^2\right)^{1/2} \end{aligned} \tag{9.22}$$

We split our argument into two cases.

1. Assume first that $\|g\| > 2|||y|||$. Then, by using (9.21), we obtain $|||\eta y_0^{**}||| < \frac{1}{4} \leq \frac{1}{2}\|g\|$ which implies that

$$|(g, y \pm \eta y_0^{**})| = \max\{\|g\|, |||y \pm \eta y_0^{**}|||\} = \|g\| = |(g, y)|. \tag{9.23}$$

2. Otherwise, $\|g\| \leq 2|||y|||$. It can be easily seen, using again (9.21), that $|||y||| \geq \frac{1}{4}$. This, in turn, implies that $|||y \pm \eta y_0^{**}||| = |||y|||$. Then we get, again, that

$$|(g, y \pm \eta y_0^{**})| = \max\{\|g\|, |||y \pm \eta y_0^{**}|||\} = \max\{\|g\|, |||y|||\} = |(g, y)|. \tag{9.24}$$

It follows from (9.22), (9.23) in case (1) and (9.24) in case (2), that $(g, y)$ is an unpreserved extreme point of $B_{(X,\nu)}$. The same remark regarding the uniqueness of the point $(0, y_0^{**})$ and the "uniformity" of the non-extremality applies, as above. □

**Corollary 9.5** *For any infinite set $\gamma$, the space $c_0(\gamma)$ has an equivalent norm with property* (M).

**Acknowledgments** We thank Prof. M. López Pellicer, who suggested to survey on this topic as a result of lecturing at the First Meeting in Topology and Functional Analysis, to honor Prof. J. Kąkol on the occasion of his 60th birthday. Elche, Alicante, 2013. We thank also *Ministerio de Economía y Competitividad* and FEDER for its support under project MTM2011-25377 (A. J. Guirao), and project MTM2011-22417 (V. Montesinos).

## References

1. Beneker, P., Wiegerinck, J.: Strongly exposed points in uniform algebras. Proc. Am. Math. Soc. **127**(5), 1567–1570 (1999)
2. Benyamini, Y., Lindenstrauss, J.: Geometric nonlinear functional analysis. Vol. 1. American Math. Soc. Colloquium Publications 48 (2000)
3. Bourgain, J.: A geometric characterization of the RNP in Banach spaces. Compositio Math. **36**, 3–6 (1978)
4. Day, M.M.: Strict convexity and smoothness of normed spaces. Trans. Amer. Math. Soc. **78**, 516–528 (1955)
5. Deville, R., Godefroy, G., Zizler, V.: Smoothness and Renormings in Banach Spaces. Pitman Monographs, vol. 64, Longman, London (1993)
6. Durier, R., Papini, P.L.: Polyhedral norms in an infinite dimensional space. Rocky Mt. J. Math. **23**(3), 863–875 (1993)
7. Fabian, M., Habala, P., Hájek, P., Montesinos, V., Zizler, V.: Banach Space Theory. The Basis for Linear and Nonlinear Analysis. CMS Books in Mathematics. Springer, New York (2011)
8. Figiel, T., Johnson, W.B., Pełczyński, A.: Some approximation properties of Banach spaces and Banach lattices. Israel J. Math. **183**, 199–231 (2011)
9. Fisher, S.: The convex hull of the finite Blaschke products. Bull. Amer. Math. Soc. **74**, 1128–1129 (1968)
10. Fonf, V.P.: One property of Lindenstrauss-Phelps spaces. Funktional Anal. i. Prilozhen **13**, 79–80 (1977)
11. Fonf, V.P.: Polyhedral Banach spaces. Math. Notes Acad. Sci. USSR **30**, 809–813 (1981)
12. Fonf, V.P., Lindenstrauss, J.: On quotients of polyhedral spaces. preprint
13. Fonf, V.P., Lindenstrauss, J., Phelps, R.R.: Infinite dimensional convexity. In: Johnson, B., Lindenstrauss, J. (eds.) Handbook of Banach Spaces I, pp. 599–670. Elsevier, Amsterdam (2001)
14. Fonf, V.P., Pallarés, A.J., Smith, R.J., Troyanski, S.: Polyhedral norms on non-separable Banach spaces. J. Funct. Anal. **255**(2), 449–470 (2008)
15. Fonf, V.P., Pallarés, A.J., Smith, R.J., Troyanski, S.: Polyhedrality in pieces. J. Funct. Anal. **266**(1), 247–264 (2014)
16. Fonf, V.P., Pallarés, A.J., Troyanski, S.: Isomorphically polyhedral Banach spaces. Zh. Mat. Fiz. Anal. Geom. **9**(1), 108–113 (2013)
17. Fonf, V.P., Veselý, L.: Infinite-dimensional polyhedrality. Canad. J. Math. **56**(3), 472–494 (2004)
18. Gleit, A., McGuigan, R.: A note on polyhedral Banach spaces. Proc. Amer. Math. Soc. **33**, 398–404 (1972)
19. Godefroy, G.: Boundaries of convex sets and interpolation sets. Math. Ann. **277**, 173–184 (1987)
20. Godefroy, G.: The Banach space $c_0$. Extr. Math. **16**(1), 1–25 (2001)
21. Godun, B.V.: Preserved extreme points. Funct. Anal. i Prilozhen **19**, 75–76 (1985)
22. Guirao, A.J., Montesinos, V., Zizler, V.: A note on extreme points of $C^\infty$-smooth balls in polyhedral spaces. To appear
23. Hájek, P.: Smooth norms that depend locally on a finite number of coordinates. Proc. Amer. Math. Soc. **123**, 3817–3821 (1995)

24. Hájek, P., Montesinos, V., Zizler, V.: Geometry and Gâteaux smoothness in separable Banach spaces. Oper. Matrices **6**(2), 201–232 (2012)
25. Hájek, P., Montesinos, V., Vanderwerff, J., Zizler, V.: Biorthogonal Systems in Banach Spaces. CMS Books in Mathematics. Springer, New York (2008)
26. Hoffman, K.: Banach spaces of analytic functions. Prentice-Hall Series in Modern Analysis. Prentice-Hall, Englewood Cliffs (1962)
27. Huff, R.E., Morris, P.D.: Dual spaces with the Krein-Milman property have the Radon-Nikodým property. Proc. Amer. Math. Soc. **49**, 104–108 (1975)
28. Klee, V.: Some characterizations of convex polyhedra. Acta Math. **102**, 79–107 (1959)
29. Klee, V.: Polyhedral sections of convex bodies. Acta Math. **103**, 243–267 (1960)
30. Kuo, T.H.: On conjugate Banach spaces with the Radon-Nikodým property. Pacific J. Math. **59**(2), 497–503 (1975)
31. Lajara, S., Pallarés, A.J., Troyanski, S.: Bidual renormings of Banach spaces. J. Math. Anal. Appl. **350**, 630–639 (2009)
32. Lin, B.L., Lin, P.K., Troyanski, S.: Characterizations of denting points. Proc. Amer. Math. Soc. **102**, 526–528 (1988)
33. Morris, P.: Dissapearance of extreme points. Proc. Amer. Math. Soc. **88**(2), 244–246 (1983)
34. Mortini, R., von Renteln, M.: Strong extreme points and ideals in uniform algebras. Arch. Math. (Basel) **52**(5), 465–470 (1989)
35. Nygaard, O., Werner, D.: Slices in the unit ball of a uniform algebra. Arch. Math. (Basel) **76**(6), 441–444 (2001)
36. Pechanec, J., Whitfield, J.H.M., Zizler, V.: Norms locally dependent on finitely many coordinates. An. Acad. Brasil Ci. **53**, 415–417 (1981)
37. Phelps, R.R.: Extreme points in function algebras. Duke Math. J. **32**, 267–277 (1965)
38. Phelps, R.R.: Extreme points of polar convex sets. Proc. Amer. Math. Soc. **12**, 291–296 (1961)
39. Rao, T.S.S.R.K.: Points of weak-norm continuity in the unit ball of Banach spaces. J. Math. Anal. Appl. **265**(1), 128–134 (2002)
40. Rudin, W.: Real and complex analysis. McGraw-Hill Book Co., New York (1987)
41. Schachermayer, W., Sersouri, A., Werner, E.: Moduli of non-dentability and the Radon-Nikodým property in Banach spaces. Israel J. Math. **65**, 225–257 (1989)
42. Sine, R.: On a paper of Phelps. Proc. Am. Math. Soc. **18**, 484–486 (1967)
43. Stegall, C.: Vorlesungen aus dem Fachbereich Mathematik der Universität Essen. Heft **10**, 1–61 (1983)
44. Valdivia, M.: On a class of Banach spaces. Studia Math. **60**, 11–30 (1977)

# Chapter 10
# Cantor Sets, Bernoulli Shifts and Linear Dynamics

**Salud Bartoll, Félix Martínez-Giménez, Marina Murillo-Arcila and Alfredo Peris**

**Abstract** Our purpose is to review some recent results on the interplay between the symbolic dynamics on Cantor sets and linear dynamics. More precisely, we will give some methods that allow the existence of strong mixing measures invariant for certain operators on Fréchet spaces, which are based on Bernoulli shifts on Cantor spaces. Also, concerning topological dynamics, we will show some consequences for the specification properties.

**Keywords** Hypercyclic operators · Specification properties · Strong mixing measures · Weighted shifts

**MSC 2010 Classification:** 47A16, 47B37, 37A25

## 10.1 Introduction

Our framework is the study of the dynamics of a linear operator $T : X \to X$ on a metrizable and complete topological vector space (in short, $F$-space) $X$. Moreover, we will assume that $X$ is separable.

---

S. Bartoll · F. Martínez-Giménez · M. Murillo-Arcila · A. Peris (✉)
IUMPA, Universitat Politècnica de València, 46022 Valencia, Spain
e-mail: aperis@mat.upv.es

S. Bartoll
e-mail: sbartoll@mat.upv.es

F. Martínez-Giménez
e-mail: fmartinez@mat.upv.es

M. Murillo-Arcila
e-mail: mamuar1@upv.es

J. C. Ferrando and M. López-Pellicer (eds.), *Descriptive Topology and Functional Analysis*, Springer Proceedings in Mathematics & Statistics 80, DOI: 10.1007/978-3-319-05224-3_10, 
© Springer International Publishing Switzerland 2014

We recall that $T$ is said to be *hypercyclic* if there is a vector $x$ in $X$ such that its *orbit* $\mathrm{Orb}(x, T) = \{x, Tx, T^2x, \dots\}$ is dense in $X$. The recent books [8, 20] contain the theory and most of the recent advances on hypercyclicity and linear dynamics.

Here we want to focus on some measure-theoretic properties and notions from topological dynamics. In recent years the study of the (chaotic) dynamics of linear operators has experienced a great development. This review article pretends to focus on the interplay between the dynamics of the Bernoulli shift on the Cantor set and the dynamics of certain operators. More precisely, we will focus on the strong specification property, which is a concept from topological dynamics, and the existence of strong mixing measures, a concern in measure-theoretic dynamics.

We also recall that a continuous map on a separable metric space is said to be chaotic in the sense of Devaney if it is topologically transitive (i.e., within our framework, it admits points with dense orbit) and the set of periodic points is dense. A notion of chaos (in the topological sense) stronger than Devaney's definition is the so called specification property. It was introduced by Bowen [12] and several versions of this property (see [27]) have been studied for the dynamics on compact metric spaces. We follow here the approach given in [4]. Some recent works on the specification properties are [21, 23, 24]. A continuous map $f : X \to X$ on a compact metric space $(X, d)$ has the strong specification property if, for any $\delta > 0$, there is a positive integer $N_\delta$ such that for each integer $s \geq 2$, for any set $\{y_1, \dots, y_s\} \subset X$, and for every tuple of integers $0 = j_1 \leq k_1 < j_2 \leq k_2 < \cdots < j_s \leq k_s$ satisfying $j_{r+1} - k_r \geq N_\delta$ for $r = 1, \dots, s-1$, there is a point $x \in X$ such that, for each $r \leq s$ and for all $i \in \mathbb{N} \cup \{0\}$ with $j_r \leq i \leq k_r$, the following conditions hold:

$$d(f^i(x), f^i(y_r)) < \delta,$$
$$f^n(x) = x, \quad \text{where } n = N_\delta + k_s.$$

Compact dynamical systems with the specification property are mixing and Devaney chaotic, among other basic properties (see, e.g., [14]). We will consider a notion of the strong specification property for operators as it was introduced in [2].

With respect to the measure-theoretic properties, we recall that ergodic theory was introduced in the dynamics of linear operators by Flytzanis [16] and Rudnicki [25]. It was only in recent years that it deserved special attention thanks to the work of Bayart and Grivaux [5, 6]. The papers [1, 7, 9, 13, 18, 26] contain recent advances on the subject.

The notion of frequently hypercyclicity was introduced by Bayart and Grivaux [6] as a way to measure the frequency of hitting times in an arbitrary non-empty open set for a dense orbit. In [6] a first version of a Frequent Hypercyclicity Criterion was given. We will work with the corresponding formulation of Bonilla and Grosse-Erdmann [11] for operators on separable $F$-spaces. Another (probabilistic) version of it was given by Grivaux [17].

After a section containing some preliminaries and basic notions, we will present the results on topological dynamics concerning the strong specification property. The last section deals with the measure-theoretic dynamics of operators with the existence of strong mixing probability measures with full support. In both cases the

Frequent Hypercyclicity Criterion will play a key role. At the end it will allow us to reduce the problem to the Bernoulli shift on countable unions of Cantor sets. The particular case of weighted shifts on sequence $F$-spaces is also presented, since we can go a bit further in this context and they serve as a typical test ground for linear dynamics.

## 10.2 Notation and Preliminaries

From now on, $T : X \to X$ will be an operator defined on a separable $F$-space $X$. We introduce first the necessary notions from topological dynamics for operators.

**Definition 10.1** An operator $T : X \to X$ on a separable $F$-space $X$ has the strong specification property (SSP) if there exists an increasing sequence $(K_m)_m$ of $T$-invariant compact sets with $0 \in K_1$ and $\overline{\cup_{m\in\mathbb{N}} K_m} = X$ such that for each $m \in \mathbb{N}$ the map $T|_{K_m}$ has the SSP.

Shifts on sequence spaces will be a matter of study in this paper. By a *sequence space* we mean a topological vector space $X$ which is continuously included in $\omega$, the countable product of the scalar field $\mathbb{K}$. A *sequence $F$-space* is a sequence space that is also an $F$-space. Given a sequence $w = (w_n)_n$ of positive weights, the associated *unilateral* (respectively, *bilateral*) *weighted backward shift* $B_w : \mathbb{K}^{\mathbb{N}} \to \mathbb{K}^{\mathbb{N}}$ is defined by $B_w(x_1, x_2, \dots) = (w_2x_2, w_3x_3, \dots)$ (respectively, $B_w : \mathbb{K}^{\mathbb{Z}} \to \mathbb{K}^{\mathbb{Z}}$ is defined by $B_w(\dots, x_{-1}, x_0, x_1, \dots) = (\dots, w_0x_0, w_1x_1, w_2x_2, \dots)$). If a sequence $F$-space $X$ is invariant under certain weighted backward shift $T$, then $T$ is also continuous on $X$ by the closed graph theorem. In [20, Chap. 4] contains more details about dynamical properties of weighted shifts on Fréchet sequence spaces.

We recall that a series $\sum_n x_n$ in $X$ *converges unconditionally* if it converges and, for any 0-neighbourhood $U$ in $X$, there exists some $N \in \mathbb{N}$ such that $\sum_{n\in F} x_n \in U$ for every finite set $F \subset \{N, N+1, N+2, \dots\}$.

The results on Devaney chaos for shift operators given in [20] (we refer the reader to [19] for the original results) remain valid for a unilateral (respectively, bilateral) weighted backward shift $T = B_w : X \to X$ on a sequence $F$-space $X$ in which the canonical unit vectors $(e_n)_{n\in\mathbb{N}}$ (respectively, $(e_n)_{n\in\mathbb{Z}}$) form an unconditional basis. In particular, $B_w$ is chaotic if, and only if, $\sum_{n=1}^{\infty} \left(\prod_{\nu=1}^{n} w_\nu\right)^{-1} e_n$ (respectively, $\sum_{n=-\infty}^{0} \left(\prod_{\nu=n+1}^{0} w_\nu\right) e_n + \sum_{n=1}^{\infty} \left(\prod_{\nu=1}^{n} w_\nu\right)^{-1} e_n$) converges unconditionally.

For a weight sequence $(v_i)_i$ the following Banach sequence spaces are considered

$$\ell^p(v) := \left\{ (x_i)_i \in \mathbb{K}^{\mathbb{N}} : \|x\| := \left( \sum_{i=1}^{\infty} |x_i|^p v_i \right)^{1/p} < \infty \right\}, \quad 1 \le p < \infty,$$

$$c_0(v) := \left\{ (x_i)_i \in \mathbb{K}^{\mathbb{N}} : \lim_{i\to\infty} |x_i| v_i = 0, \ \|x\| := \sup_i |x_i| v_i \right\}.$$

In this situation, the required condition to have $B_w : \ell^p(v) \to \ell^p(v)$ bounded is

$$\sup_{i\in\mathbb{N}} \left|w_{i+1}^p\right| \frac{v_i}{v_{i+1}} < \infty,$$

condition that will always be assumed to hold.

For a weight sequence $(v_i)_{i\in\mathbb{Z}}$ indexed on the set of integers, we consider the Banach sequence spaces

$$\ell^p(v,\mathbb{Z}) := \left\{(x_i)_i \in \mathbb{K}^{\mathbb{Z}} : \|x\| := \left(\sum_{i=-\infty}^{\infty} |x_i|^p\, v_i\right)^{1/p} < \infty\right\}, \quad 1 \le p < \infty,$$

$$c_0(v,\mathbb{Z}) := \left\{(x_i)_i \in \mathbb{K}^{\mathbb{Z}} : \lim_{|i|\to\infty} |x_i|\, v_i = 0,\ \|x\| := \sup_i |x_i|\, v_i\right\}.$$

The condition that characterizes that $B_w : \ell^p(v,\mathbb{Z}) \to \ell^p(v,\mathbb{Z})$ is bounded is

$$\sup_{i\in\mathbb{Z}} \left|w_{i+1}^p\right| \frac{v_i}{v_{i+1}} < \infty.$$

Finally, for measure-theoretic dynamics, let $(X, \mathfrak{B}, \mu)$ be a probability space, where $X$ is a topological space and $\mathfrak{B}$ denotes the $\sigma$-algebra of Borel subsets of $X$. We say that a Borel probability measure $\mu$ has *full support* if for all non-empty open set $U \subset X$ we have $\mu(U) > 0$. A measurable map $T : (X, \mathfrak{B}, \mu) \to (X, \mathfrak{B}, \mu)$ is called a *measure-preserving* transformation if $\mu(T^{-1}(A)) = \mu(A)$ for all $A \in \mathfrak{B}$). $T$ is *ergodic* if $T^{-1}(A) = A$ for certain $A \in \mathfrak{B}$ necessarily implies that $\mu(A)(1-\mu(A)) = 0$. $T$ is said to be *strongly mixing* with respect to $\mu$ if

$$\lim_{n\to\infty} \mu(A \cap T^{-n}(B)) = \mu(A)\mu(B) \qquad (A, B \in \mathfrak{B}),$$

and it is *exact* if given $A \in \bigcap_{n=0}^{\infty} T^{-n}\mathfrak{B}$ then either $\mu(A)(1-\mu(A)) = 0$. We refer to [15, 28] for a detailed account on these properties.

Given $A \subset \mathbb{N}$, its *lower density* is defined by

$$\underline{\text{dens}}(A) = \liminf_{n\to\infty} \frac{\text{card}(A \cap [1,n])}{n}.$$

An operator $T$ is *frequently hypercyclic* (see [6]) if there exists $x \in X$ such that for every non-empty open subset $U$ of $X$, the set $\{n \in \mathbb{N}\,;\ T^n x \in U\}$ has positive lower density.

The following version of the so called Frequently Hypercyclicity Criterion given by Bonilla and Grosse-Erdmann [11] (see also [20, Theorem 9.9 and Remark 9.10]) gives a sufficient condition for frequent hypercyclicity.

**Theorem 10.1** [11] *Let $T$ be an operator on a separable $F$-space $X$. If there is a dense subset $X_0$ of $X$ and a sequence of maps $S_n : X_0 \to X$ such that, for each $x \in X_0$:*

(i) *$\sum_{n=0}^{\infty} T^n x$ converges unconditionally.*
(ii) *$\sum_{n=0}^{\infty} S_n x$ converges unconditionally.*
(iii) *$T^n S_n x = x$ and $T^m S_n x = S_{n-m} x$ if $n > m$.*

*Then the operator $T$ is frequently hypercyclic.*

## 10.3 Cantor Sets and the Strong Specification Property

This section is devoted to a dynamical property in the topological sense, namely, the SSP. We first study the dynamics on certain Cantor sets of the backward shift on a sequence $F$-space $X$, by following an approach slightly different from the one in [2]. Later in this section we present a more general argument that yields the SSP for operators on $F$-spaces satisfying the Frequent Hypercyclicity Criterion.

**Theorem 10.2** *Let $B_w : X \to X$ be a unilateral weighted backward shift on a sequence $F$-space $X$ in which $(e_n)_{n\in\mathbb{N}}$ is an unconditional basis. Then the following conditions are equivalent:*

(i) *$B_w$ is chaotic.*
(ii) *The series*

$$\sum_{n=1}^{\infty} \Big(\prod_{\nu=1}^{n} w_\nu\Big)^{-1} e_n$$

*converges in $X$.*
(iii) *$B_w$ has a nontrivial periodic point.*
(iv) *$B_w$ has the SSP.*

*Proof* The equivalence of (i)–(ii)–(iii) was given in [19] (See also [20, Chapt. 4]). Obviously, (iv) implies (iii). For the converse, we fix a countable set $M = \{z_n \,;\, n \in \mathbb{N}\}$ of pairwise different scalars which form a dense set in $\mathbb{K}$ with $z_1 = 0$. Let $(U_n)_n$ be a basis of balanced open 0-neighbourhoods in $X$ such that $U_{n+1} + U_{n+1} \subset U_n$, $n \in \mathbb{N}$.

Since $\sum_{n=1}^{\infty} \left(\prod_{\nu=1}^{n} w_\nu\right)^{-1} e_n$ converges, there exists an increasing sequence of positive integers $(N_n)_n$ with $N_{n+2} - N_{n+1} > N_{n+1} - N_n$ for all $n \in \mathbb{N}$ such that

$$\sum_{k>N_n} \alpha_k \Big(\prod_{\nu=1}^{k} w_\nu\Big)^{-1} e_k \in U_{n+1}, \text{ if } \alpha_k \in \{z_1, \ldots, z_n\}, \text{ for each } n \in \mathbb{N}. \quad (10.1)$$

We define $A_m = \{z_1, \dots z_m\}$ for $m \in \mathbb{N}$. $A_m^{\mathbb{N}}$ is a compact space when endowed with the product topology inherited from $M^{\mathbb{N}}$, $m \in \mathbb{N}$. Now we define the map $\Phi : \bigcup_{m=1}^{\infty} A_m^{\mathbb{N}} \to X$ given by

$$\Phi((\alpha_k)_{k\in\mathbb{N}}) = \sum_{k=1}^{\infty} \alpha_k \Big(\prod_{\nu=1}^{k} w_\nu\Big)^{-1} e_k.$$

$\Phi$ is well-defined and $\Phi|_{A_m^{\mathbb{N}}}$ is continuous for each $m \in \mathbb{N}$ by (10.1). We have that $K_m := \Phi(A_m^{\mathbb{N}})$ is a compact Cantor subset of $X$, invariant under $B_w$, and such that $B_w|_{K_m}$ is conjugated to $\sigma|_{A_m^{\mathbb{N}}}$ via $\Phi$, for each $m \in \mathbb{N}$. It is well-known that $\sigma|_{A_m^{\mathbb{N}}}$ satisfies the SSP (see, e.g., [27]), and by conjugacy we obtain that $B_w|_{K_m}$ satisfies the SSP too, for every $m \in \mathbb{N}$. Finally, by density of $M$ in the scalar field $\mathbb{K}$ we conclude that $\bigcup_{m\in\mathbb{N}} K_m$ is dense in $X$, therefore $B_w$ satisfies the SSP. □

A similar argument, by considering the bilateral shift on the sets $A_m^{\mathbb{Z}}$, $m \in \mathbb{N}$, and the results of [19] (See also of [20, Chap. 4]), yields the analogous characterizations for bilateral weighted shifts on sequence $F$-spaces.

**Theorem 10.3** *Let $B_w : X \to X$ be a bilateral weighted backward shift on a sequence $F$-space $X$ in which $(e_n)_{n\in\mathbb{Z}}$ is an unconditional basis. Then the following conditions are equivalent:*

(i) *$B_w$ is chaotic.*
(ii) *The series*

$$\sum_{n=-\infty}^{0} \Big( \prod_{\nu=n+1}^{0} w_\nu\Big) e_n + \sum_{n=1}^{\infty} \Big(\prod_{\nu=1}^{n} w_\nu\Big)^{-1} e_n$$

*converges in $X$.*
(iii) *$B_w$ has a nontrivial periodic point.*
(iv) *$B_w$ has the SSP.*

Particular cases on which the above results are interesting are the (unilateral and bilateral) weighted shifts on (weighted or not) $\ell^p$-spaces and $c_0$ (see [2]). Moreover, Bayart and Ruzsa [10] recently proved that weighted shift operators on $\ell^p$, $1 \le p < \infty$, are frequently hypercyclic if, and only if, they are Devaney chaotic. This fact adds a new equivalence for $\ell^p$-spaces.

**Corollary 10.1** *For a bounded weighted backward shift operator $B_w$ defined on $X = \ell^p(v)$, $1 \le p < \infty$, (respectively, on $X = c_0(v)$) the following conditions are equivalent:*

(i) $\displaystyle\sum_{i=1}^{\infty} \frac{v_i}{\prod_{j=1}^{i} \left|w_j^p\right|} < \infty$*(respectively,* $\displaystyle\lim_{i\to\infty} \frac{v_i}{\prod_{j=1}^{i} |w_j|} = 0$).
(ii) *$B_w$ has SSP.*
(iii) *$B_w$ is Devaney chaotic.*

*Moreover, for $X = \ell^p(v)$, $1 \le p < \infty$, the above items are equivalent to*

(iv) *$B_w$ is frequently hypercyclic.*

**Corollary 10.2** *For a bounded bilateral weighted backward shift operator $B_w$ defined on $\ell^p(v,\mathbb{Z})$, $1 \le p < \infty$, (respectively, on $c_0(v,\mathbb{Z})$) the following conditions are equivalent:*

(i) $\displaystyle\sum_{i=1}^{\infty} \frac{v_i}{\prod_{j=1}^{i} \left|w_j^p\right|} < \infty$ *and* $\sum_{i=1}^{\infty} \prod_{j=0}^{-i+1} \left|w_j^p\right| v_{-i} < \infty$

*(respectively,* $\displaystyle\lim_{i\to\infty} \frac{v_i}{\prod_{j=1}^{i} |w_j|} = \lim_{i\to\infty} \prod_{j=0}^{-i+1} |w_j| \, v_{-i} = 0$).

(ii) *$B_w$ has SSP.*
(iii) *$B_w$ is Devaney chaotic.*

*Moreover, for $X = \ell^p(v,\mathbb{Z})$, $1 \le p < \infty$, the above items are equivalent to*

(iv) *$B_w$ is frequently hypercyclic.*

In the previous results we deduced the SSP from the dynamics on certain invariant Cantor sets. Something similar can be done under the more general assumptions that the operator satisfies the Frequent Hypercyclicity Criterion (see [3]). The idea now is to work with certain invariant factors of Cantor sets. We will take the general version of the Frequent Hypercyclicity Criterion given in [20, Remark 9.10].

**Theorem 10.4** *Let $T$ be an operator on a separable $F$-space $X$. If there is a dense subset $X_0$ of $X$ and a sequence of maps $S_n : X_0 \to X$ such that, for each $x \in X_0$:*

(i) *$\sum_{n=0}^{\infty} T^n x$ converges unconditionally.*
(ii) *$\sum_{n=0}^{\infty} S_n x$ converges unconditionally.*
(iii) *$T^n S_n x = x$ and $T^m S_n x = S_{n-m} x$ if $n > m$.*

*Then the operator $T$ satisfies the SSP.*

*Proof* We suppose that $X_0 = \{x_n \,;\, n \in \mathbb{N}\}$ with $x_1 = 0$ and $S_n 0 = 0$ for all $n \in \mathbb{N}$. Let $(U_n)_n$ be a basis of balanced open 0-neighbourhoods in $X$ such that $U_{n+1} + U_{n+1} \subset U_n$, $n \in \mathbb{N}$. By (i) and (ii), there exists an increasing sequence of positive integers $(N_n)_n$ with $N_{n+2} - N_{n+1} > N_{n+1} - N_n$ for all $n \in \mathbb{N}$ such that

$$\sum_{k>N_n} T^k x_{m_k} \in U_{n+1} \text{ and } \sum_{k>N_n} S_k x_{m_k} \in U_{n+1}, \text{ if } m_k \in \{1,\ldots,n\}, \text{ for each } n \in \mathbb{N}. \tag{10.2}$$

We let $B_m = \{1,\ldots,m\}$ and define the map $\Phi : \bigcup_{m=1}^{\infty} B_m^{\mathbb{Z}} \to X$ given by

$$\Phi((n_k)_{k\in\mathbb{Z}}) = \sum_{k<0} S_{-k} x_{n_k} + x_{n_0} + \sum_{k>0} T^k x_{n_k}.$$

$\Phi$ is well-defined and $\Phi|_{B_m^{\mathbb{Z}}}$ is continuous for each $m \in \mathbb{N}$ by (10.2). As in Theorem 10.2 we have that $K_m := \Phi(B_m^{\mathbb{Z}})$ is a compact subset of $X$, invariant under the operator $T$, and such that $T|_{K_m}$ is conjugated to $\sigma^{-1}|_{B_m^{\mathbb{Z}}}$ via $\Phi$, for each $m \in \mathbb{N}$. Again, by conjugacy, we obtain that $T|_{K_m}$ satisfies the SSP too, for every $m \in \mathbb{N}$ and, since $\bigcup_{m\in\mathbb{N}} K_m$ is dense in $X$ because it contains $X_0$, we conclude that $T$ satisfies the SSP. □

## 10.4 Mixing Measures and Bernoulli Shifts

For the existence of strong mixing measures with full support, certain Cantor subsets of $\mathbb{N}^N$, with either $N = \mathbb{N}$ or $N = \mathbb{Z}$, will be needed. This time we precise some finer adjustments that involve Cantor sets larger than $A^N$ with $A$ finite. Actually they will be of the form $C = \prod_{n\in N} A_n$, where the cardinalities of the finite sets $A_n$ tend to infinity as $n \to \infty$. The problem now is that these type of sets are not invariant under the shift, so actually we will use invariant sets that are countable unions of Cantor sets. The contents of this section are based on [22]. We also recall that, recently, Bayart and Matheron gave very general conditions expressed on eigenvector fields associated to unimodular eigenvalues under which an operator $T$ admits a $T$-invariant mixing measure [9].

In all the cases that we treat in this section the idea is to construct a model probability space $(Z, \overline{\mu})$ and a (Borel) measurable map $\Phi : Z \to X$, where:

(a) $Z \subset \mathbb{N}^N$ is such that $\sigma(Z) = Z$ for the Bernoulli shift (either unilateral or bilateral).
(b) $\overline{\mu}$ is a $\sigma$-invariant strongly mixing measure.
(c) $Y := \Phi(Z)$ is a $T$-invariant dense subset of $X$.
(d) The operator $T$ is either (quasi)conjugated to $\sigma$ or to $\sigma^{-1}$ through $\Phi$.

This way, the measure $\overline{\mu}$ induces by (quasi)conjugacy a Borel probability measure $\mu$ on $X$ that is $T$-invariant and strongly mixing.

### The Model Probability Space $(Z, \overline{\mu})$

We will construct $Z \subset \mathbb{N}^N$, invariant under the shift, where $N = \mathbb{N}$ or $N = \mathbb{Z}$. In all the cases we are given certain increasing sequence of positive integers $(N_n)_n$ with $N_{n+2} - N_{n+1} > N_{n+1} - N_n$. We define the compact space $L = \prod_{k\in N} A_k$ where

$$A_k = \{1, \dots m\} \text{ if } N_m < |k| \leq N_{m+1},\ m \in \mathbb{N}, \text{ and } A_k = \{1\}, \text{ if } |k| \leq N_1.$$

Let $L(s) := \sigma^s(L)$, $s \in N \cup \{0\}$. $L(s)$ is a subspace of $\mathbb{N}^N$, $s \in N \cup \{0\}$.

In $\mathbb{N}^N$ we define the product probability measure $\overline{\mu} = \bigotimes_{k\in N} \overline{\mu_k}$, where $\overline{\mu_k}(\{n\}) = p_n$ for all $n \in \mathbb{N}$ and $\overline{\mu_k}(\mathbb{N}) = \sum_{n=1}^{\infty} p_n = 1$, $k \in N$. The numbers $p_n \in ]0,1[$, $n \in \mathbb{N}$, are so that, if

$$\beta_j := \left(\sum_{i=1}^{j} p_i\right)^{N_{j+1}-N_j}, \quad j \in \mathbb{N}, \quad \text{then } \prod_{j=1}^{\infty} \beta_j > 0.$$

We define $Z = \bigcup_{s\in N} L(s)$, which is a countable union of Cantor sets, invariant under the shift, and satisfies

$$\begin{aligned}\overline{\mu}(Z) \geq \overline{\mu}(L) &= \prod_{|k|\leq N_1} \overline{\mu}_k(\{1\}) \prod_{l=1}^{\infty} \left( \prod_{N_l<|k|\leq N_{l+1}} \overline{\mu}_k(\{1,\ldots,l\})\right) \\ &\geq p_1^{2N_1+1} (\prod_{l=1}^{\infty} \beta_l)^2 > 0.\end{aligned}$$

By [28] we know that $\overline{\mu}$ is a $\sigma$-invariant strongly mixing Borel probability measure on $\mathbb{N}^N$. Since $\sigma(Z) = Z$, it has positive measure, and every strong mixing measure is ergodic, we necessarily have that $\overline{\mu}(Z) = 1$. Even more, in the case $N = \mathbb{N}$ it is known (see [28, Sect. 4.12]) that $\overline{\mu}$ is a $\sigma$-invariant exact Borel probability measure.

Once we have defined the model, as in the previous section we first analyze the situation for unilateral shifts.

**Theorem 10.5** *Let $B_w : X \to X$ be a unilateral weighted backward shift on a sequence $F$-space $X$ in which $(e_n)_{n\in\mathbb{N}}$ is an unconditional basis. If $B_w$ is chaotic then there exists a $T$-invariant exact Borel probability measure on $X$ with full support.*

*Proof* As in the proof of Theorem 10.2, we fix a countable set $M = \{z_n; n \in \mathbb{N}\}$ of pairwise different scalars which form a dense set in $\mathbb{K}$ with $z_1 = 0$, and a basis $(U_n)_n$ of balanced open 0-neighbourhoods in $X$ such that $U_{n+1} + U_{n+1} \subset U_n$, $n \in \mathbb{N}$.

Again, although not so immediate, the fact that $\sum_{n=1}^{\infty} \left(\prod_{\nu=1}^{n} w_\nu\right)^{-1} e_n$ converges, implies the existence of an increasing sequence of positive integers $(N_n)_n$ with $N_{n+2} - N_{n+1} > N_{n+1} - N_n$ for all $n \in \mathbb{N}$ such that

$$\sum_{k>N_n} \alpha_k \Big(\prod_{\nu=1}^{k} w_\nu\Big)^{-1} e_k \in U_{n+1}, \text{ if } \alpha_k \in \{z_1,\ldots,z_{2m}\}, \text{ for } N_m < k \leq N_{m+1},\ m \geq n. \tag{10.3}$$

We set $Z = \bigcup_{s\geq 0} L(s) \subset \mathbb{N}^{\mathbb{N}}$ the $\sigma$-invariant set and the exact probability measure $\overline{\mu}$ considered in the model above, and we define the map $\Phi : Z \to X$ given by

$$\Phi((n(k))_{k\in\mathbb{N}}) = \sum_{k=1}^{\infty} \alpha_{n(k)} \Big(\prod_{\nu=1}^{k} w_\nu\Big)^{-1} e_k.$$

$\Phi$ is well-defined and $\Phi|_{L(s)}$ is continuous for each $s \geq 0$ by (10.3). We also have that $Y := \Phi(Z)$ is a countable union of Cantor subsets of $X$, invariant under $B_w$, and such that $B_w|_Y$ is conjugated to $\sigma|_Z$ via $\Phi$. Since we know that $\overline{\mu}$ is exact on $Z$, the measure $\mu(A) = \overline{\mu}(\Phi^{-1}(A))$, $A \in \mathfrak{B}(X)$, is well-defined on $X$, it is $B_w$-invariant and exact, so it only remains to show that it has full support. Indeed, given a non-empty open set $U$ in $X$, we pick $y = \Phi((n(k))_k) \in Y$ satisfying $y + U_n \subset U$. Thus

$$\mu(U) \geq \mu(\{x = y + \sum_{k>N_n} \frac{\alpha_k}{\prod_{\nu=1}^k w_\nu} e_k \; ; \; \alpha_k \in \{z_1, \ldots, z_{2m}\},$$

$$\text{for } N_m < k \leq N_{m+1}, \; m \geq n\})$$

$$\geq \prod_{k=1}^{N_n} \overline{\mu}_k(\{n(k)\}) \prod_{l=n}^{\infty} \left( \prod_{N_l<k\leq N_{l+1}} \overline{\mu}_k(\{1, \ldots, 2l\}) \right) > \prod_{k=1}^{N_n} p_{n(k)} (\prod_{l=n}^{\infty} \beta_l)^2 > 0,$$

by (10.3) and by the selection of the sequence $(p_n)_n$ in the model. Therefore we conclude the result. □

As in the previous section, a general result can be obtained for operators satisfying the Frequent Hypercyclicity Criterion.

**Theorem 10.6** *Let $T$ be an operator on a separable $F$-space $X$. If there is a dense subset $X_0$ of $X$ and a sequence of maps $S_n : X_0 \to X$ such that, for each $x \in X_0$:*

(i) *$\sum_{n=0}^{\infty} T^n x$ converges unconditionally.*
(ii) *$\sum_{n=0}^{\infty} S_n x$ converges unconditionally.*
(iii) *$T^n S_n x = x$ and $T^m S_n x = S_{n-m} x$ if $n > m$.*

*Then there is a $T$-invariant strongly mixing Borel probability measure $\mu$ on $X$ with full support.*

*Proof* As in Theorem 10.4 we suppose that $X_0 = \{x_n \; ; \; n \in \mathbb{N}\}$ with $x_1 = 0$ and $S_n 0 = 0$ for all $n \in \mathbb{N}$, and $(U_n)_n$ is a basis of balanced open 0-neighbourhoods in $X$ such that $U_{n+1} + U_{n+1} \subset U_n$, $n \in \mathbb{N}$. Again, with a more subtle argument, by (i) and (ii) we have the existence of an increasing sequence of positive integers $(N_n)_n$ with $N_{n+2} - N_{n+1} > N_{n+1} - N_n$ for all $n \in \mathbb{N}$ such that

$$\sum_{k>N_n} T^k x_{m_k} \in U_{n+1} \text{ and } \sum_{k>N_n} S_k x_{m_k} \in U_{n+1},$$

$$\text{if } m_k \leq 2l, \text{ for } N_l < k \leq N_{l+1}, \; l \geq n. \tag{10.4}$$

Actually, this is a consequence of the completeness of $X$ and the fact that, for each 0-neighbourhood $U$ and for all $l \in \mathbb{N}$, there is $N \in \mathbb{N}$ such that $\sum_{k\in F} T^k x \in U$ and $\sum_{k\in F} S_k x \in U$ for any finite subset $F \subset ]N, +\infty[$ and for each $x \in \{x_1, \ldots, x_{2l}\}$.

We fix now our model space $(Z, \overline{\mu})$ with $Z = \bigcup_{s\in\mathbb{Z}} L(s) \subset \mathbb{N}^{\mathbb{Z}}$ and the measure $\overline{\mu}$ associated with the increasing sequence $(N_n)_n$.

1. **The map $\Phi$**
   We define the map $\Phi : Z \to X$ by

$$\Phi((n(k))_{k\in\mathbb{Z}}) = \sum_{k<0} S_{-k}x_{n(k)} + x_{n(0)} + \sum_{k>0} T^k x_{n(k)}. \tag{10.5}$$

   $\Phi$ is well-defined since, given $(n(k))_{k\in\mathbb{Z}} \in L(s)$, and for $l \geq |s|$, we have $n(k) \leq 2l$ if $N_l < |k| \leq N_{l+1}$, which shows the convergence of the series in (10.5) by (10.4).
   $\Phi|_{L(s)}$ is also continuous for each $s \in \mathbb{Z}$. Indeed, let $(\gamma_j)_j$ be a sequence of elements of $L(s)$ that converges to $\gamma \in L(s)$ and fix any $n \in \mathbb{N}$ with $n > |s|$. We will find $n_0 \in \mathbb{N}$ such that $\Phi(\gamma_j) - \Phi(\gamma) \in U_n$ for $n \geq n_0$. To do this, by definition of the topology of $L(s)$ there exists $n_0 \in \mathbb{N}$ such that

$$\gamma(k)_j = \gamma(k) \text{ if } |k| \leq N_{n+1} \text{ and } j \geq n_0.$$

   By (10.4) we have

$$\Phi(\gamma_j) - \Phi(\gamma) = \sum_{k<-N_{n+1}} S_{-k}(x_{\gamma(k)_j} - x_{\gamma(k)}) + \sum_{k>N_{n+1}} T^k(x_{\gamma(k)_j} - x_{\gamma(k)}) \in U_n$$

   for all $j \geq n_0$. This shows the continuity of $\Phi : L(s) \to X$ for every $s \in \mathbb{Z}$.
   The map $\Phi : Z \to X$ is then measurable (i.e., $\Phi^{-1}(A) \in \mathfrak{B}(Z)$ for every $A \in \mathfrak{B}(X)$).
2. **The measure $\mu$ on $X$**
   $Y := \Phi(Z)$ is a $T$-invariant Borel subset of $X$ because it is a countable union of compact sets and $\Phi\sigma^{-1} = T\Phi$. As before, the measure $\mu$ on $X$ is defined by $\mu(A) = \overline{\mu}(\Phi^{-1}(A))$ for all $A \in \mathfrak{B}(X)$, which is well-defined, $T$-invariant, and a strongly mixing Borel probability measure. As in Theorem 10.5, it only remains to show that $\mu$ has full support. Indeed, given a non-empty open set $U$ in $X$, we fix $n \in \mathbb{N}$ such that $x_n + U_n \subset U$. Therefore,

$$\begin{aligned}\mu(U) &\geq \mu\Big(\{x = x_n + \sum_{k>N_n} T^k x_{m(k)} + \sum_{k>N_n} S_k x_{m(-k)}\ ;\ m(k) \leq 2l \\ &\qquad\qquad \text{for } N_l < |k| \leq N_{l+1},\ l \geq n\}\Big) \\ &\geq \overline{\mu}_0(\{n\}) \prod_{0<|k|\leq N_n} \overline{\mu}_k(\{1\}) \prod_{l=n}^{\infty} \Bigg( \prod_{N_l<|k|\leq N_{l+1}} \overline{\mu}_k(\{1,\dots,2l\}) \Bigg) \\ &> p_n p_1^{2N_n} (\prod_{l=n}^{\infty} \beta_l)^2 > 0,\end{aligned}$$

   and we obtain that $\mu$ has full support. □

As a consequence, since every chaotic bilateral shift on a sequence $F$-space in which the natural basis $(e_n)_{n\in\mathbb{Z}}$ is an unconditional basis satisfies the Frequent Hypercyclicity Criterion (see, e.g., [20, Chap. 9]), we obtain strong mixing measures for these shifts.

**Corollary 10.3** *Let $T : X \to X$ be a chaotic bilateral weighted shift on a sequence $F$-space $X$ in which $(e_n)_{n\in\mathbb{Z}}$ is an unconditional basis. Then there exists a $T$-invariant strongly mixing Borel probability measure on $X$ with full support.*

**Acknowledgments** This work is supported in part by MICINN and FEDER, Projects MTM2010-14909 and MTM2013-47093-P. The third author is supported by a grant from the FPU Program of MEC. The second and fourth authors were partially supported by GVA, Project PROMETEOII/2013/013.

## References

1. Badea, C., Grivaux, S.: Unimodular eigenvalues, uniformly distributed sequences and linear dynamics. Adv. Math. **211**, 766–793 (2007)
2. Bartoll, S., Martínez-Giménez, F., Peris, A.: The specification property for backward shifts. J. Difference Equ. Appl. **18**, 599–605 (2012)
3. Bartoll, S., Martínez-Giménez, F., Peris, A.: The specification property in linear dynamics. Preprint
4. Bauer, W., Sigmund, K.: Topological dynamics of transformations induced on the space of probability measures. Monatsh. Math. **79**, 81–92 (1975)
5. Bayart, F., Grivaux, S.: Hypercyclicity and unimodular point spectrum. J. Funct. Anal. **226**, 281–300 (2005)
6. Bayart, F., Grivaux, S.: Frequently hypercyclic operators. Trans. Amer. Math. Soc. **358**, 5083–5117 (2006)
7. Bayart, F., Grivaux, S.: Invariant Gaussian measures for operators on Banach spaces and linear dynamics. Proc. Lond. Math. Soc. **3**(94), 181–210 (2007)
8. Bayart, F., Matheron, É.: Dynamics of Linear Operators. Cambridge University Press, Cambridge (2009)
9. Bayart, F., Matheron, É.: Mixing operators and small subsets of the circle. Preprint, arXiv:1112.1289v1
10. Bayart, F., Ruzsa, I.Z.: Frequently hypercyclic weighted shifts. Ergodic Theory Dynam. Systems (to appear)
11. Bonilla, A., Grosse-Erdmann, K.G.: Frequently hypercyclic operators and vectors. Ergodic Theory Dynam. Systems **27**, 383–404 (2007). [Erratum: Ergodic Theory Dynam. Systems **29**, 1993–1994 (2009)]
12. Bowen, R.: Topological entropy and axiom A. In: Global Analysis (Proceedings of Symposium on Pure Mathematics, vol. XIV, Berkeley, California, 1968). American Mathematical Societ, Providence, R.I., 23–41 (1970)
13. De la Rosa, M., Frerick, L., Grivaux, S., Peris, A.: Frequent hypercyclicity, chaos, and unconditional Schauder decompositions. Israel J. Math. **190**, 389–399 (2012)
14. Denker, M., Grillenberger, C., Sigmund, K.: Ergodic Theory on Compact Spaces. Springe, Berlin (1976)
15. Einsiedler, M., Ward, T.: Ergodic Theory with a View Towards Number Theory. Graduate Texts in Mathematics 259. Springer, London (2011)
16. Flytzanis, E.: Unimodular eigenvalues and linear chaos in Hilbert spaces. Geom. Funct. Anal. **5**, 1–13 (1995)

17. Grivaux, S.: A probabilistic version of the frequent hypercyclicity criterion. Studia Math. **176**, 279–290 (2006)
18. Grivaux, S.: A new class of frequently hypercyclic operators. Indiana Univ. Math. J. **60**, 1177–1201 (2011)
19. Grosse-Erdmann, K.G.: Hypercyclic and chaotic weighted shifts. Studia Math. **139**, 47–68 (2000)
20. Grosse-Erdmann, K.G., Peris Manguillot, A.: Linear Chaos. Universitext. Springer, London (2011)
21. Lampart, M., Oprocha, P.: Shift spaces, $\omega$-chaos and specification property. Topology Appl. **156**, 2979–2985 (2009)
22. Murillo-Arcila, M., Peris, A.: Strong mixing measures for linear operators and frequent hypercyclicity. J. Math. Anal. Appl. **398**, 462–465 (2013)
23. Oprocha, P.: Specification properties and dense distributional chaos. Discrete Contin. Dyn. Syst. **17**, 821–833 (2007)
24. Oprocha, P., Štefánková, M.: Specification property and distributional chaos almost everywhere. Proc. Amer. Math. Soc. **136**, 3931–3940 (2008)
25. Rudnicki, R.: Gaussian measure-preserving linear transformations. Univ. Iagel. Acta Math. **30**, 105–112 (1993)
26. Rudnicki, R.: Chaoticity and invariant measures for a cell population model. J. Math. Anal. Appl. **393**, 151–165 (2012)
27. Sigmund, K.: On dynamical systems with the specification property. Trans. Amer. Math. Soc. **190**, 285–299 (1974)
28. Walters, P.: An Introduction to Ergodic Theory. Springer, New York (1982)

# Chapter 11
# Some Non-linear Geometrical Properties of Banach Spaces

**Domingo García and Manuel Maestre**

**Abstract** In this survey we report on very recent results about some non-linear geometrical properties of many classes of real and complex Banach spaces and uniform algebras, including the ball algebra $\mathscr{A}_u(B_X)$ of all uniformly continuous functions on the closed unit ball and holomorphic on the open unit ball of a complex Banach space $X$. These geometrical properties are: Polynomial numerical index, Polynomial Daugavet property and Bishop-Phelp-Bollobás property for multilinear mappings.

**Keywords** Disk algebra · Numerical index · Daugavet property · Banach space

**MSC 2000 Classification:** Primary 46B20, Secondary 46G25, 47A12

## 11.1 Polynomial Numerical Index

Throughout this survey, all the Banach spaces considered will be real or complex unless the scalar field is specified and all polynomials will be continuous. Let $X$ be a Banach space. By $B_X$ we denote the closed unit ball and by $S_X$ the unit sphere of $X$. Given $k \geq 0$, a mapping $P : X \to Y$ between two Banach spaces $X$ and $Y$ is called a continuous $k$-homogeneous polynomial if there exists a continuous $k$-linear mapping $A : X^k \to Y$ such that $P(x) = A(x, \ldots, x)$ for every $x$ in $X$. We denote by $\mathscr{P}\left({}^k X; X\right)$ the space of all $k$-homogeneous polynomials from $X$ to $X$, and by $\mathscr{P}\left({}^k X\right)$ the space of all $k$-homogeneous scalar polynomials. If $k = 0$,

D. García · M. Maestre (✉)
Departamento de Análisis Matemático, Universidad de Valencia, Doctor Moliner 50, 46100 Burjasot (Valencia), Spain
e-mail: manuel.maestre@uv.es

D. García
e-mail: domingo.garcia@uv.es

J. C. Ferrando and M. López-Pellicer (eds.), *Descriptive Topology and Functional Analysis*, Springer Proceedings in Mathematics & Statistics 80, DOI: 10.1007/978-3-319-05224-3_11, 
© Springer International Publishing Switzerland 2014

$\mathscr{P}\left({}^{0}X; X\right)$ identify with the space of constant functions, i.e. $\mathscr{P}\left({}^{0}X; X\right) \equiv X$; for $k = 1$, $\mathscr{P}\left({}^{1}X; X\right)$ is equal to $L(X)$, the algebra of all bounded linear operators on $X$. We say that $P : X \longrightarrow X$ is a polynomial on $X$, writing $P \in \mathscr{P}(X; X)$, if $P$ is a finite sum of homogeneous polynomials from $X$ to $X$. We will use the notation $\mathscr{P}(X)$ to denote the space of all finite sums of homogeneous scalar polynomials. Let us recall that $\mathscr{P}(X; X)$ is a normed space if we endowed it with the norm

$$\|P\| := \sup\{\|P(x)\| \ : \ x \in B_X\},$$

and that $\mathscr{P}({}^{k}X; X)$ and $\mathscr{P}({}^{k}X)$ are Banach spaces. Therefore, $\mathscr{P}(X; X)$ embeds isometrically into $\ell_\infty(B_X, X)$, the Banach space of all bounded functions from $B_X$ to $X$ endowed with the supremum norm. We will write $\ell_\infty(B_X)$ when only scalar valued functions are considered. Analogous definitions will be used changing $B_X$ to $S_X$.

For a Banach space $X$, we write $\Pi(X)$ to denote the subset of $X \times X^*$ given by

$$\Pi(X) := \left\{(x, x^*) \ : \ x \in S_X, \ x^* \in S_{X^*}, \ x^*(x) = 1\right\}.$$

Given a bounded function $\Phi : S_X \to X$, its *numerical range* is defined by

$$V(\Phi) := \{x^*(\Phi(x)) \ : \ (x, x^*) \in \Pi(X)\}$$

and its *numerical radius* is defined by

$$v(\Phi) := \sup\{|\lambda| \ : \ \lambda \in V(\Phi)\}.$$

Let us comment that for a bounded function $\Phi : \Omega \to X$, where $S_X \subset \Omega \subset X$, the above definitions are applied by just considering $V(\Phi) := V(\Phi|_{S_X})$.

For $k \in \mathbb{N}$, we define

$$n^{(k)}(X) = \inf\{v(P) \ : \ P \in \mathscr{P}({}^{k}X; X), \ \|P\| = 1\},$$

and call it the *polynomial numerical index of order k of X* [34]. The case $k = 1$ is what is usually called the *numerical index* of $X$. Let us observe that the numerical index is the greatest constant of equivalence (whenever exists) between the numerical radius and the usual norm in the Banach algebra $L(X)$ of all bounded linear operators on the space. It was extended to the polynomial case in [34] by considering homogeneous polynomials instead of linear operators.

The concept of the numerical range was first considered by Toeplitz [84] in 1918 for matrices. In the 1960s, F. Bauer [20] and G. Lumer (this last author in 1968) in [48] gave the definition of the *numerical index* of a Banach space. They gave a theory of the numerical range of bounded linear operators on Banach spaces, extended to arbitrary continuous functions from the unit sphere of a Banach space into the space by F. F. Bonsall, B. E. Cain, and H. Schneider. The classical references for these early studies are the books by Bonsall and Duncan [26, 27].

In the last years it has been a very active area of research the study of the numerical index of a Banach space. For the linear case, the excellent 2006 survey by Kadetse et al. [59] contains a very complete report of these advances until that time. The survey is also very relevant since it raised many open problems (at that time) and afterward many results has been obtained and the reader should give a look at least at [5, 12, 13, 18, 19, 25, 29, 30, 49, 50, 57, 58, 70, 71, 73–75]. Some of these papers goal was to solve open problems of the list proposed in the mentioned [59]. But in this survey we are going to concentrate in the non-linear situation, i.e. in the description of the results obtained about the $k$-polynomial numerical index, when $k > 1$.

Let us go to the polynomial case. The easiest examples are $n^{(k)}(\mathbb{R}) = 1$ and $n^{(k)}(\mathbb{C}) = 1$ for every $k \in \mathbb{N}$. In the complex case, one has $n^{(k)}(C(K)) = 1$ for every $k \in \mathbb{N}$ (see below), while the situation for $L_1(\mu)$ spaces is different since, for instance, $n^{(2)}(\ell_1) \leq \frac{1}{2}$ in the complex case. This shows that the polynomial numerical index distinguishes between $L$-spaces and $M$-spaces in the complex case, and this is not possible with the (linear) numerical index. In the real case, the situation is even more different from the linear case. For instance, a very relevant result is the following by H. J. Lee in 2008.

**Theorem 11.1** ([66, Theorem 2.7]) *The only finite-dimensional real Banach space $X$ with $n^{(2)}(X) = 1$ is $X = \mathbb{R}$.*

Since then, this author together with others has given several improvements of this result. Actually in 2009, J. Kim and H.J. Lee obtained the following theorem.

**Theorem 11.2** ([62, Theorem 4.6]) *Let $X$ be an $n$-dimensional complex Banach space with an absolute norm and let $k$ be an integer greater than or equal to 2. Then $n^{(k)}(X) = 1$ if and only if $X$ is isometric to $\ell_\infty^n$.*

Here $\ell_p^n$ stands for $(\mathbb{K}^n, \|.\|_p)$ for $1 \leq p \leq \infty$ and $\mathbb{K} = \mathbb{R}$ or $\mathbb{K} = \mathbb{C}$.

For infinite-dimensional Banach spaces, H.J. Lee, M. Martín and J. Merí obtained in 2011 the next theorem.

**Theorem 11.3** ([68, Theorem 4.2]) *Let $\Lambda$ be a non-empty set and let $X$ be a linear subspace of $\mathbb{C}^\Lambda$ with absolute norm such that $n^{(2)}(X) = 1$.*

*(a) If $X$ has the Radon-Nikodým property, then $\Lambda$ is finite and $X = \ell_\infty^m$ for some $m \in \mathbb{N}$.*

*(b) If $X$ is an Asplund space, then $X = c_0(\Lambda)$.*

This theorem has the following real version.

**Theorem 11.4** ([68, Theorem 4.3]) *Let $\Lambda$ be a non-empty set and let $X$ be a linear subspace of $\mathbb{R}^\Lambda$ with absolute norm such that $n^{(2)}(X) = 1$. If $X$ has the Radon-Nikodým property or if $X$ is an Asplund space, then $X = \mathbb{R}$.*

Finally we state their last improvement for separable Banach spaces.

**Theorem 11.5** ([67, Theorem 3.1]) *Let $X$ be a Banach space with 1-unconditional basis and $n^{(2)}(X) = 1$. If $X$ is a complex Banach space, then either $X = c_0$ or there exists $m \in \mathbb{N}$ such that $X = \ell_\infty^m$. If $X$ is real, then $X = \mathbb{R}$.*

For non-separable Banach spaces they have another result for Banach spaces of sequences such that their Köthe dual $X'$ is norming.

**Theorem 11.6** ([67, Corollary 3.4]) *Let $X$ be an infinite-dimensional Banach sequence space with absolute norm and such that $X'$ is norming for $X$. In the complex case if $n^{(2)}(X) = 1$, then $c_0 \subset X \subset \ell_\infty$ isometrically. In the real case we always have $n^{(2)}(X) < 1$.*

The real spaces $c_0$, $\ell_1$ and $\ell_\infty$ have polynomial numerical index of order 2 equal to $1/2$. The inequality

$$n^{(k+1)}(X) \leq n^{(k)}(X)$$

holds for every real or complex Banach space $X$ and every $k \in \mathbb{N}$.

It is known that

$$\{n(X) : \ X \text{ real Banach space}\} = [0, 1]$$

with $n(H) = 0$ for any real Hilbert space $H$. Actually, Rosenthal [80] proved that the two-dimensional real Hilbert space is the only one space of that dimension with numerical index zero. In a precise way.

**Theorem 11.7** ([80]) *Let $X$ be a two dimensional real space with $n(X) = 0$. Then, $X$ is isometrically isomorphic to the two dimensional real Euclidean space.*

One elementary proof can be found in [52, Corollary 2.9]. Above inequality gives that $n^{(k)}(H) = 0$ for every $k \in \mathbb{N}$ in the real case. On the other hand in this last paper is it obtained the following fact.

**Proposition 11.1** [52, Corollary 2.8] *If $X$ is a two-dimensional real Banach space with $n^{(2)}(X) = 0$, then $n(X) = 0$.*

As a consequence the real Hilbert space is again the only two-dimensional real Banach space $X$ such that $n^{(2)}(X) = 0$. This is not true anymore for $k \geq 3$. see [52, Theorem 2.10].

In the complex case the situation changes dramatically. We have

$$\{n(X) : \ X \text{ complex Banach space}\} = [e^{-1}, 1].$$

Finally, it follows from a result by Harris [55, Theorem 1] of 1971 that

$$1 \geq n^{(k)}(X) \geq k^{\frac{k}{1-k}}$$

for every complex Banach space $X$ and every $k \geq 2$. Actually, also from Harris' result, the constant $k^{\frac{k}{1-k}}$ is sharp, in the sense that given $k \geq 1$, there exists a two dimensional complex Banach space $X$ such that $n^{(k)}(X) = k^{\frac{k}{1-k}}$.

The situation as stated already above is very different for the real case. Actually it is possible to have $k$-polynomial numerical index positive until some given $k_0$ and zero from $k_0 + 1$ on. There are many examples in the literature. Some of them are given in [52, Example 2.1]:

Let $1 < p < \infty$,

(a) If $p$ is an even number and $k \in \mathbb{N}$, then $n^{(k)}(\ell_p^2) = 0$ if $k \geq p - 1$ and $n^{(k)}(\ell_p^2) > 0$ if $k < p - 1$.
(b) If $p$ is not an even number, then $n^{(k)}(\ell_p^2) > 0$ for every $k \in \mathbb{N}$.

As a corollary [52, Corollary 2.2] it is proven that if $p$ is an even number and $d \geq 2$ an integer, then, $n^{(p-1)}(\ell_p) = n^{(p-1)}(\ell_p^d) = 0$.

Actually in the same paper is given a method to find real Banach spaces, finite-dimensional or not, for which from some index on, the polynomial numerical radius is zero.

**Proposition 11.2** [52, Proposition 2.5] *Let $k$ be a positive integer and let $(X, \|\cdot\|)$ be a real Banach space of dimension greater than one. If the mapping $x \longmapsto \|x\|^{2k}$ is a $2k$-homogeneous polynomial, then $n^{(2k-1)}(X) = 0$.*

In [70, 76] it is proven that the numerical index of a real or complex Banach space $X$ is well preserved for some vector valued Banach spaces as

$$n\big(C(K, X)\big) = n(X).$$

All proofs on linear numerical index case, depend upon some convexity arguments that are going to fail for polynomial numerical index. In the real case, they are not true. For instance,

$$n^{(2)}(\mathbb{R}) = 1 \text{ while } n^{(2)}(c_0) = n^{(2)}(c) = n^{(2)}(\ell_\infty) = 1/2.$$

These results were obtained by Kim et al. in [63]. They were consequence of the following general result.

A Banach space $X$ is said to be *lush* if for every $x, y \in S_X$ and for every $\varepsilon > 0$ there is a slice

$$S = S(B_X, x^*, \varepsilon) = \{x \in B_X \ : \ \operatorname{Re} x^*(x) > 1 - \varepsilon\}, \ x^* \in S_{X^*}$$

such that $x \in S$ and $dist(y, aconv(S)) < \varepsilon$. A real Banach space is said to be *CL-space* if $B_X$ is the absolutely convex hull of every maximal proper face. Finally a closed subspace of a $C(K)$ is *C-rich* if for every nonempty open subset $U$ of $K$ and every $\varepsilon > 0$, there is a positive function $h$ of norm 1 with support inside $U$ such that the distance from $h$ to $X$ is less than $\varepsilon$.

**Theorem 11.8** ([63, Theorem 2.1, Corollary 2.4]) *Let X be a real lush space, a real CL-space or C-rich subspace of a $C(K)$ and $k \geq 1$. We have*

$$\frac{2^k}{2 + M_k(2^k - 2)} \leq n^{(k)}(X)$$

*where* $M_k = \sum_{j=1}^{k} \frac{j^k}{j!(k-j)!}$.

**Theorem 11.9** ([63, Corollary 2.5]) *Let X denote any of the real Banach spaces* $\ell_\infty^m$, $\ell_1^m$ *($m \geq 2$),* $c_0$, $\ell_1$, *and* $\ell_\infty$. *Then*

$$n^{(2)}(X) = \frac{1}{2} \text{ and } \frac{2^k}{2 + M_k(2^k - 2)} \leq n^{(k)}(X) \leq \frac{2}{k}\left(\frac{k-2}{k}\right)^{\frac{k-2}{k}} \quad (k \geq 3).$$

For real spaces $C(K)$ we have the following.

**Theorem 11.10** ([53, theorem 2.1]) *Let K be a compact Hausdorff space with at least two points. Then*

$$n^{(2)}(C(K)) = \frac{1}{2} \quad \text{and} \quad n^{(2k)}(C(K)) \leq \frac{1}{2^k} \ (k \geq 2)$$

*hold in the real case.*

The same inequalities remain true for locally compact Hausdorff topological spaces $L$ with at least two points.

**Theorem 11.11** ([53, Corollary 2.2]) *Let L be a locally compact Hausdorff space L with at least two points. Then*

$$n^{(2)}(C_0(L)) = \frac{1}{2} \quad \text{and} \quad n^{(2k)}(C_0(L)) \leq \frac{1}{2^k} \ (k \geq 2)$$

*hold in the real case.*

What can be said for the polynomial index in the complex case?

Given a compact Hausdorff space $K$, we denote by $C_w(K, X))$ the space of all weakly continuous functions from $K$ to $X$, and by $C_{w^*}(K, X^*)$ the space of all weakly* continuous functions from $K$ into $X^*$. For a locally compact Hausdorff space $L$, we denote by $C_0(L, X)$ the Banach space of all continuous functions from $L$ into $X$ vanishing at infinity and for a completely regular Hausdorff space $\Omega$, $C_b(\Omega, X)$ stands for the Banach space of all bounded an continuous functions from $\Omega$ to $X$, all of them endowed with the supremum norm.

If $Z$ is any of the above spaces, $\Gamma$ is either $K$, $L$ or $\Omega$, and $t \in \Gamma$, we write $\delta_t(f) = f(t)$ for every $f \in Z$.

As already stated for $C(K; X)$ in [70, 76] it is shown that the numerical index of a real or complex Banach space $X$ is well preserved in some of the above mentioned $X$-valued function spaces, concretely

$$n\big(C_w(K, X)\big) = n\big(C(K, X)\big) = n(X).$$

We have already pointed out that the proofs of these results depend upon some convexity arguments and cannot be extended to the polynomial numerical index and that, in general, the results fails for real Banach spaces. However, in the complex case the convexity arguments can be replaced sometimes by the maximum modulus theorem to extend most of the above equalities to the polynomial setting.

**Theorem 11.12** ([37, Corollary 3.3]) *If $X$ is a complex Banach space, $L$ is a locally compact Hausdorff space, $\Omega$ is a completely regular Hausdorff space, and $K$ is a compact Hausdorff space, then*

$$n^{(k)}(C_0(L, X)) = n^{(k)}(C_b(\Omega, X)) = n^{(k)}(C(K, X)) = n^{(k)}(X)$$

*for every $k \in \mathbb{N}$.*

Since $n^{(k)}(\mathbb{C}) = 1$, a direct consequence of above result is that

$$n^{(k)}(C_0(L)) = n^{(k)}(C_b(\Omega)) = n^{(k)}(C(K)) = 1$$

for every $k \in \mathbb{N}$, every locally compact Hausdorff space $L$, any completely regular Hausdorff space $\Omega$, and any compact Hausdorff space $K$. For the more exotic spaces, paradoxically what always it is true is a lower bound.

**Theorem 11.13** ([37, Theorem 3.3]) *Let $X$ be a complex Banach space and $k \in \mathbb{N}$.*

*(a) If $K$ is a compact Hausdorff space, then $n^{(k)}(C_w(K, X)) \geq n^{(k)}(X)$.*
*(b) Let $K$ be a compact Hausdorff space. If $X$ is an Asplund space or $K$ has a dense subset of isolated points, then $n^{(k)}(C_{w^*}(K, X^*)) \geq n^{(k)}(X^*)$.*
*(c) For a compact Hausdorff space $K$ and an open and dense subset $U$ of $K$, we write*

$$Y = \big\{f \in C(K) \ : \ f(K \setminus U) = 0\big\}.$$

*Then, for every subspace $X$ of $C(K)$ containing $Y$, one has $n^{(k)}(X) = 1$ for every $k \in \mathbb{N}$.*

In [78, Theorem 2.3], Martín and Villena showed that $n\big(L_\infty(\mu, X)\big) = n(X)$ for every $\sigma$-finite measure $\mu$ and every Banach space $X$. As usual, given a $\sigma$-finite measure space $(\Omega, \Sigma, \mu)$, we denote by $L_\infty(\mu, X)$ the Banach space of all (equivalence classes of) essentially bounded Böchner-measurable functions from $\Omega$ into $X$, endowed with the essential supremum norm (see [45, 46]).

**Theorem 11.14** ([37, Theorem 4.1]) *Let $(\Omega, \Sigma, \mu)$ be a $\sigma$-finite measure space and let $X$ be a complex Banach space. Then*

$$n^{(k)}\left(L_\infty(\mu, X)\right) = n^{(k)}(X)$$

*for every $k \in \mathbb{N}$.*

If $X$ is a real Banach space. Then in [37, Proposition 4.2] only one inequality is shown,

$$n^{(k)}\left(L_\infty(\mu, X)\right) \leq n^{(k)}(X)$$

for every $k \in \mathbb{N}$.

As already remarked for real Banach spaces there is no a regular pattern on the behavior of polynomial numerical index. But even so it is obtained that for the most representative example of this situation the polynomial numerical index is positive [37, Example 4.3]. The real space $L_\infty[0,1]$ satisfies

$$n^{(2)}\big(L_\infty[0,1]\big) = \frac{1}{2}, \qquad \frac{2^k}{2 + M_k(2^k - 2)} \leq n^{(k)}\big(L_\infty[0,1]\big) \leq \frac{2}{k}\left(\frac{k-2}{k}\right)^{\frac{k-2}{2}} \qquad (k \geq 3)$$

where $M_k = \sum_{j=1}^{k} \frac{j^k}{j!\,(k-j)!}$.

Other example of the strange behavior of numerical index of Banach spaces is that for a positive measure $\mu$ and the spaces $L_p(\mu)$ for $1 < p < \infty$ and $p \neq 2$, it was unknown until 2011 whether their numerical index was positive or not. This was solved in the positive by M. Martín and J. Merí and M.M. Popov.

**Theorem 11.15** ([73, Corollary 3]) *Let $1 < p < \infty$, $p \neq 2$ and let $\mu$ be a positive measure. Then, in the real case, one has*

$$n(L_p(\mu)) \geq \frac{M_p}{12e},$$

*where* $M_p = \max_{t \geq 1} \frac{|t^{p-1} - t^p|}{1 + t^p} > 0$.

Also very recently is has been obtained next result for 2-numerical index.

**Theorem 11.16** ([53, Theorem 5]) *Let $(\Omega, \Sigma, \mu)$ be a measure space so that the dimension of $L_1(\mu)$ is greater than or equal to 2. Then the following hold*

$$n^{(2)}\big(L_1(\mu)\big) = \frac{1}{2} \text{ in the real case and } \frac{1}{3} \leq n^{(2)}\big(L_1(\mu)\big) \leq \frac{1}{2} \text{ in the complex case.}$$

On the other hand, most of the above results about polynomial numerical index, but also to another geometrical properties of the space of real or complex continuous functions as the (polynomial) Daugavet property [38, 61, 86, 87] use at one point or another the classical Urysohn Lemma, that states that a Hausdorff topological space is normal if and only if given two disjoint closed subsets can be separated by a continuous function with values on [0, 1] and taking the value 1 on one closed set and 0 on the other set. Hence many of those results could not be extended to a uniform algebra (i.e. to a closed subalgebra of a complex $C(K)$ that separates points) since the Urysohn Lemma cannot be true, in general, if we ask the function to be in a given uniform algebra, for example its most representative case, the disk algebra $\mathscr{A}(\mathbb{D})$ of continuous functions on the closed unit disk $\overline{\mathbb{D}}$ and holomorphic in the open unit disk $\mathbb{D}$ of $\mathbb{C}$. But in 2013, Cascales et al. in [31, Lemmas 2.8 and 2.12] have constructed Urysohn type Lemmas in order to extend some results on the Bishop-Phelps-Bollobás property to uniform algebras, and the functions obtained in those Lemmas satisfy completely the "spirit" of "separating" the closed sets.

For a unital uniform algebra $A$ of $C(K)$, if

$$S = \{x^* \in A^* : \|x^*\| = 1, x^*(\mathbf{1}) = 1\},$$

then the set $\Gamma_0(A)$ of all $t \in K$ such that $\delta_t$ is an extreme point of $S$ is a boundary for $A$ that is called the *Choquet boundary of* $A$.

**Lemma 11.1** [31, Lemma 2.2] *Let $A \subset C(K)$ be a unital uniform algebra for some compact Hausdorff space $K$ and $\Gamma_0 = \Gamma_0(A)$. Then, for every open set $U \subset K$ with $U \cap \Gamma_0 \neq \phi$ and $0 < \varepsilon < 1$, there exist $f \in A$ and $t_0 \in U \cap \Gamma_0$ such that $f(t_0) = \|f\|_\infty = 1$, $|f(t)| < \varepsilon$ for every $t \in K \backslash U$ and $f(K) \subset R_\varepsilon = \{z \in \mathbb{C} : |\operatorname{Re}(z) - 1/2| + (1/\sqrt{\varepsilon})|\operatorname{Im}(z)| \leq 1/2\}$. In particular,*

$$|f(t)| + (1-\varepsilon)|1 - f(t)| \leq 1, \ \ for all\ t \in K.$$

Given an algebra $A$, by $m(A)$ we denote its spectrum, i.e. the set of non-zero linear forms from $A$ to $\mathbb{K}$ that are also multiplicative (and then automatically continuous). It is well known that $A \subset C(K)$ is a uniform algebra if and only if the Gelfand transform $\widehat{\ }: A \longrightarrow \widehat{A} \subset C(m(A), w(A^*, A))$ defined by $\widehat{a}(\phi) = \phi(a)$ for every $\phi \in m(A)$ and $a \in A$ is an injective isometry. Moreover, if $A$ is a unital uniform algebra, then $\widehat{A}$ is also a unital uniform algebra.

For a non-unital uniform algebra $B \subset C(S)$, that is, a uniform algebra without $\mathbf{1}$, unless otherwise stated in the remainder of this paper, we let $K = (m(B), w^*)$ and identify $B$ with a subalgebra of $C(m(B))$ by using the Gelfand transform. We denote by $A = \{c\mathbf{1} + f : c \in \mathbb{C}, f \in B\}$ the $\|\cdot\|_\infty$-closed subalgebra generated by $B \cup \{\mathbf{1}\}$. Consider the Choquet boundary $\Gamma_0(A)$. Since $B$ is a maximal ideal of $A$, the Gelfand-Mazur theorem ensures that there exists $v \in K$ such that $B = \{f \in A : \delta_v = 0\}$. Denote $\Gamma_0 = \Gamma_0(A) \backslash \{v\}$, then $\Gamma_0$ is a boundary for $B$.

**Lemma 11.2** *([31, Lemma 2.12]) Let $B$ be a non-unital uniform algebra and consider $K = m(B)$. Then, for every open set $U \subset K$ with $U \cap \Gamma_0 \neq \phi$ and $0 < \varepsilon < 1$, there exist $f \in B$ and $t_0 \in U \cap \Gamma_0$ such that $f(t_0) = \|f\|_\infty = 1$, $|f(t)| < \varepsilon$ for every $t \in K \backslash U$ and*

$$|f(t)| + (1-\varepsilon)|1 - f(t)| \leq 1, \ \ for all\ t \in K.$$

Now we are going to show, for a uniform algebra $A$, positive theorems about the polynomial numerical index extending many of the results for $C(K)$ or $C(K, X)$ to $A^X$ (see below for the definition). For all of them this Urysohn type Lemma is the key to succeed. Also, in the next section will do the same for the polynomial Daugavet property and finally, in the last section, to prove results on the approximate hyperplane series property for uniform algebras. If $A$ is a uniform algebra, then $A^X$ is defined to be a subspace of $C(K, X)$ such that

$$A^X = \{f \in C(K, X) \ : \ x^* \circ f \in A \ \text{ for all } \ x^* \in X^*\}.$$

In the case that $A$ is the whole $C(K)$, it is clear that $A^X = C(K, X)$.

Given $f \in A$ and $x \in X$, we define $f \otimes x \in C(K, X)$ by $(f \otimes x)(t) = f(t)x$ for $t \in K$. We write

$$A \otimes X = \{f \otimes x \ ; \ f \in A, \ x \in X\}.$$

From the definition of $A^X$ we note that $A \otimes X \subset A^X$. Given two complex Banach spaces $X$ and $Y$, by $\mathscr{A}_u(B_X, Y)$ we denote the space of all uniformly continuous mappings on the closed unit ball $\overline{B}_X$ and holomorphic on the open unit ball $B_X$ of $X$ with values in $Y$. If $X$ is finite-dimensional since the continuous mappings on the closed unit ball $\overline{B}_X$ are uniformly continuous then we will use the standard notation $\mathscr{A}(B_X, Y)$ instead of $\mathscr{A}_u(B_X, Y)$.

In [36, Theorem 2.1] it is shown that if $A$ is a uniform algebra, then $n^{(k)}(A^X) \geq n^{(k)}(X)$ for every $k \geq 1$. The equality can be obtained with a natural additional property.

**Theorem 11.17** ([36, Theorem 2.2]) *Let $A$ be a uniform algebra and $X$ be a complex Banach space. Assume that $A^X$ has the following property: For every $P \in \mathscr{P}(^k X; X)$ and $t \in K$, $Q : A^X \longrightarrow C(K, X)$ where $Q(f)(t) = P(f(t))$ satisfies that $Q(f) \in A^X$ for every $f \in A^X$. Then $n^{(k)}(A^X) = n^{(k)}(X)$.*

That happens in the following relevant cases.

**Corollary 11.1** [36, Corollary 2.3] *For any complex Banach space $X$, and $k \geq 1$, the following hold:*

1. *For a uniform algebra $A$ we have $n^{(k)}(A) = 1$ for every $k \geq 1$.*
2. *$n^{(k)}(\mathscr{A}(\mathbb{D}^n, X)) = n^{(k)}(X)$ for every $n \in \mathbb{N}$.*
3. *$n^{(k)}\big(\mathscr{A}_u(B_X)\big) = 1$ for every $k \geq 1$.*

For every real or complex Banach space $X$ and every $P \in \mathscr{P}(X; X)$, one has $\|P\| = \|\overline{P}\|$ and $v(P) = v(\overline{P})$, where $\overline{P} \in \mathscr{P}(X^{**}; X^{**})$ is the Aron-Berner extension of $P$ (see [35, Theorem 3.1]). It follows that

$$n^{(k)}(X^{**}) \leq n^{(k)}(X),$$

for every $k \in \mathbb{N}$. It is a very classical result that in the linear case (i.e. k $= 1$) $n(X^*) \leq n(X)$. Hence above inequality is obvious in this case. But the situation in the non-linear case is very different. Before describing this situation let us point out that until 2007 it was unknown if a Banach space (real or complex) could be found such that $n(X^*) < n(X)$. This was solved completely in the negative in the important paper by Boyko et al. [29, Examples 3.1, 3.3]. They built an example of real Banach space such that

$$n(X) = 1 \text{ and } n(X^*) = 0.$$

Also of a complex Banach space such that

$$n(X) = 1 \text{ and } n(X^*) = 1/e.$$

For the polynomial case the situation changes dramatically, for example in [52, Example 2.4] it is shown that the reflexive real Banach space $X = \ell_4^2$ satisfies $n^{(k)}(X) = 0$ and $n^{(k)}(X^*) > 0$ for all $k \geq 3$. For the complex case more is said in [37, Example 3.6]. There is found a complex Banach space $X$ so that

$$n^{(k)}(X) = 1 \quad \text{and} \quad n^{(k)}(X^{**}) = k^{\frac{k}{1-k}}$$

for every $k \geq 2$, i.e. the polynomial numerical index of $X$ is the greatest possible, while that of $X^{**}$ is the smallest polynomial numerical index possible. By $\mathscr{A}_{w^*u}(B_{Y^*}, X)$ it is denoted the space of all $w^*$-uniformly continuous mappings on the closed unit ball and holomorphic on the open unit ball of the dual $Y^*$ of a complex Banach space $Y$ with values in $X$ and by $\mathscr{A}_{wu}(B_Y, X)$ we denote the space of all weakly uniformly continuous mappings on the closed unit ball and holomorphic on the open unit ball of a complex Banach space $Y$ with values in $X$.

**Corollary 11.2** [36, Corollary 2.4] *Let $X$ and $Y$ be complex Banach spaces, $n^{(k)}\big(\mathscr{A}_{w^*u}(B_{Y^*}, X)\big) = n^{(k)}(X)$ for every $k \geq 1$ and $n^{(k)}\big(\mathscr{A}_{wu}(B_Y, X)\big) = n^{(k)}(X)$ for every $k \geq 1$.*

If we consider elements of $\mathscr{A}_u(B_X, X)$ instead of continuous $k$-homogeneous polynomials, we can define, as done in [66], the analytic numerical index of $X$ by

$$n_a(X) = \inf\{v(f) \; : \; f \in \mathscr{A}_u(B_X, X), \; \|f\| = 1\}.$$

Since the space $\mathscr{P}(X; X)$ of all continuous polynomials from $X$ into $X$ is dense in $\mathscr{A}_u(B_X, X)$ we have that

$$n_a(X) = \inf\{v(P) \; : \; P \in \mathscr{P}(X; X), \; \|P\| = 1\}.$$

Thus $n_a(X)$ can be considered the "non-homogeneous polynomial numerical index of $X$". It is immediate that $n_a(X) \leq n^{(k)}(X)$ for every $k \in \mathbb{N}$. But here even in the complex case is possible for a complex Banach space $X$ to have $n_a(X) = 0$. Indeed, in [37, Example 5.2] is given an example of a complex Banach space such that

$$n_a(X) = 1 \quad \text{but} \quad n_a(X^{**}) = 0.$$

On the other hand in 2009 it was obtained next theorem by J. Kim and H.J. Lee.

**Theorem 11.18** ([62, Corollary 2.5]) *Let $X$ be an $n$-dimensional complex Banach space with an absolute norm. Then $n_a(X) = 1$ if and only if $X$ is isometric to $\ell_\infty^n$.*

Suppose that $A$ is a uniform algebra, next result resumes what is known about the analytic numerical index next.

**Theorem 11.19** ([36, Corollary 2.3]) *For any complex Banach space $X$, and $k \geq 1$, the following hold. For a uniform algebra $A$, we have Then $n_a(A^X) \geq n_a(X)$. In particular $n_a(A) = 1$. Moreover, $n_a(\mathscr{A}(\mathbb{D}^n, X)) = n_a(X)$ for every $n \in \mathbb{N}$, $n_a\big(\mathscr{A}_u(B_X)\big) = 1$, $n_a\big(\mathscr{A}_{w^*u}(B_{Y^*}, X)\big) = n_a(X)$ and $n_a\big(\mathscr{A}_{wu}(B_Y, X)\big) = n_a(X)$.*

## 11.2 Polynomial Daugavet Property

The aim of this section is to survey on very recent developments about the $k$-order Daugavet property and the $k$-order alternative Daugavet property.

In 1963, Daugavet [44] showed that every compact linear operator $T$ on $C[0,1]$ satisfies

$$\|\mathrm{Id} + T\| = 1 + \|T\|,$$

a norm equality which has currently become known as the Daugavet equation. Over the years, the validity of the above equality has been established for many classes of operators on many Banach spaces. For instance, weakly compact linear operators on $C(K)$, $K$ perfect, and $L_1(\mu)$, $\mu$ atomless, satisfy Daugavet equation (see [85] for an elementary approach). We refer the reader to the books [1, 2] and the papers [3, 11, 21, 59, 61, 83, 87] for more information and background. It is also a remarkable result given in 1970 by Duncan et al. [48] that, for every compact Hausdorff space $K$ and every bounded linear operator $T$ on $C(K)$, the equality

$$\max_{\omega \in \mathbb{T}} \|\mathrm{Id} + \omega T\| = 1 + \|T\|$$

holds, where we use $\mathbb{T}$ to denote $\{-1, 1\}$ for the real case and the torus for the complex case. The above norm equality is now known as the alternative Daugavet equation [75], and it is satisfied by all bounded linear operators on $C(K)$ and $L_1(\mu)$, $K$ and $\mu$ arbitrary. We refer the reader to [48, 59, 71, 75] and references there in for background.

The numerical range of an operator, described in the previous section, is intimately related to the Daugavet and alternative Daugavet equations. In the seventies, Harris [55, 56] showed that a good setting to deal with numerical ranges is the space of bounded uniformly continuous functions on the unit sphere of a Banach space with values in the space (see also Rodríguez-Palacios [79]).

Generalizing the linear case, it is said that $\Phi \in \ell_\infty(B_X, X)$ satisfies the *Daugavet equation* if the norm equality

$$\|\mathrm{Id} + \Phi\| = 1 + \|\Phi\| \tag{DE}$$

holds, and we say that $\Phi$ satisfies the *alternative Daugavet equation* if there exists $\omega \in \mathbb{T}$ such that $\omega\,\Phi$ satisfies (DE) or, equivalently, if

$$\max_{|\omega|=1} \|\mathrm{Id} + \omega\,\Phi\| = 1 + \|\Phi\|. \tag{ADE}$$

It is clear that (DE) implies (ADE) but, in general, they are not the same ($-\mathrm{Id}$ always satisfies (ADE) but never (DE)).

If (DE) (resp. (ADE)) happens for all rank-one $P \in \mathscr{P}(X; X)$, it is said that $X$ has *the polynomial Daugavet property* (resp. *alternative polynomial Daugavet property*). Analogously, we say that a Banach space $X$ has the *$k$-order Daugavet property* ($k$-DP in short) if all rank-one $k$-homogeneous polynomials satisfy (DE). When $k = 1$, we simply say that $X$ has the *Daugavet property*. Finally, $X$ has the *$k$-order alternative Daugavet property* ($k$-ADP in short) if all rank-one $k$-homogeneous polynomials satisfy (ADE), and we use the name alternative Daugavet property in the linear case. These polynomial counterparts extension of the original Daugavet equation were introduced in [38].

If $X$ is a Banach space, by $C_u(B_X, X)$ we denote the Banach space of all uniformly continuous $X$ valued functions on $B_X$ endowed with the supremum norm. Note that since $B_X$ is convex and bounded, then every function in $C_u(B_X, X)$ is also bounded. If $X$ is a complex Banach space, we denote by $\mathscr{A}_\infty(B_X)$ (resp. $\mathscr{A}_\infty(B_X, X)$) the Banach space of all complex valued (resp. $X$ valued) functions on $B_X$ which are holomorphic in the open unit ball, and bounded and continuous in $B_X$. $\mathscr{A}_\infty(B_X, X)$ embeds isometrically into $\ell_\infty(S_X, X)$.

Next theorem is a generalization of [61, Lemma 2.2 and Theorem 2.3], where only bounded linear operators were considered.

The following notation is an extension to Banach spaces of the one given in the first section for uniform algebras. If $X$ is a Banach space and $\mathscr{X}$ is a subspace of $\ell_\infty(B_X)$, we write $\mathscr{X}^X$ to denote the space of all functions $\Phi : B_X \longrightarrow X$ such that $x^* \circ \Phi \in \mathscr{X}$ for every $x^* \in X^*$. The simplest elements of $\mathscr{X}^X$ are the tensors of functions of $\mathscr{X}$ and elements of $X$, namely, for $\varphi \in \mathscr{X}$ and $x_0 \in X$, we denote by $\varphi \otimes x_0$ the element of $\mathscr{X}^X$ given by $[\varphi \otimes x_0](x) = \varphi(x)x_0$ for every $x \in B_X$. We will use the result below for the cases $\mathscr{X} = \mathscr{P}(X)$ and $\mathscr{X} = \mathscr{P}\left({}^k X\right)$.

**Theorem 11.20** ([38, Theorem 1.1]) *Let $X$ be a Banach space and let $\mathscr{X}$ be a subspace of $\ell_\infty(B_X)$. Then the following are equivalent:*

*(i) For every $\varphi \in \mathscr{X}$ and every $x_0 \in X$, $\varphi \otimes x_0$ satisfies* (DE).
*(ii) For every $\varphi \in S_{\mathscr{X}}$, every $x_0 \in S_X$, and every $\varepsilon > 0$, there exist $\omega \in \mathbb{T}$ and $y \in B_X$ such that*

$$\mathrm{Re}\,\omega\varphi(y) > 1 - \varepsilon \quad \textit{and} \quad \|x_0 + \omega y\| > 2 - \varepsilon.$$

*(iii) Every $\Phi \in \mathscr{X}^X$ whose image is relatively weakly compact satisfies* (DE).

Let us point out that a consequence of above theorem is that a Banach space $X$ has the polynomial Daugavet property if and only if every weakly compact polynomial

$P : X \to X$ satisfies the Daugavet equation. Also that a Banach space $X$ has the $k$-order Daugavet property if and only if every weakly compact $k$-homogeneous polynomial $P : X \to X$ satisfies the Daugavet equation.

For the alternative Daugavet equation we can state an analogous result.

**Theorem 11.21** ([38, Corollary 1.2]) *Let $X$ be a Banach space and let $\mathscr{Z}$ be a subspace of $\ell_\infty(B_X)$. Then the following are equivalent:*

*(i) For every $\varphi \in \mathscr{Z}$ and every $x_0 \in X$, $\varphi \otimes x_0$ satisfies* (ADE).
*(ii) For every $\varphi \in S_{\mathscr{Z}}$, every $x_0 \in S_X$, and every $\varepsilon > 0$, there exist $\omega_1, \omega_2 \in \mathbb{T}$ and $y \in B_X$ such that*

$$\operatorname{Re} \omega_1 \varphi(y) > 1 - \varepsilon \quad \text{and} \quad \|x_0 + \omega_2 y\| > 2 - \varepsilon.$$

*(iii) For every $\varphi \in S_{\mathscr{Z}}$, every $x_0 \in S_X$, and every $\varepsilon > 0$, there exist $\omega \in \mathbb{T}$ and $y \in B_X$ such that*

$$|\varphi(y)| > 1 - \varepsilon \quad \text{and} \quad \|x_0 + \omega y\| > 2 - \varepsilon.$$

*(iv) Every $\Phi \in \mathscr{Z}^X$ whose image is relatively weakly compact satisfies* (ADE).

Next proposition shows the relationship between Daugavet equation and numerical range.

**Proposition 11.3** [38, Proposition 1.3] *Let $X$ be a Banach space and let $\Phi$ be an element of $C_u(B_X, X)$. Then we have:*

*(a) $\Phi$ satisfies* (DE) *if and only if* $\|\Phi\| = \sup \operatorname{Re} V(\Phi)$.
*(b) $\Phi$ satisfies* (ADE) *if and only if* $\|\Phi\| = v(\Phi)$.

Neither $\mathbb{R}$ nor $\mathbb{C}$ have the property of that every (weakly compact) polynomial on them satisfies (DE). Actually this a property only fulfilled by some infinite-dimensional Banach spaces. But every polynomial on $\mathbb{C}$ satisfies (ADE).

For polynomials we have the following characterization of when all weakly compact polynomials satisfy (DE).

**Theorem 11.22** ([38, Corollary 2.2]) *Let $X$ be a Banach space. Then the following are equivalent:*

*(i) For every $P \in \mathscr{P}(X)$ and every $x_0 \in X$, the polynomial $P \otimes x_0$ satisfies* (DE).
*(ii) For every $P \in \mathscr{P}(X)$ with $\|P\| = 1$, every $x_0 \in S_X$, and every $\varepsilon > 0$, there exist $\omega \in \mathbb{T}$ and $y \in B_X$ such that*

$$\operatorname{Re} \omega P(y) > 1 - \varepsilon \quad \text{and} \quad \|x_0 + \omega y\| > 2 - \varepsilon.$$

*(iii) Every weakly compact $P \in \mathscr{P}(X; X)$ satisfies* (DE).

We have already defined C-rich subspaces of a $C(K)$. Now we extend that definition to a completely regular Hausdorff topological space $\Omega$. Let $X$ be a Banach space, it is said that a subspace $\mathscr{F}$ of $C_b(\Omega, X)$, the Banach space of all bounded $X$ valued continuous functions on $\Omega$ endowed with the supremum norm, is *$C_b$-rich* if for every open subset $U$ of $\Omega$, every $x \in X$ and every $\varepsilon > 0$ there exists a continuous function $\varphi : \Omega \longrightarrow [0, 1]$ of norm one with support included in $U$ such that the distance of $\varphi \otimes x$ to $\mathscr{F}$ is less than $\varepsilon$. Examples of $C_b$-rich subspaces of $C_b(\Omega, X)$ spaces appear in [60].

**Theorem 11.23** ([38, theorem 2.4]) *Let $\Omega$ be a completely regular Hausdorff topological space without isolated points, let $X$ be a Banach space and let $\mathscr{F}$ be a $C_b$-rich subspace of $C_b(\Omega, X)$. Then every weakly compact polynomial from $\mathscr{F}$ into itself satisfies the Daugavet equation.*

We are going to present here, for the most important case $\mathscr{F} = C_b(\Omega, X)$, a proof different of the one published at [38] that it is self contained and does not depend on a result by Bogdanowicz [23] on weak continuity of polynomials on the ball of $c_0$.

We need a technical preliminary result.

**Lemma 11.3** *Given a Banach space $X$ over $\mathbb{K}$ and $k$ a positive integer, there exists a constant $D(k)$ such that, for every $P \in \mathscr{P}\left({}^kX\right)$ and every $x_1, \ldots, x_n \in X$, we have*

$$|P(x_1)| + \cdots + |P(x_n)| \leq D(k)\|P\| \sup_{\alpha_j \in \mathbb{K},\ |\alpha_j| \leq 1} \|\alpha_1 x_1 + \cdots + \alpha_n x_n\|^k$$

*Moreover, $D(k) = 1$ for $\mathbb{K} = \mathbb{C}$ and $D(k) \leq \frac{(4k)^k}{k!}$ for $\mathbb{K} = \mathbb{R}$.*

*Proof* If $\mathbb{K} = \mathbb{C}$, the inequality was proved by probabilistic methods by Aron and Glovecnik [17] (for a proof see also [47, (1.55) in p. 64]).

If $\mathbb{K} = \mathbb{R}$, we consider the complexification $Y = X + iX$ of $X$, which is a complex Banach space when endowed with the norm $\|x + iy\| = \sup_{\theta \in [0, 2\pi]} \|(\cos\theta)x + (\sin\theta)y\|$, and we consider $Q : Y \longrightarrow \mathbb{C}$ the complexification of $P$ given by

$$Q(x + iy) := \sum_{j=0}^{k} \binom{k}{j} i^j \widehat{P}(x^{(k-j)}, y^{(j)}),$$

where $\widehat{P}$ is the symmetric $k$-linear form associated with $P$. Hence, by the polarization constants, we have $\|Q\| \leq 2^k \frac{k^k}{k!} \|P\|$. By the complex case

$$\begin{aligned} |P(x_1)| + \cdots + |P(x_n)| &= Q(x_1)| + \cdots + |Q(x_n)| \\ &\leq \|Q\| \sup_{\alpha_j \in \mathbb{C},\ |\alpha_j| \leq 1} \|\alpha_1 x_1 + \cdots + \alpha_n x_n\|^k \\ &\leq \frac{(4k)^k}{k!} \|P\| \sup_{\lambda_j \in \mathbb{R},\ |\lambda_j| \leq 1} \|\lambda_1 x_1 + \cdots + \lambda_n x_n\|^k. \end{aligned}$$ □

*Proof (Proof of Theorem 11.23)* Let $P \in \mathscr{P}(C_b(\Omega, X))$ such that $\|P\| = 1$, $P = \sum_{i=0}^{m} P_i$, with $P_i \in \mathscr{P}\left({}^i C_b(\Omega, X)\right)$ and $f_0 \in S_{C_b(\Omega,X)}$. Let $M > 0$ such that $\|P_i\| \leq M$ for every $0 \leq i \leq m$ (in the complex case, by the Cauchy inequalities, we have $\|P_i\| \leq \|P\| = 1$ for every $0 \leq i \leq m$ and we can take $M = 1$). Since $P$ is uniformly continuous when restricted to any bounded subset of $C_b(\Omega, X)$, given $\varepsilon > 0$, there exists $0 < \delta < \varepsilon/2$ such that

$$|P(f) - P(g)| < \frac{\varepsilon}{2}, \tag{11.1}$$

for all $f, g \in C_b(\Omega, X)$ satisfying $\|f - g\| < \delta$, $\|f\| \leq 1$, $\|g\| \leq 1$. We fix $n \in \mathbb{N}$ such that

$$\frac{1}{n} < \frac{\delta}{2Mm9^m}\left(\frac{m!}{m^m}\right)^2, \tag{11.2}$$

we take $h \in C_b(\Omega, X)$ with $\|h\| \leq 1$ such that $|P(h)| > 1 - \frac{\varepsilon}{4}$, and we also take $\omega \in \mathbb{T}$ such that

$$\operatorname{Re} \omega P(h) = |P(h)| > 1 - \frac{\varepsilon}{4}. \tag{11.3}$$

Clearly, by the maximum modulus theorem, $h$ can be consider in $S_{C_b(\Omega,X)}$ in the complex case. Since the open set $U = \{t \in \Omega : \|f_0(t)\| > 1 - \delta\}$ is not empty and $\Omega$ is perfect, there exist $U_1, \ldots, U_n$ non empty open subsets of $U$ pairwise disjoint. As $\Omega$ is completely regular and Hausdorff, we can find non empty open subsets $V_j$ of $U_j$ and continuous functions $\eta_j : \Omega \longrightarrow [0, 1]$ with

$$\operatorname{supp} \eta_j \subset U_j \quad \text{and} \quad \eta_j(t) = 1 \quad (t \in V_j)$$

for all $j = 1, \ldots, n$. For each $j$ we take a $t_j \in V_j$ and a continuous function $\phi_j : \Omega \longrightarrow [0, 1]$ such that

$$\operatorname{supp} \phi_j \subset V_j \quad \text{and} \quad \phi_j(t_j) = 1$$

for all $j = 1, \ldots, n$. We have that

$$\left\| \sum_{j=1}^{n} \alpha_j(-h\eta_j + \omega^{-1} f_0(t_j)\phi_j) \right\| \leq 2,$$

for all $\alpha_j \in \mathbb{K}$, $|\alpha_j| \leq 1$, $j = 1, \ldots, n$. On the other hand

$$\begin{aligned} P(h - h\eta_j + \omega^{-1} f_0(t_j)\phi_j) - P(h) = \sum_{i=1}^{m}\sum_{k=1}^{i} \binom{i}{k} \widehat{P_i}(h^{(i-k)}, \\ (-h\eta_j + \omega^{-1} f_0(t_j)\phi_j)^{(k)}), \end{aligned} \tag{11.4}$$

where, for each $1 \leq i \leq m$, $\widehat{P_i}$ is the symmetric $i$-linear form associated with $P_i$. For each $1 \leq k \leq i$ the map

$$f \longmapsto \widehat{P_i}(h^{(i-k)}, f^{(k)}) \qquad \big(f \in C_b(\Omega, X)\big)$$

is a $k$-homogeneous continuous polynomial on $C_b(\Omega, X)$ of norm less than or equal to $\frac{k^k}{k!}\|P_i\|\|h\|^{i-k} \leq M\frac{k^k}{k!}$. If we apply Lemma 11.3, we obtain

$$\begin{aligned}
&\sum_{j=1}^{n}\sum_{i=1}^{m}\sum_{k=1}^{i}\binom{i}{k}\left|\widehat{P_i}(h^{(i-k)}, (-h\eta_j + \omega^{-1}f_0(t_j)\phi_j)^{(k)})\right| \\
&\quad = \sum_{i=1}^{m}\sum_{k=1}^{i}\binom{i}{k}\sum_{j=1}^{n}\left|\widehat{P_i}(h^{(i-k)}, (-h\eta_j + \omega^{-1}f_0(t_j)\phi_j)^{(k)})\right| \\
&\quad \leq \sum_{i=1}^{m}\sum_{k=1}^{i}\binom{i}{k}D(k)\frac{k^k}{k!}M \sup_{\alpha_j\in\mathbb{K},\ |\alpha_j|\leq 1}\left\|\sum_{j=1}^{n}\alpha_j(-h\eta_j + \omega^{-1}f_0(t_j)\phi_j)\right\|^k \\
&\quad \leq \sum_{i=1}^{m}\sum_{k=1}^{i}\binom{i}{k}D(k)\frac{k^k}{k!}2^k M \leq Mm\left(\frac{m^m}{m!}\right)^2 9^m.
\end{aligned}$$

Thus, by (11.2), there exists a $j_0 \in \{1, \ldots, n\}$ such that

$$\sum_{i=1}^{m}\sum_{k=1}^{i}\binom{i}{k}\left|\widehat{P_i}(h^{(i-k)}, (-h\eta_{j_0} + \omega^{-1}f_0(t_{j_0})\phi_{j_0})^{(k)})\right| < \frac{\delta}{2} < \frac{\varepsilon}{4}.$$

Now, by applying (11.4), we obtain that

$$|P(h - h\eta_{j_0} + \omega^{-1}f_0(t_{j_0})\phi_{j_0}) - P(h)| < \frac{\varepsilon}{4}.$$

Combining with (11.3), we have

$$\operatorname{Re}\,\omega P(h - h\eta_{j_0} + \omega^{-1}f_0(t_{j_0})\phi_{j_0}) > 1 - \frac{\varepsilon}{2}. \tag{11.5}$$

Since $\operatorname{supp}(h - h\eta_{j_0}) \cap \operatorname{supp}(\omega^{-1}f_0(t_{j_0})\phi_{j_0}) = \emptyset$, we get

$$1 \geq \|h - h\eta_{j_0} + \omega^{-1}f_0(t_{j_0})\phi_{j_0}\| \geq |f_0(t_{j_0})| > 1 - \delta.$$

If we define

$$g = \frac{h - h\eta_{j_0} + \omega^{-1}f_0(t_{j_0})\phi_{j_0}}{\|h - h\eta_{j_0} + \omega^{-1}f_0(t_{j_0})\phi_{j_0}\|} \in C_b(\Omega, X),$$

we will have $\|g\| = 1$ and

$$\|g - (h - h\eta_{j_0} + \omega^{-1} f_0(t_{j_0})\phi_{j_0})\| = 1 - \|(h - h\eta_{j_0} + \omega^{-1} f_0(t_{j_0})\phi_{j_0})\| < \delta.$$

Hence, by (11.1) and (11.5)

$$\text{Re } \omega P(g) > 1 - \varepsilon, \tag{11.6}$$

and

$$\begin{aligned} \|f_0 + \omega g\| &\geq |f_0(t_{j_0}) + \omega g(t_{j_0})| \\ &= \left| f_0(t_{j_0}) + \frac{f_0(t_{j_0})}{\|h - h\eta_{j_0} + \omega^{-1} f_0(t_{j_0})\phi_{j_0}\|} \right| \geq 2|f_0(t_{j_0})| > 2 - \varepsilon. \end{aligned} \tag{11.7}$$

By Theorem 11.2 we have that every weakly compact polynomial from $C_b(\Omega, X)$ into itself satisfies the Daugavet equation. □

The density of the space of polynomials on a complex Banach space $Z$ in $\mathscr{A}_u(B_Z, Z)$, gives us the following consequence.

**Corollary 11.3** [38, Corollary 2.6] *Let $\Omega$ be a completely regular Hausdorff topological space without isolated points and let $X$ be a complex Banach space. Then, every weakly compact $\Phi$ in $\mathscr{A}_u\big(B_{C_b(\Omega,X)}, C_b(\Omega, X)\big)$ satisfies the Daugavet equation.*

We now state the main results known about the rest of the classical spaces that we have already dealt with related to the polynomial numerical index.

**Theorem 11.24** ([37, Theorem 6.5 and Corollary 6.8]) *Let $X$ be a real or complex Banach space. We have:*

*(a) $L_\infty(\mu, X)$ has the polynomial Daugavet property only if $\mu$ is an atomless $\sigma$-finite measure or $X$ has the polynomial Daugavet property.*
*(b) If $K$ is a perfect compact Hausdorff topological space, then $C_w(K, X)$ and $C_{w^*}(K, X^*)$ have the polynomial Daugavet property.*

For complex Banach spaces (a) above has a counterpart for $C_w(K, X)$ and $C_{w^*}(K, X^*)$.

**Proposition 11.4** [37, Proposition 6.10] *Let $X$ be a complex Banach space, $K$ a compact space, $L$ a locally compact space and $\Omega$ a completely regular space.*

*(a) $C(K, X)$ has the polynomial Daugavet property if and only if $X$ does or $K$ is perfect.*
*(b) $C_w(K, X)$ has the polynomial Daugavet property if and only if $X$ does or $K$ is perfect.*

(c) *$C_0(L, X)$ has the polynomial Daugavet property if and only if X does or L is perfect.*
(d) *$C_b(\Omega, X)$ has the polynomial Daugavet property if and only if X does or $\Omega$ is perfect.*
(e) *If K contains a dense subset of isolated points and $X^*$ has the polynomial Daugavet property, then $C_{w^*}(K, X^*)$ has the polynomial Daugavet property.*

**Corollary 11.4** [37, Corollary 6.6] *Let X be a complex Banach space. Let Y denote any of the complex spaces $C_w(K, X)$, $C_{w^*}(K, X^*)$ for K compact and perfect, or $C_0(L, X)$ for L locally compact and perfect, or $C_b(\Omega, X)$ for $\Omega$ completely regular without isolated points, or $L_\infty(\mu, X)$ with $\mu$ an atomless $\sigma$-finite measure. Then every weakly compact $\Phi \in \mathscr{A}_u(B_Y, Y)$ satisfies the Daugavet equation.*

If a Hausdorff compact set $K$ has at least two point and at least one is isolated, then the Daugavet property does not hold anymore. Even so the alternative Daugavet property remains true.

**Corollary 11.5** [38, Corollary 2.10] *Let K be a compact Hausdorff space. Then every polynomial from the complex space $C(K)$ into itself satisfies* (ADE).

Finally we state the most recent results in this area.

**Theorem 11.25** ([72, Theorem 2.6 and Theorem 3.3]) *Let $(\Omega, \Sigma, \mu)$ be an atomless measure space and X any Banach space. Then the space $L_1(\mu, X)$ has the polynomial Daugavet property.*

Next theorem shows that a result by Werner [86, Theorem 3.3] on the Daugavet equation is actually true for the polynomial Daugavet equation.

**Theorem 11.26** ([36, Corollary 2.9]) *If A is a uniform algebra whose Choquet boundary has no isolated points $A^X$ has the polynomial Daugavet property for any complex Banach space X.*

If we assume that $X$ has the polynomial Daugavet property, then $A^X$ has this property too.

**Theorem 11.27** ([36, Theorem 2.11]) *Let A be a uniform algebra and X be a complex Banach space with the polynomial Daugavet property. Then $A^X$ has the polynomial Daugavet property.*

Since given $U$ either the Euclidean ball in $\mathbb{C}^n$ or the polydisk $\mathbb{D}^n$, the Choquet boundary of $\mathscr{A}(U)$ does not have isolated points, we have the following corollary, that extends a result by Wojtaszczyk [88, Remark after Corollary 3].

**Theorem 11.28** ([36, Corollary 2.10]) *If U is either the Euclidean ball in $\mathbb{C}^n$ or the polydisk $\mathbb{D}^n$, then $\mathscr{A}(U, X)$ has the polynomial Daugavet property for every complex Banach space X. In particular, $\mathscr{A}(U)$ has the polynomial Daugavet property.*

Now we study the $k$-order Daugavet property and the $k$-order alternative Daugavet property as there exist spaces (for example the trivial case $\mathbb{R}$), for which there exist weakly compact polynomials that do not satisfy (ADE) and nevertheless every weakly compact $k$-homogeneous polynomial satisfies (ADE) ($k \geq 2$).

The following result gives us a surprising behavior of the Daugavet equation and the alternative Daugavet equation for $k$-homogeneous polynomials with $k > 1$.

**Proposition 11.5** [38, Corollary 3.3] *Let $X$ be a Banach space and let $k$ be an integer, $k \geq 2$.*

*(a) If $X$ is complex, then the $k$-DP and the $k$-ADP are equivalent.*
*(b) If $X$ is real and $k$ is even, then the $k$-DP and the $k$-ADP are equivalent.*

On the other hand [38, Example 3.14] the real spaces $c_0$, $c$, and $\ell_\infty$ do not have the $k$-ADP for any $k \geq 2$.

Now, we will study the relationship between the $k$-DP and the $k$-ADP for different values of the integer $k$.

**Proposition 11.6** [38, Corollaries 6 and 7] *Let $X$ be a Banach space and let $k$ be a positive integer. If $X$ has the $(k+1)$-ADP, then $X$ has the $k$-ADP. Let $X$ be a complex Banach space and let $k$ be a positive integer, $k \geq 2$. If $X$ has the $(k+1)$-DP, then $X$ has the $k$-DP.*

But the above result is not valid for $k = 1$. Since the complex space $c_0$ has the 2-DP, but it does not have the 1-DP. Also the above result is false in the real case, since the real space $\mathbb{R}$ has the $2m$-DP for every $m \in \mathbb{N}$, but it does not have the $(2m-1)$-DP for any $m \in \mathbb{N}$.

In [39, pp. 141] it is shown a weakly compact 2-homogeneous polynomial $P \in \mathcal{P}\left({}^2\ell_1; \ell_1\right)$ (either in the real case and in the complex case) which satisfies $v(P) \leq \frac{1}{2}\|P\|$. Hence, The space $\ell_1$ does not have the $k$-ADP for any $k \geq 2$.

In [38, Example 3.14] it is proven that the real spaces $c_0$, $c$, $\ell_\infty$, $\ell_\infty^n$ for every $n \geq 2$ do not have the $k$-ADP for any $k \geq 2$. Actually, [38, Example 3.15] shows that the real $C(K)$ when $K$ is a compact set with at least two isolated points do not have the $k$-ADP for any $k \geq 2$.

Finally J. Kim and H.J. Lee obtained the following result about the $k$-DP.

**Theorem 11.29** ([62, Corollary 3.5]) *Let $X$ be a finite-dimensional Banach space with an absolute norm. The following are equivalent:*

1. *$X$ has the alternative polynomial Daugavet property.*
2. *$X$ has the $k$-DP for some $k \geq 2$.*
3. *$X$ has the $k$-DP for every $k \geq 2$.*
4. *$n^{(k)}(X) = 1$ for some $k \geq 2$.*
5. *$n^{(k)}(X) = 1$ for every $k \geq 2$.*
6. *$n_a(X) = 1$.*
7. *$X = \ell_\infty^n$.*

## 11.3 The Bishop-Phelps-Bollobás Property for Multilinear Mappings

The Bishop-Phelps Theorem states the denseness of the set of norm attaining functionals in the (topological) dual of any Banach space [22]. Bollobás proved a quantitative version of this result, known nowadays as the Bishop-Phelps-Bollobás Theorem [24], which is very useful to study numerical ranges of operators (see for instance [26]). The Bishop-Phelps-Bollobás Theorem can be stated as follows:

*Let $X$ be a Banach space and $0 < \varepsilon < 1$. Given $x \in B_X$ and $x^* \in S_{X^*}$ with $|1 - x^*(x)| < \frac{\varepsilon^2}{4}$, there are elements $y \in S_X$ and $y^* \in S_{X^*}$ such that $y^*(y) = 1$, $\|y - x\| < \varepsilon$ and $\|y^* - x^*\| < \varepsilon$.*

The study of both results in the vector valued case has attracted the interest of many authors. In his pioneering work [69] Lindenstrauss studied versions of the Bishop-Phelps Theorem for operators. He gave the first counterexample of Banach spaces $X$ and $Y$ such that the subset of norm attaining operators between $X$ and $Y$ is not dense in $\mathscr{L}(X, Y)$. He also provided either isomorphic or isometric properties of the Banach spaces $X$ and $Y$ to obtain positive results. Bourgain generalized one of these results proving that for each Banach space $X$ with the Radon-Nikodým property, the mentioned density holds for every Banach space $Y$ [28]. The excellent survey [4] contains the most relevant achievements in the field until 2006 (see also [7, 10, 16]).

For versions of the Bishop-Phelps-Bollobás Theorem the situation is quite different. For instance, let us mention that the Radon-Nikodým property on $X$ does not imply a version of the Bishop-Phelps-Bollobás Theorem for operators from $X$ into any Banach space $Y$. Even in the case of $\ell_1$ this result fails. The next property was introduced in [6].

**Definition 11.1** [6, *Definition 1.1*] Let $X$ and $Y$ be Banach spaces. The pair $(X, Y)$ satisfies the *Bishop-Phelps-Bollobás property for operators* if given $\varepsilon > 0$, there are $\eta(\varepsilon) > 0$ and $\beta(\varepsilon) > 0$ with $\lim_{t \to 0} \beta(t) = 0$ such that for any $T \in S_{\mathscr{L}(X,Y)}$, if $x_0 \in S_X$ is such that $\|Tx_0\| > 1 - \eta(\varepsilon)$, then there exist a point $u_0 \in S_X$ and an operator $S \in S_{\mathscr{L}(X,Y)}$ that satisfy the following conditions:

$$\|Su_0\| = 1, \quad \|u_0 - x_0\| < \beta(\varepsilon) \quad \text{and} \quad \|S - T\| < \varepsilon.$$

In this case, we also say that the space $\mathscr{L}(X, Y)$ has the Bishop-Phelps-Bollobás property.

Here we are going to concentrate only in the non-linear case and we also study the applications of the Urysohn type Lemmas to the AHSP of $A^X$ whenever $A$ is a uniform algebra. This property was introduced in [6, Definition 3.1] to characterize a Banach space $X$ such that the pair $(\ell_1, X)$ has the Bishop-Phelps-Bollobás property for operators.

A Banach space X is said to have the AHSP if for every $\varepsilon > 0$ there exist $\gamma(\varepsilon) > 0$ and $\eta(\varepsilon) > 0$ with $\lim_{\varepsilon \to 0^+} \gamma(\varepsilon) = 0$ such that for every sequence $(x_k)_{k=1}^{\infty} \subset B_X$ and for every convex series $\sum_{k=1}^{\infty} \alpha_k$ satisfying

$$\Big\| \sum_{k=1}^{\infty} \alpha_k x_k \Big\| > 1 - \eta(\varepsilon)$$

there exist a subset $A \subset \mathbb{N}$, a subset $\{z_k : k \in A\} \subset S_X$ and $x^* \in S_{X^*}$ such that:

(i) $\sum_{k \in A} \alpha_k > 1 - \gamma(\varepsilon)$.
(ii) $\|z_k - x_k\| < \varepsilon$ for all $k \in A$.
(iii) $x^*(z_k) = 1$ for all $k \in A$.

It is easy to check that the AHSP holds if it is satisfied just for every finite convex combination instead of every infinite convex series. The following Banach spaces were shown to have the AHSP (see [6, 41]): (a) a finite-dimensional space, (b) a real or complex space $L_1(\mu)$ for a $\sigma$-finite measure $\mu$, (c) a uniformly convex space, (d) a lush space, and (see [9]) (e) the space of continuous mappings $C(K, X)$ defined on a compact Hausdorff topological space $K$ with values in a uniformly convex Banach space $X$.

**Corollary 11.6** [36, Corollary 2.14] *Suppose that A is a uniform algebra.*

1. *If X is a lush complex Banach space, then $A^X$ has the AHSP. In particular, A has the AHSP.*
2. *If A is unital and $A^X$ is lush, then the complex Banach space X has the AHSP.*

Now we generalize this corollary.

**Theorem 11.30** ([36, Theorem 2.15]) *Suppose that A is a uniform algebra.*

1. *If X is a complex Banach space that has the AHSP, then $A^X$ has the AHSP.*
2. *If A is unital and $A^X$ has the AHSP, then the complex Banach space X has the AHSP.*

**Corollary 11.7** [36, Corollary 2.17] *If X is complex Banach space that has the AHSP, then for any complex Banach space Y:*

1. *$\mathcal{A}_u(B_Y)$ has the AHSP.*
2. *$\mathcal{A}_{w^*u}(B_{Y^*}, X)$ has the AHSP.*
3. *$\mathcal{A}_{wu}(B_Y, X)$ has the AHSP.*

If $\mu$ is a $\sigma$-finite measure and $m$ is the Lebesgue measure on the unit interval, it was shown in [15] that the Bishop-Phelps-Bollobás Theorem holds for operators from $L_1(\mu)$ into $L_\infty(m)$. In the case of bilinear forms the parallel problem was initiated by Choi and Song [42], there are only a few results and the answers are quite different from the operator case.

For two Banach spaces $X$ and $Y$, by using the usual identification of the continuous bilinear forms on $X \times Y$ and the space $\mathcal{L}(X, Y^*)$ it holds that a version of the Bishop-Phelps-Bollobás Theorem for bilinear forms on $X \times Y$ implies the parallel result for the space $\mathcal{L}(X, Y^*)$. The converse is no longer true even for $X = Y = \ell_1$ in view of [42] and [6, Theorem 4.1].

**Theorem 11.31** ([42, Theorem 2]) *The Bishop-Phelps-Bollobás Theorem fails for bilinear forms on $\ell_1 \times \ell_1$.*

Let $X, Y$ be real or complex Banach spaces. By $\mathcal{L}(^2X \times Y)$ we denote the space of continuous bilinear forms on $X \times Y$.

**Definition 11.2** [8, 42] The pair $(X, Y)$ has the *Bishop-Phelps-Bollobás property for bilinear forms* (*BPBP* for bilinear forms), if given $\varepsilon > 0$ there exist $\beta(\varepsilon) > 0$ and $\eta(\varepsilon) > 0$ with $\lim_{\varepsilon \to 0^+} \beta(\varepsilon) = 0$ such that for any $A \in S_{\mathcal{L}(^2X \times Y)}$, if $(x_0, y_0) \in S_X \times S_Y$ is such that $|A(x_0, y_0)| > 1 - \eta(\varepsilon)$, then there exist $(u_0, v_0) \in S_X \times S_Y$ and $B \in S_{\mathcal{L}(^2X \times Y)}$ satisfying the following conditions:

$$|B(u_0, v_0)| = 1, \ \|u_0 - x_0\| < \beta(\varepsilon), \ \|v_0 - y_0\| < \beta(\varepsilon) \text{ and } \|B - A\| < \varepsilon.$$

Acosta et al. [8, Theorem 3.6] characterized the Banach spaces $X$ such that the pair $(\ell_1, X)$ has the *BPBP* for bilinear forms. They introduced the approximate hyperplane series property (*AHSP*) for a pair $(X, X^*)$, closely related to the AHSP for the dual $X^*$, and showed that $(\ell_1, X)$ has the *BPBP* for bilinear forms if and only if the pair $(X, X^*)$ has the *AHSP*.

**Definition 11.3** [8, *Definition 3.1]* For a Banach space X, the pair $(X, X^*)$ has the *approximate hyperplane series property* (*AHSP*) if for every $\varepsilon > 0$ there exist $0 < \delta, \eta < \varepsilon$ such that for every convex series $\sum_n \alpha_n$ and for every sequence $(x_k^*) \subset S_{X^*}$ and $x_0 \in S_X$ with

$$\operatorname{Re} \sum_{n=1}^{\infty} \alpha_n x_n^*(x_0) > 1 - \eta$$

there exist a subset $A \subset \mathbb{N}$, $\{z_k^* : k \in A\} \subset S_{X^*}$ and $z_0 \in S_X$ satisfying:

(1) $\sum_{k \in A} \alpha_k > 1 - \delta$.
(2) $\|z_0 - x_0\| < \varepsilon$ and $\|z_k^* - x_k^*\| < \varepsilon$ for all $k \in A$.
(3) $z_k^*(z_0) = 1$ for all $k \in A$.

It is easy to check that we may assume in Definition 11.3 that the sequence $(x_k^*)$ is contained in $B_{X^*}$.

Here is the characterization.

**Theorem 11.32** ([8, Theorem 3.6]) *Let X be a Banach space. Then the pair $(\ell_1, X)$ has the Bishop-Phelps-Bollobás property for bilinear forms if and only if the pair $(X, X^*)$ satisfies the AHSP.*

Clearly the AHSP for $(X, X^*)$ implies the AHSP for $X^*$ and, by using Theorem 11.32, the converse is not true taking, for instance, the space $X = \ell_1$ (see [6, Theorem 4.1])

Banach spaces $X$ such that $(X, X^*)$ has the AHSP were provided in [8, Propositions 4.1, 4.2, 4.4, 4.7].

**Theorem 11.33** *The following spaces $X$ satisfy that the pair $(X, X^*)$ has the AHSP:*

1. *Every uniformly smooth Banach space $X$.*
2. *Every finite-dimensional normed space $X$.*
3. *The space $X = C_0(\Omega)$, of real or complex valued and continuous functions on $\Omega$ vanishing at infinity, for every locally compact Hausdorff topological space $\Omega$.*
4. *The space $X = K(H)$ of compact operators on any Hilbert space $H$.*

That was all about the space $\ell_1$ but it is natural to ask about the space $L_1(\mu)$, for any measure $\mu$. In [9] it was shown the following.

**Theorem 11.34** ([9, Theorem 2.4]) *Let $Y$ be a Banach space and suppose that $L_1(\mu)$ is infinite-dimensional. If the pair $(L_1(\mu), Y)$ has the* BPBP *for bilinear forms, then $(Y, Y^*)$ has the* AHSP.

If $L_1(\mu)$ is finite-dimensional and $N = \dim L_1(\mu)$, by looking carefully the proof of Theorem 11.34 and [8, Theorem 3.6], it can be obtained that the pair $(L_1(\mu), Y)$ has the *BPBP* for bilinear forms if and only if $(Y, Y^*)$ has the *AHSP* only for finite sums of $N$ elements (instead of any convex series). In case that $L_1(\mu)$ is infinite-dimensional, a characterization for Asplund spaces $Y$ and under a mild assumption on the measure $\mu$ is also obtained in [9]. The argument used to prove this characterization is inspired by the proof of [40, Theorem 2.2].

**Theorem 11.35** ([9, Theorem 2.6]) *Let $\mu$ be a $\sigma$-finite measure such that $L_1(\mu)$ is infinite-dimensional and $Y$ an Asplund space. Then the pair $(L_1(\mu), Y)$ has the* BPBP *for bilinear forms if and only if $(Y, Y^*)$ has the* AHSP.

**Corollary 11.8** [9, Corollary 2.7] *Let $\mu$ be a $\sigma$-finite measure.*

*(1) Assume that $X$ is a finite-dimensional normed space. Then $(L_1(\mu), X)$ has the* BPBP *for bilinear forms. As a consequence, $(X, L_\infty(\mu))$ has the* BPBP *for operators.*
*(2) $(L_1(\mu), c_0)$ has the* BPBP *for bilinear forms, so $(c_0, L_\infty(\mu))$ and $(L_1(\mu), \ell_1)$ have the* BPBP *for operators.*
*(3) Assume that* $\dim L_1(\mu) = \infty$ *and $X$ is a smooth Banach space. Then $(L_1(\mu), X)$ has the* BPBP *for bilinear forms if and only if $X$ is uniformly smooth.*
*(4) If $\mu_1$ and $\mu_2$ are measures such that $L_1(\mu_1)$ and $L_1(\mu_2)$ are infinite-dimensional spaces, the pair $(L_1(\mu_1), L_1(\mu_2))$ fails the* BPBP *for bilinear forms.*

The statement (4) in the previous corollary generalizes the fact that $(\ell_1, \ell_1)$ does not have the *BPBP* for bilinear forms [42]. It is worth to remark that Choi [33] proved that there is no Bishop-Phelps Theorem for bilinear forms on $L_1[0, 1] \times L_1[0, 1]$.

Afterwards Saleh [82] showed that the set of norm attaining bilinear forms on $L_1(\mu)$ is dense in the set of all bounded bilinear forms on $L_1(\mu)$ if and only if $\mu$ is purely atomic (see also [81]).

Related to the assertions (1) and (2) above, it is also worthwhile to notice that Aron, Cascales and Kozhushkina [14] studied the *BPBP* for operators for the case $Y = C_0(L)$ (where $L$ is a locally compact Hausdorff space) and showed that the pair $(X, C_0(L))$ has the *BPBP* if $X$ is Asplund, and so $(c_0, C_0(L))$ has the *BPBP* for operators. Kim [64] also proved that $(c_0, Y)$ has the *BPBP* for operators for every uniformly convex space $Y$.

Now we are going to give a direct proof of the statement (4) in the previous corollary which is not included in another paper.

**Theorem 11.36** *If $\mu$ and $\nu$ are measures such that $L_1(\mu)$ and $L_1(\nu)$ are infinite-dimensional spaces, $\mathcal{L}(^2L_1(\mu) \times L_1(\nu))$ does not satisfy the Bishop-Phelps-Bollobás Theorem.*

*Proof* Let denote by $(\Omega_1, \mathcal{A}_1, \mu)$ and $(\Omega_2, \mathcal{A}_2, \nu)$ the measure spaces involved.

Since $L_1(\mu)$ and $L_1(\nu)$ are infinite-dimensional spaces, there are sequences of sets $\{E_n\}$ and $\{F_n\}$ satisfying the following conditions:

$$E_n \in \mathcal{A}_1, 0 < \mu(E_n) < \infty, \ \forall n, \ E_n \cap E_m = \varnothing \ \text{ for } \ n \neq m,$$

and

$$F_n \in \mathcal{A}_2, 0 < \nu(F_n) < \infty, \ \forall n, \ F_n \cap F_m = \varnothing \ \text{ for } \ n \neq m.$$

Now we define the bilinear form on $L_1(\mu) \times L_1(\nu)$ given by

$$A(f, g) := \sum_{i=1}^{\infty} \Big( \int_{E_i} f \, d\mu \Big) \sum_{\substack{j=1 \\ j \neq i}}^{\infty} \Big( \int_{F_j} g \, d\nu \Big) \quad (f \in L_1(\mu), g \in L_1(\nu)) \, .$$

We put $E_0 = \Omega_1 \setminus \Big( \bigcup_{n=1}^{\infty} E_n \Big)$ and $F_0 = \Omega_2 \setminus \Big( \bigcup_{n=1}^{\infty} F_n \Big)$. If we take

$$f_0 = \sum_{i=1}^{2n^2} \frac{1}{2n^2} \frac{\chi_{E_i}}{\mu(E_i)} \, , \quad g_0 = \sum_{i=1}^{2n^2} \frac{1}{2n^2} \frac{\chi_{F_i}}{\nu(F_i)} \quad (n \ \text{ fixed}, \quad n \in \mathbb{N}) \, ,$$

we have that $A(f_0, g_0) = 1 - \frac{1}{2n^2}$. Thus $\|A\| = 1$.

Suppose that there are $B \in \mathcal{L}(^2L_1(\mu) \times L_1(\nu))$ and $(\overline{f}, \overline{g}) \in S_{L_1(\mu)} \times S_{L_1(\nu)}$ such that $1 = \|B\| = |B(\overline{f}, \overline{g})|$.

Now, we write

$$a_n = \int_{E_n} |\overline{f}| \, \chi_{E_n} \, d\mu \, , \quad b_n = \int_{F_n} |\overline{g}| \, \chi_{F_n} \, d\nu \, , \quad \text{for all} \quad n = 0, 1, \ldots \, .$$

Thus

$$1 = |B(\overline{f}, \overline{g})| = |\sum_{n=0}^{\infty}\sum_{m=0}^{\infty} B(\overline{f}\,\chi_{E_n}, \overline{g}\,\chi_{F_m})|$$
$$\leq \sum_{n=0,m=0}^{\infty} |B(\overline{f}\,\chi_{E_n}, \overline{g}\,\chi_{F_m})| \leq \sum_{n=0,m=0}^{\infty} a_n b_m = 1.$$

Hence
$$|B(\overline{f}\,\chi_{E_n}, \overline{g}\,\chi_{F_m})| = a_n b_m \quad \text{for all} \quad n, m = 0, 1, 2, \ldots.$$

Take $L = \{n \in \mathbb{N} : a_n > 0\}$ and $M = \{m \in \mathbb{N} : b_m > 0\}$.
So for all $(n, m) \in L \times M$ we have that

$$\text{(1)} \qquad \left| B\left(\frac{1}{a_n}\overline{f}\,\chi_{E_n}, \frac{1}{b_m}\overline{g}\,\chi_{F_m}\right)\right| = 1\,.$$

We claim that if $\|A - B\| < 1$ then either $\|f_0 - \overline{f}\| \geq \frac{1}{2}$ or $\|g_0 - \overline{g}\| \geq \frac{1}{2}$.

First, we see that $0 \notin L$. If this were not the case, then $A(\frac{1}{a_0}\overline{f}\,\chi_{E_0}, \frac{1}{b_m}\overline{g}\,\chi_{F_m}) = 0$, for all $m \in M$. Therefore, by (1), $\|A - B\| = 1$ and this is a contradiction. Analogously, $0 \notin M$.

Let us prove that $L \bigcap M = \emptyset$. If $n \in L \bigcap M$, then by (1)

$$\left| B\left(\frac{1}{a_n}\overline{f}\,\chi_{E_n}, \frac{1}{b_n}\overline{g}\,\chi_{F_n}\right)\right| = 1\,.$$

Hence $\|A - B\| = 1$ .

Finally, if $\|f_0 - \overline{f}\| < \frac{1}{2}$ then $\operatorname{card}(\{1, \ldots, 2n^2\} \bigcap L) > n^2$. To see this, assume $\operatorname{card}(\{1, \ldots, 2n^2\} \bigcap L) \leq n^2$. Thus

$$\|f_0 - \overline{f}\| \geq \int_{\bigcup_{i \in \{1,\ldots,2n^2\}\setminus L} E_i} |f_0|\, d\mu \geq \frac{n^2}{2n^2} = \frac{1}{2}\,,$$

a contradiction.

Since $L \bigcap M = \emptyset$ we have that $\operatorname{card}(\{1, \ldots, 2n^2\} \bigcap M) < n^2$. Therefore $\|g_0 - \overline{g}\| > \frac{1}{2}$. The claim follows and the proof is completed. □

What about the $n$-linear case and the $n$-homogeneous polynomial case?

In [8, Theorem 2.2] it was shown the Bishop-Phelps-Bollobás Theorem holds for $n$-linear mappings defined on uniformly convex Banach spaces.

Let $X_1, \ldots, X_n$ and $Y$ be either real or complex Banach spaces. As usual, $\mathcal{L}(^n X_1 \times \cdots \times X_n, Y)$ will be the set of continuous $n$-linear mappings from $X_1 \times \cdots \times X_n$ into $Y$. It is said that the space $\mathcal{L}(^n X_1 \times \cdots \times X_n, Y)$ has the *Bishop-Phelps-Bollobás property* (*BPBP*) if the following condition is satisfied:

Given $\varepsilon > 0$ there exist $\beta(\varepsilon) > 0$ and $\eta(\varepsilon) > 0$ with $\lim_{\varepsilon \to 0^+} \beta(\varepsilon) = 0$ such that if $\|A(x_1, ..., x_n)\| > 1 - \eta(\varepsilon)$ for $A \in S_{\mathscr{L}(^n X_1 \times \cdots \times X_n, Y)}$ and $(x_1, ..., x_n) \in S_{X_1} \times \cdots \times S_{X_n}$, then there exist both an $n$-linear mapping $B \in S_{\mathscr{L}(^n X_1 \times \cdots \times X_n, Y)}$ and $(u_1, ..., u_n) \in S_{X_1} \times \cdots \times S_{X_n}$ such that

$$\|B(u_1, ..., u_n)\| = 1, \|B - A\| < \varepsilon, \text{ and } \|u_i - x_i\| < \beta(\varepsilon) \text{ for all } 1 \leq i \leq n.$$

**Theorem 11.37** ([8, Theorem 2.2]) *Let $X_1, \ldots, X_n, Y$ be Banach spaces and assume that every $X_j$ is uniformly convex with modulus of convexity $\delta_j$, $1 \leq j \leq n$. Given $\varepsilon > 0$, if $0 < \eta < \min\{\delta_j(\varepsilon) : 1 \leq j \leq n\}\frac{\varepsilon}{8+2\varepsilon}$, then for every $A \in S_{\mathscr{L}(^n X_1 \times \cdots \times X_n, Y)}$ and every $x_0 \in S$ such that $\|A(x_0)\| > 1 - \eta$, there are a point $z_0 \in S$ and $B \in S_{\mathscr{L}(^n X_1 \times \cdots \times X_n, Y)}$ satisfying the following conditions:*

$$\|B(z_0)\| = 1, \quad \|z_0 - x_0\|_\infty \leq \varepsilon \text{ and } \|B - A\| < \varepsilon.$$

*Moreover if $A$ belongs to some linear subspace of $\mathscr{L}(^n X_1 \times \cdots \times X_n, Y)$ containing the finite-type $n$-linear mappings, then $B$ belongs to the same subspace.*

The following consequence for operators is immediate.

**Corollary 11.9** [8, Corollary 2.3] *Let $X$ and $Y$ be Banach spaces, and assume that $X$ is uniformly convex. Then given $\varepsilon > 0$, there is $\eta > 0$ such that for every $R \in S_{\mathscr{L}(X,Y)}$ and $x_0 \in S_X$ such that $\|R(x_0)\| > 1 - \eta$, then there exist a point $u_0 \in S_X$ and $T \in S_{\mathscr{L}(X,Y)}$ satisfying the following conditions:*

$$\|T(u_0)\| = 1, \quad \|u_0 - x_0\| \leq \varepsilon \text{ and } \|T - R\| < \varepsilon.$$

*Actually any $0 < \eta < \frac{\varepsilon\delta(\varepsilon)}{8+2\varepsilon}$ satisfies the above condition.*

*In addition if $R$ belongs to some linear space $M \subset \mathscr{L}(X, Y)$ containing the finite-rank operators, then $T$ also belongs to $M$.*

As we already mentioned the BPBP for operators from $X$ into $Y^*$ is satisfied if the pair $(X, Y)$ satisfies the BPBP for bilinear forms and the converse is not true. However in [8] (independently in [43]) is provided a class of Banach spaces for which the converse holds.

**Proposition 11.7** [8, Proposition 2.4] *Let $X$ and $Y$ be Banach spaces and assume that $Y$ is uniformly convex. If the pair $(X, Y^*)$ has the Bishop-Phelps-Bollobás property for operators then $(X, Y)$ satisfies the Bishop-Phelps-Bollobás property for bilinear forms.*

**Corollary 11.10** [8, Corollary 2.6] *If $Y$ is a uniformly convex Banach space and $Y^*$ has the AHSP, then the pair $(\ell_1, Y)$ has the Bishop-Phelps-Bollobás property for bilinear forms.*

The previous statement is a consequence of the fact that the pair $(\ell_1, Y^*)$ has the BPBP for operators only when $Y^*$ has the AHSP (see [6, Theorem 4.1]) and

Proposition 11.7. Examples of classes of spaces having the AHSP can be found in [6, 32]. For instance, it is known that the finite-dimensional spaces, $C(K)$, $L_1(\mu)$ and uniform convex Banach spaces have this property. Every almost CL-space satisfies the AHSP (see for instance [77] for the definitions and also [32]). Furthermore every lush space has the AHSP (see [41]). Also spaces whose dual norm satisfies some uniform condition of smoothness (USSD) at some boundary have this property (see [32, 51]).

It is said that $\mathscr{P}(^n X, Y)$ has the *Bishop-Phelps-Bollobás property* (*BPBP*) when the following condition is satisfied: Given $\varepsilon > 0$ there exist $\beta(\varepsilon) > 0$ and $\eta(\varepsilon) > 0$ with $\lim_{\varepsilon \to 0^+} \beta(\varepsilon) = 0$ such that if $\|P x_0\| > 1 - \eta(\varepsilon)$ for $P \in S_{\mathscr{P}(^n X, Y)}$ and $x_0 \in S_X$, then there exist both $Q \in S_{\mathscr{P}(^n X, Y)}$ and $u_0 \in S_X$ such that

$$\|Q u_0\| = 1, \quad \|u_0 - x_0\| < \beta(\varepsilon) \quad \text{and} \quad \|Q - P\| < \varepsilon.$$

Modifying an argument given in [65] it was shown in [9] the following result for the space of homogeneous polynomials.

**Theorem 11.38** ([9, Theorem 3.1]) *Let $X$ be a uniformly convex Banach space. Then $\mathscr{P}(^n X, Y)$ has the* BPBP *for every Banach space $Y$.*

It is unknown if the analogous result holds for symmetric $n$-linear forms on a uniformly convex space.

Now we recall an isometric condition, called *property $\beta$* that was introduced by Lindenstrauss [69].

**Definition 11.4** A Banach space $Y$ is said to have *property $\beta$* (of Lindenstrauss) if there are two sets $\{y_i : i \in I\} \subset S_Y$, $\{y_i^* : i \in I\} \subset S_{Y^*}$ and $0 \leq \rho < 1$ such that the following conditions hold:

(1) $y_i^*(y_i) = 1, \forall i \in I$.
(2) $|y_i^*(y_j)| \leq \rho < 1$ if $i, j \in I, i \neq j$.
(3) $\|y\| = \sup_{i \in I}\{|y_i^*(y)|\}$, for all $y \in Y$.

The spaces $c_0$ and $\ell_\infty$ satisfy the above property. In both cases $\rho = 0$.

Given a family $\mathscr{F}$ of mappings defined on a Banach space and bounded on its unit ball, by $\mathrm{NA}(\mathscr{F})$ we will denote the subset of norm attaining elements of $\mathscr{F}$. In case that $Y$ has property $\beta$, the following results are shown in [41]:

(1) If $\mathrm{NA}(\mathscr{L}(^n X))$ is dense in $\mathscr{L}(^n X)$, then $\mathrm{NA}(\mathscr{L}(^n X, Y))$ is dense in $\mathscr{L}(^n X, Y)$.
(2) If $\mathrm{NA}(\mathscr{L}_s(^n X))$ is dense in $\mathscr{L}_s(^n X)$, then $\mathrm{NA}(\mathscr{L}_s(^n X, Y))$ is dense in $\mathscr{L}_s(^n X, Y)$, where $\mathscr{L}_s(^n X, Y)$ stands for the symmetric $n$-linear mappings from $X$ into $Y$.
(3) If $\mathrm{NA}(\mathscr{P}(^n X))$ is dense in $\mathscr{P}(^n X)$, then $\mathrm{NA}(\mathscr{P}(^n X, Y))$ is dense in $\mathscr{P}(^n X, Y)$.

In [9] it was shown similar results for the Bishop-Phelps-Bollobás property.

**Proposition 11.8** [9, Proposition 3.3] *Suppose that the Banach space $Y$ has property $\beta$.*
(1) *If $\mathscr{L}(^n X_1 \times \cdots \times X_n)$ has the* BPBP, *then $\mathscr{L}(^n X_1 \times \cdots \times X_n, Y)$ has the* BPBP.
(2) *If $\mathscr{P}(^n X)$ has the* BPBP, *then $\mathscr{P}(^n X, Y)$ has the* BPBP.

An analogous argument to the used in the proof of the above result also works for the symmetric case. However we do not know of any infinite-dimensional Banach space $X$ such that the Bishop-Phelps-Bollobás property is satisfied for the symmetric $n$-linear forms on $X$. The only positive result known is the following given in 2014.

**Theorem 11.39** ([54, Theorems 3.2 and 3.4]) *The BPBP property holds for symmetric bilinear forms on any real or complex Hilbert space.*

**Acknowledgments** The authors were supported by MICINN Project MTM2011-22417 and Prometeo II/2013/013.

## References


1. Abramovich, Y.A., Aliprantis, C.D.: An Invitation to Operator Theory. Graduate Texts in Mathematics, vol. 50. AMS, Providence (2002)
2. Abramovich, Y.A., Aliprantis, C.D.: Problems in Operator Theory. Graduate Texts in Mathematics, vol. 51. AMS, Providence (2002)
3. Abramovich, Y.A., Aliprantis, C.D., Burkinshaw, O.: The Daugavet equation in uniformly convex Banach spaces. J. Funct. Anal. **97**, 215–230 (1991)
4. Acosta, M.D.: Denseness of norm attaining mappings. Rev. R. Acad. Cien. Ser. A. Mat. (RACSAM) 100, 9–30 (2006)
5. Acosta, M.D.: Boundaries for spaces of holomorphic functions on $C(K)$. Publ. Res. I. Math. Sci. **42**, 27–44 (2006)
6. Acosta, M.D., Aron, R.M., García, D., Maestre, M.: The Bishop-Phelps-Bollobás Theorem for operators. J. Funct. Anal. **254**, 2780–2799 (2008)
7. Acosta, M.D., Becerra-Guerrero, J., García, D., Kim, S.K., Maestre, M.: The Bishop-Phelps-Bollobás property and its finite dimensional approach (preprint)
8. Acosta, M.D., Becerra-Guerrero, J., García, D., Maestre, M.: The Bishop-Phelps-Bollobás theorem for bilinear forms. Trans. Amer. Math. Soc. **365–11**, 5911–5932 (2013)
9. Acosta, M.D., Becerra-Guerrero, J., Choi, Y.S., García, D., Kim, S.K., Lee, H.J., Maestre, M.: The Bishop-Phelps-Bollobás property for bilinear forms and polynomials. J. Math. Soc. Japan **66**, 957–979 (2014)
10. Acosta, M.D., García, D., Maestre, M.: A multilinear Lindenstrauss theorem. J. Funct. Anal. **235**, 122–136 (2006)
11. Acosta, M.D., Kamińska, A.: Weak neighborhoods and the Daugavet property of the interpolation spaces $L^1 + L^\infty$ and $L^1 \cap L^\infty$. Indiana Univ. Math. J. **57**, 77–96 (2008)
12. Acosta, M.D., Kim, S.G.: Numerical boundaries for some classical Banach spaces. J. Math. Anal. Appl. **350**, 694–707 (2009)
13. Aksoy, A.G., Lewicki, G.: Limit theorems for the numerical index. J. Math. Anal. Appl. **398**, 296–302 (2013)
14. Aron, R.M., Cascales, B., Kozhushkina, O.: The Bishop-Phelps-Bollobás theorem and Asplund operators. Proc. Amer. Math. Soc. **139**(10), 3553–3560 (2011)
15. Aron, R.M., Choi, Y.S., García, D., Maestre, M.: The Bishop-Phelps-Bollobás Theorem for $\mathscr{L}(L_1(\mu), L_\infty[0, 1])$. Adv. Math. **228**, 617–628 (2011)
16. Aron, R.M., García, D., Maestre, M.: On norm attaining polynomials. Publ. Res. Inst. Math. Sci. **39**, 165–172 (2003)
17. Aron, R.M., Globevnik, J.: Analytic functions on $c_0$. Rev. Mat. Univ. Complt. Madrid **2**, 27–33 (1989)
18. Avilés, A., Kadets, V.M., Martín, M., Merí, J., Shepelska, V.: Slicely countably determined Banach spaces. C. R. Math. Acad. Sci. Paris. **347**, 1277–1280 (2009)

19. Avilés, A., Kadets, V.M., Martín, M., Merí, J., Shepelska, V.: Slicely countably determined Banach spaces. Trans. Amer. Math. Soc. **362**, 4871–4900 (2010)
20. Bauer, F.L.: On the field of values subordinate to a norm. Numer. Math. **4**, 103–111 (1962)
21. Guerrero, Becerra: J., Rodríguez-Palacios, A.: Banach spaces with the Daugavet property, and the centralizer. J. Funct. Anal. **254**, 2294–2302 (2008)
22. Bishop, E., Phelps, R.R.: A proof that every Banach space is subreflexive. Bull. Amer. Math. Soc. **67**, 97–98 (1961)
23. Bogdanowicz, W.: On the weak continuity of the polynomial functionals defined on the space $c_0$. Bull. Acad. Polon. Sci. Cl. III. **5**, 243–246 (in Russian) (1957)
24. Bollobás, B.: An extension to the theorem of Bishop and Phelps. Bull. London Math. Soc. **2**, 181–182 (1970)
25. Bonsall, F.F., Cain, B.E., Schneider, H.: The numerical range of a continuos mapping of a normed space. Aequationes Math. **2**, 86–93 (1968)
26. Bonsall, F.F., Duncan, J.: Numerical Ranges of Operators on Normed Spaces and of Elements of Normed Algebras. London Mathematical Society Lecture Note Series 2, Cambridge (1971)
27. Bonsall, F.F., Duncan, J.: Numerical Ranges II. London Mathematical Society Lecture Note Series 10, Cambridge (1973)
28. Bourgain, J.: Dentability and the Bishop-Phelps property. Israel J. Math. **28**, 265–271 (1977)
29. Boyko, K., Kadets, V.M., Martín, M., Werner, D.: Numerical index of Banach spaces and duality. Math. Proc. Cambridge Phil. Soc. **142**, 93–102 (2007)
30. Boyko, K., Kadets, V.M., Martín, M., Werner, D.: Properties of lush spaces and applications to Banach spaces with numerical index 1. Studia Math. **190**, 117–133 (2009)
31. Cascales, B., Guirao, A.J., Kadets, V.M.: A Bishop-Phelps-Bollobás type theorem for uniform algebras. Adv. Math. **240**, 370–382 (2013)
32. Cheng, L., Dai, D., Dong, Y.: A sharp operator version of the Bishop-Phelps Theorem for operators from $\ell_1$ to CL-spaces. Proc. Amer. Math. Soc. **141**, 867–872 (2013)
33. Choi, Y.S.: Norm attaining bilinear forms on $L^1[0, 1]$. J. Math. Anal. Appl. **211**, 295–300 (1997)
34. Choi, Y.S., García, D., Kim, S.G., Maestre, M.: The polynomial numerical index of a Banach space. Proc. Edinb. Math. Soc. **49**, 32–52 (2006)
35. Choi, Y.S., García, D., Kim, S.G., Maestre, M.: Composition, numerical range and Aron-Berner extension. Math. Scand. **103**, 97–110 (2008)
36. Choi, Y.S., García, D., Kim, S.K., Maestre, M.: Some geometric properties on disk algebra. J. Math. Anal. Appl. **409**, 147–157 (2014)
37. Choi, Y.S., García, D., Maestre, M., Martín, M.: The polynomical numerical index for some complex vector-valued function spaces. Q. J. Math. **59**, 455–474 (2008)
38. Choi, Y.S., García, D., Maestre, M., Martín, M.: The Daugavet equation for polynomials. Studia Math. **178**, 63–82 (2007)
39. Choi, Y.S., Kim, S.G.: Norm or numerical radius attaining multilinear mappings and polynomials. J. London Math. Soc. **54**, 135–147 (1996)
40. Choi, Y.S., Kim, S.K.: The Bishop-Phelps-Bollobás theorem for operators from $L_1(\mu)$ to Banach spaces with the Radon-Nikodým property. J. Funct. Anal. **261**, 1446–1456 (2011)
41. Choi, Y.S., Kim, S.K.: The Bishop-Phelps-Bollobás property and lush spaces. J. Math. Anal. Appl. **390**, 549–555 (2012)
42. Choi, Y.S., Song, H.G.: The Bishop-Phelps-Bollobás theorem fails for bilinear forms on $l_1 \times l_1$. J. Math. Anal. Appl. **360**, 752–753 (2009)
43. Dai, D.: The Bishop-Phelps-Bollobás Theorem for bilinear mappings (preprint)
44. Daugavet, I.K.: On a property of completely continuous operators in the space $C$. Uspekhi Mat. Nauk **18**, 157–158 (1963)
45. Diestel, J.: Sequences and Series in Banach Space. Graduate Texts in Mathematics, vol. 92. Springer, New York (1984)
46. Diestel, J., Uhl, J.J.: Vector Measures. Mathematical Surveys 15. AMS, Providence (1977)
47. Dineen, S.: Complex Analysis on Infinite Dimensional Spaces. Springer Monographs in Mathematics, Springer, London (1999)

48. Duncan, J., McGregor, C.M., Pryce, J.D., White, A.J.: The numerical index of a normed space. J. London Math. Soc. **2**, 481–488 (1970)
49. Ed-dari, E.: On the numerical index of Banach spaces. Linear Algebra Appl. **403**, 86–96 (2005)
50. Faried, N., Ibrahim, F.A., Bakery, A.A.: The numerical index of the $L_{(p)}$ space of dimension two. Int. Math. Forum **6**, 2101–2108 (2011)
51. Franchetti, C., Payá, R.: Banach spaces with strongly differentiable norm. Boll. U.M.I. **7**, 45–70 (1993)
52. García, D., Grecu, B., Maestre, M., Martín, M., Merí, J.: Two-dimensional Banach spaces with Polynomial numerical index zero. Linear Algebra Appl. **430**, 2488–2500 (2009)
53. García, D., Grecu, B., Maestre, M., Martín, M., Merí, J.: Polynomial numerical indices of $C(K)$ and $L_1(\mu)$. Proc. Amer. Math. Soc. **142**(4), 1229–1235 (2014)
54. García, D., Lee, H.J., Maestre, M.: The Bishop-Phelps-Bollobás property for Hermitian forms on Hilbert spaces. Q. J. Math. **65**, 201–209 (2014)
55. Harris, L.A.: The numerical range of holomorphic functions in Banach spaces. Amer. J. Math. **93**, 1005–1019 (1971)
56. Harris, L.A.: The numerical range of functions and best approximation. Proc. Camb. Phil. Soc. **76**, 133–141 (1974)
57. Kadets, V.M., Martín, M., Merí, J., Shepelska, V.: Lushness, numerical index one and duality. J. Math. Anal. Appl. **357**, 15–24 (2009)
58. Kadets, V.M., Martín, M., Merí, J., Werner, D.: Lushness, numerical index 1 and the Daugavet property in rearrangement invariant spaces. Canad. J. Math. **65**, 331–348 (2013)
59. Kadets, V.M., Martín, M., Payá, R.: Recent progress and open questions on the numerical index of Banach spaces. Rev. R. Acad. Cien. Serie A Mat. (RACSAM) **100**, 155–182 (2006)
60. Kadets, V.M., Popov, M.M.: The Daugavet property for narrow operators in rich subspaces of $C[0, 1]$ and $L_1[0, 1]$. St. Petersburg Math. J. **8**, 571–584 (1997)
61. Kadets, V.M., Shvidkoy, R.V., Sirotkin, G.G., Werner, D.: Banach spaces with the Daugavet property. Trans. Amer. Math. Soc. **352**, 855–873 (2000)
62. Kim, J., Lee, H.J.: Strong peak points and strongly norm attaining points with applications to denseness and polynomial numerical indices. J. Funct. Anal. **257**, 931–947 (2009)
63. Kim, S.G., Martín, M., Merí, J.: On the polynomial numerical index of the real spaces $c_0$, $\ell_1$ and $\ell_\infty$. J. Math. Anal. Appl. **337**, 98–06 (2008)
64. Kim, S.K.: The Bishop-Phelps-Bollobás Theorem for operators from $c_0$ to uniformly convex spaces. Israel J. Math. **197**, 425–435 (2013)
65. Kim, S.K., Lee, H.J.: Uniform convexity and Bishop-Phelps-Bollobás property. Can. J. Math. **66**, 373–386 (2014)
66. Lee, H.J.: Banach spaces with polynomial numerical index 1. Bull. London Math. Soc. **40**, 193–198 (2008)
67. Lee, H.J., Martín, M.: Polynomial numerical indices of Banach spaces with 1-unconditional bases. Linear Algebra Appl. **437**, 2001–2008 (2012)
68. Lee, H.J., Martín, M., Merí, J.: Polynomial numerical indices of Banach spaces with absolute norm. Linear Algebra Appl. **435**, 400–408 (2011)
69. Lindenstrauss, J.: On operators which attain their norm. Israel J. Math. **1**, 139–148 (1963)
70. López, G., Martín, M., Merí, J.: Numerical Index of Banach spaces of weakly or weakly-star continuous functions. Rocky Mountain J. Math. **38**, 213–223 (2008)
71. Martín, M.: The alternative Daugavet property of $C^*$-algebras and $JB^*$-triples. Math. Nachr. **281**, 376–385 (2008)
72. Martín, M., Merí, J., Popov, M.M.: The polynomial Daugavet property for atomless $L_1(\mu)$-spaces. Arch. Math. (Basel) **94**, 383–389 (2010)
73. Martín, M., Merí, J., Popov, M.M.: On the numerical index of real $L_p(\mu)$-spaces. Israel J. Math. **184**, 183–192 (2011)
74. Martín, M., Merí, J., Popov, M.M., Randrianantoanina, B.: Numerical index of absolute sums of Banach spaces. J. Math. Anal. Appl. **375**, 207–22 (2011)
75. Martín, M., Oikhberg, T.: The alternative Daugavet property. J. Math. Anal. Appl. **294**, 158–180 (2004)

76. Martín, M., Payá, R.: Numerical index of vector-valued function spaces. Studia Math. **142**, 269–280 (2000)
77. Martín, M., Payá, R.: On CL-spaces and almost CL-spaces. Ark. Mat. **42**, 107–118 (2004)
78. Martín, M., Villena, A.: Numerical index and the Daugavet property for $L_\infty(\mu, X)$. Proc. Edinb. Math. Soc. **46**(2), 415–420 (2003)
79. Rodríguez-Palacios, A.: Numerical ranges of uniformly continuous functions on the unit sphere of a Banach space. J. Math. Anal. Appl. **297**, 472–476 (2004)
80. Rosenthal, H.: The Lie algebra of a Banach space. Lecture Notes in Mathematics, vol. 1166. Springer, Berlin (1985)
81. Saleh, Y.: Norm attaining bilinear forms on $L_1(\mu)$. Int. J. Math. & Math. Sci. **23**, 833–837 (2000)
82. Saleh, Y.: Norm attaining multilinear forms on $L_1(\mu)$. Int. J. Math. & Math. Sci. 1–6 (2008)
83. Sánchez Pérez, E., Werner, D.: The $p$-Daugavet property for function spaces. Arch. Math. (Basel) 96, 565–575 (2011)
84. Toeplitz, O.: Das algebraische Analogon zu einem Satze von Fejer. Math. Z. **2**, 187–197 (1918)
85. Werner, D.: An elementary approach to the Daugavet equation. In: Kalton, N., Saab, E., Montgomery-Smith, S. (eds.) Interaction between Functional Analysis, Harmonic Analysis and Probability. Lecture Notes in Pure and Applied Mathematics, vol. 175, pp. 449–454. Dekker, New York (1996)
86. Werner, D.: The Daugavet equation for operators on function spaces. J. Funct. Anal. **143**, 117–128 (1997)
87. Werner, D.: Recent progress on the Daugavet property. Irish Math. Soc. Bull. **46**, 77–97 (2001)
88. Wojtaszczyk, P.: Some remarks on the Daugavet equation. Proc. Amer. Math. Soc. **115**, 1047–1052 (1992)

MIX
Papier aus verantwortungsvollen Quellen
Paper from responsible sources
FSC® C105338

If you have any concerns about our products,
you can contact us on
**ProductSafety@springernature.com**

In case Publisher is established outside the EU,
the EU authorized representative is:
**Springer Nature Customer Service Center GmbH**
**Europaplatz 3, 69115 Heidelberg, Germany**

Printed by Libri Plureos GmbH
in Hamburg, Germany